Fuzzy Techniques
in Pattern
Recognition

FUZZY TECHNIQUES IN PATTERN RECOGNITION

ABRAHAM KANDEL

The Florida State University, Tallahassee

A Wiley-Interscience Publication
JOHN WILEY & SONS

New York   Chichester   Brisbane   Toronto   Singapore

Copyright © 1982 by John Wiley & Sons, Inc.

All rights reserved. Published simultaneously in Canada.

Reproduction or translation of any part of this work
beyond that permitted by Section 107 or 108 of the
1976 United States Copyright Act without the permission
of the copyright owner is unlawful. Requests for
permission or further information should be addressed to
the Permissions Department, John Wiley & Sons, Inc.

Library of Congress Cataloging Publication Data:

Kandel, Abraham.
   Fuzzy techniques in pattern recognition.

   "A Wiley-Interscience publication."
   Bibliography: p.
   Includes index.
   1. Pattern perception. 2. Fuzzy sets. I. Title.
Q327.K3 1982   001.53'4   82-8359
ISBN 0-471-09136-7   AACR2

Printed in the United States of America

10 9 8 7 6 5 4 3 2 1

To my wife, Nurit

Preface

Fuzzy techniques represent the most recent aspect of the newly emerging field of pattern recognition, in itself an important component of the relatively new discipline of computer applications to science and engineering. My intent in writing this book is to cover in descriptive terms a good part of the problems in pattern recognition by use of fuzzy techniques. I am not concerned with the substantial body of research in fuzzy set theory and its applications, but restrict the discussion to those subjects having implications to the fundamental nature of the recognition process. During the past decade, there has been a considerable growth of interest in the theory and applications of fuzzy sets as well as in the field of pattern recognition. Most books published so far in the recognition area deal with either the decision-theoretic approach or the syntactic approach. This book treats the design of pattern recognition systems by use of the fuzzy approach. It may be used either (1) in a graduate-level course devoted specifically to fuzzy techniques in pattern recognition, a course planned with the idea that, after covering the basic material of this text, the class can profitably turn to the journal literature on the subject, or (2) as a reference research monograph on fuzzy set theory and its applications in the design of pattern recognition systems. The presentation is kept concise, and the comprehensive bibliography is designed to facilitate the use of the book both as a research compendium and as a graduate-level textbook.

The text is organized in five chapters. Chapter 1 presents a discussion of pattern recognition characterization and cluster analysis. The classical approach to pattern recognition is introduced and several pattern analysis techniques are briefly reviewed. In Chapter 2 the algebra of inexactness and the nature of fuzziness are treated. Fuzzy sets, different kinds of uncertainty, the theory of possibility, and the concept of fuzzy expectation are covered, as well as operations on fuzzy relations and estimated similarities. Various recognition techniques for pattern description that include syntactic and semantic methods, decision trees, fuzzy partitions, and fuzzy ISODATA are presented in Chapter 3. Chapter 4 is a treatment of inexact hierarchical clustering via the use of fuzzy relations, inexact matrices, similarity relations, and convex decompositions. Several important classes of imprecise operations are defined and their properties investigated. Chapter 5 begins with some specific applications of the previous concepts. These applications include phonetic and phonemic learning of speech, speech identification, and fuzzy filters. The final sections of the chapter treat the topics of clustering via unimodal fuzzy sets, fuzzy covariance matrix, and cluster validity. An updated bibliography listing more than 3000 publications in the field of fuzzy sets and its applications is included at the end of the book.

Some of the material contained in this book has been used in courses at The Florida State University. I am indebted to my students and peers for their help and encouragement throughout the development of this book, especially I want to express my lasting gratitude to Douglas H. Schlak for assistance in putting together the extensive bibliography, and to Pamela L. Flowers and Lawrence O. Hall for proofreading an early version of the manuscript. Portions of the material in this text are derived from many early investigations listed in the bibliography.

I would like to acknowledge my indebtedness to all those researchers and the many scientists who have contributed to the fast growing field of research known as "fuzzy set theory and its applications." It is my pleasure to particularly acknowledge the constant encouragement and consistent support of Lotfi A. Zadeh, King-Sun Fu, Madan M. Gupta, Ramesh Jain, Arnold Kaufmann, Ebrahim H. Mamdani, Ronald R. Yager, Hans J. Zimmermann, and William J. Byatt. I owe a debt of gratitude to all of them.

ABRAHAM KANDEL

Tallahassee, Florida
June 1982

Contents

ONE   PATTERN CHARACTERIZATION   1

    1.1   Cluster Analysis Theory, 1
    1.2   Need for Clustering Methods, 2
    1.3   Pattern Analysis, 4
    1.4   Similarity Measures and Clustering, 8
    1.5   Applications, 18
          Bibliography, 20

TWO   THE ALGEBRA OF INEXACTNESS   22

    2.1   Introduction, 22
    2.2   Fuzzy Sets, 23
    2.3   Inexact Algebra and Uncertainty, 43
    2.4   Toward the Theory of Possibility, 52
    2.5   Fuzzy Relations and Estimated Similarities, 59
    2.6   Fuzzy Statistics, 64

THREE   PATTERN CLASSIFICATION AND FUZZY SETS   91

    3.1   Introduction, 91
    3.2   Pattern Recognition and Cluster Analysis in a Fuzzy-Set-Theoretic Framework, 92
    3.3   Syntactic and Semantic Techniques, 94
    3.4   Recognition via Fuzzy Decision Tree, 110
    3.5   Fuzzy Partition via Ruspini, 113
    3.6   Fuzzy Isodata, 119

FOUR   INEXACT HIERARCHICAL CLASSIFICATION   130

    4.1   Introduction, 130
    4.2   Fuzzy Relations, 131
    4.3   Similarity Relations, 139

4.4 Inexact Matrices, 142
4.5 Applications to Clustering, 148
4.6 Convex Decompositions, 168

FIVE APPLICATIONS AND PERFORMANCE MEASURES 173

5.1 Phonetic and Phonemic Labeling of Speech, 173
5.2 Speech Identification, 177
5.3 Fuzzy Filters, 182
5.4 Clustering via Unimodal Fuzzy Sets, 186
5.5 Fuzzy Covariance Matrix, 193
5.6 Cluster Validity, 200

Key References in Fuzzy Pattern Recognition 209

Bibliography 212

Index 355

**Fuzzy Techniques
in Pattern
Recognition**

CHAPTER ONE

Pattern Characterization

1.1 CLUSTER ANALYSIS THEORY

It is frequently stated that the process of recognition and classification is one of the most fundamental of human activities. As a matter of fact, one of the most primitive and common activities of animals (human beings included) consists of sorting like items into groups. These groups are described by patterns and what we perform is the act of recognition of certain patterns and then classification of them into groups. The word "pattern" follows the root of the word "patron" and reflects the concept of an ideal model of a set of objects or structures. This perfect pattern brings to mind the Platonic philosophy of perfect structures; the structures of the "real world" are considered to be imperfect replicas of the ideal. This deep question of the imperfection of "real world" patterns and their classification is the subject of this work.

Without pretending to offer a general theory of pattern recognition and classification, we try to suggest a new point of view about the subject of clustering, a subject representing a mental activity in which we formulate, select, modify, and adjust our frames of reference so that we can relate to a certain structure. This is a psychophysiological process that involves a relationship between a recognizer (person) and a physical stimulus. In psychology pattern recognition is defined as the process by which "external signals arriving at the sense organs are converted into meaningful perceptual experiences" (Lindsay and Norman, 1972).[†] Following our intuitive search for computerized recognition and classification, we obviously have to quantize the structural representation of patterns. For example, when we say that

[†]References in this chapter are found at the end of the chapter.

$Q(s)$ holds, the predicate Q plays the role of a quantizer on an item s. If s is a picture, for example, representing raw information, the process of recognition and classification becomes more complex (Pavlidis, 1977) and the quantization Q might involve the transformation of the scene to a numerical array, texture identification via scene segmentation and analysis, and quantitative description of shapes and objects.

The Greek word βοτρυς means a cluster of grapes. In his excellent paper, Good (1977) proposed again the adoption of this origin of the English prefix botryo, and used the term "botryology," meaning the theory of clusters. To quote Good:

> It seems to me that the subject of clustering is now wide enough and respectable enough to deserve a name like those of other disciplines, and the existence of such a name enables one to form adjectives and so on. For example, one can use expressions such as "a botryological analysis" or "a well-known botryologist said so and so." There is another word that serves much the same purpose, namely "taxonomy," but this usually refers to biological applications whereas "botryology" is intended to refer to the entire field, provided that mathematical methods are used.

Adopting this concept, this work can be labeled as "fuzzy botryology."

1.2 NEED FOR CLUSTERING METHODS

Classification is the process of assigning an item or an observation to its proper place; the problem of cluster analysis is frequently stated as one of finding the "natural groups." In a more concrete sense, the objective is to sort a data set into categories such that the degree of "natural association" is high among members of the same category and low between members of different categories. The essence of cluster analysis might be viewed as assigning appropriate meaning to the terms "natural groups" and "natural association," where "natural" usually refers to homogeneous and "well-separated" structures.

We have used the word "theory" with relation to the above stated problem. The word "theory" has two rather different general connotations. One has the character of speculation, an hypothesis or guess; this kind of theory is the opposite of "practice" and sometimes is criticized as being "idle." The other kind of "theory" is the scientific one, an explanation (as opposed to a catalog of facts) from general principles. In mathematics there is also a technical meaning: a body of definitions, axioms, and theorems that are a systematic presentation of a subject.

Need for Clustering Methods

As a field that encompasses many diverse techniques for discovering structure within complex bodies of data, cluster analysis occupies an interesting position with regard to theory and practice.

On the one hand, the field has practical problems, of importance not only to itself, but of economic and social interest as well. Applications of pattern recognition and classification exist in many important areas such as computer-assisted medical diagnosis and treatment, multiphasic screening and analysis, image processing and scene analysis, weather prediction, process control, neurobiological signal processing, analysis of aerial photography, earth-resource analysis, sonar detection and classification, speech and fingerprint recognition, and many, many more. So, as in a physical or social science, phenomena exist in cluster analysis that require understanding and may even be controlled through that understanding.

At the same time, the fundamental entities of the field, digital computers, are themselves entirely the product of human design, and in a sense represent an idealization of the way certain applied mathematicians performed calculations. In this aspect, the field is like a combination of engineering, mathematics, and computer science in that the subject of study is under explicit control.

The last decade has witnessed considerable interest and rapid advances in both the development and research of automatic pattern recognizers and classifiers, but many questions are still open.

One responsibility of theory in cluster analysis is to explain practical matters, and another to meet conventional mathematical criteria as systematic abstraction. These two goals are not incompatible, but neither are they close together.

Exact problems can be fully stated by means of mathematics, for example, in the form of equations. Algorithms for the solution of exact problems need only be translated into a computer language and the deed is done; the mathematician has acquired a strong and untiring helper.

The situation is different with respect to the problems that we have conditionally called *inexact*, problems for which the mathematical methods have not been defined, so that there is very little to translate into machine language. The inexact tasks include a wide variety of possibilities. Human beings handle them more or less satisfactorily, but they do it subconsciously, so to speak, not knowing how they really solve them. Cluster analysis is one of these inexact problems.

The variety of applications listed above may cause us to question whether a unified approach to pattern recognition and classification is at all possible. I believe that there is a general approach to the problem but the techniques under this general approach still vary, based on the judgment as well as the imagination of the scientist or the engineer.

In earlier studies, such as the optical character recognition device that reads the code characters on bank checks, one advantage of the technique has been the simplicity of the algorithm. In the era of networking and parallel systems, we do not measure the success of an algorithm by its simplicity. Rather we will encourage a complex scheme, as long as the implementation of the algorithm is on a parallel system with k processors and with a reasonable computational complexity (linear or low order polynomial function of n, the data size). The complexity of computations is especially important, as in most computer-implemented algorithms that are applied to complex and large data sets.

Even though little or nothing about the category structure can be stated in advance, the analyst usually is informed sufficiently about the problem that he or she can distinguish between "good" and "bad" category structures when confronted with them. Hence, Anderberg (1973) asks the question: Why not enumerate all the possibilities and simply choose the most appealing?

The number of ways of sorting n observations into m groups is a Stirling number of the second kind:[†]

$$S_n^{(m)} = \frac{1}{m!} \sum_{j=0}^{m} (-1)^{m-j} \binom{m}{j} j^n$$

The problem is compounded by the fact that the number of groups is usually unknown, so that the number of possibilities is a sum of Stirling numbers. In the case of 25 observations it is

$$\sum_{k=1}^{25} S_{25}^{(k)} > 4 \times 10^{18}.$$

Hence it is generally the intent of cluster analysis techniques to emulate human efficiency and find an acceptable solution while considering only a small number of the alternatives.

1.3 PATTERN ANALYSIS

The common problem in data analysis is the lack of homogeneity among attributes, which leads to the inexact meaning of the word "cluster." Most cluster analysis methods require a measure of similarity to be defined for every pairwise combination of the entities to be clustered. When clustering

---

[†] For even the relatively tiny problem of sorting 25 observations into 5 categories, the number of possibilities is the astounding quantity $S_{25}^{(5)} = 2,436,684,974,110,751$.

Pattern Analysis

data units, the proximity of individuals is usually expressed as a distance. The clustering of variables generally involves a correlation or other such measure of association, many of which are discussed in the literature. Some have operational interpretations while others are rather difficult to describe. The measures interact with cluster analysis criteria so that some measures give identical results with one criterion and distinctly different results with another. The combined choices of variables, transformations, and similarity measures give operational meaning to the term "natural association." The choices are often made separately and should be reviewed for their composite effect to make sure the result is satisfactory.

The term "cluster" is often left undefined and taken as a primitive notion in much the same manner as "point" is treated in geometry. Such treatment is fine for theoretical discussions, but when it comes to finding clusters in real data, the term must bear a definite meaning. The choice of a clustering criterion is tantamount to defining a cluster. It may not be possible to say just what a cluster is in abstract terms, but it can always be defined constructively through statement of the criterion and an implementing algorithm.

Many criteria for clustering have been proposed and used. In some problems there is a natural choice, while in others almost any criterion might have status as a candidate. It should not be necessary to choose only a single criterion because clustering the data set several times with different criteria is a good way to reveal various facets of the structure. On the other hand, the expense of using cluster analysis prohibits trying out everything that is available.

In spite of the lack of a complete theory of recognition and classification, extensive study of these problems has led to some excellent treatments of the subject in the nonfuzzy way. Many of the books and papers in the brief reference list at the end of this chapter provide a systematically treated overview of classical (nonfuzzy) pattern recognition and classification. The purpose of this work is to give a systematic discussion of some important principles of pattern recognition and classification that involve fuzzy set theory.

The pattern space is essentially the domain that is defined by the data observed by a sensory device. A way of characterizing an object q is to assign to it the values of a finite set of parameters considered relevant to the object. A column vector $\mathbf{x}$ in the pattern space will have scalar elements

$$\mathbf{x} = (x_1, \ldots, x_n)^t \mid t = \text{transpose},$$

where each x_i, $1 \leq i \leq n$, represents the particular value associated with the ith dimension, or the value associated with feature i of object q. Namely, object q is represented by pattern $\mathbf{x}$ where

$$\mathbf{x} = f(q) = (f_1(q),\ldots,f_n(q)) = (x_1,\ldots,x_n)^t$$

and f_i, $1 \leq i \leq n$, is the measurement procedure associated with feature i. $f_i(q)$ is obviously the feature value. The first problem deals with feature extraction; this is the selection of a small set of measurement procedures $\{f_i\}_{i=1}^{n}$ and/or a set of primitives into which q is decomposed in order to formulate the mathematical description x. The other problem is that of cluster analysis (also referred to as classification theory categorization), which is concerned with the problem of partitioning a given set of entities into homogeneous and well-separated subsets.

The concepts of homogeneity and of separation can be made precise when a precise measure of dissimilarity between the entities is given. In his extensive "Review of Classification," Cormak (1971) notes that, "A classification, as usually understood, allocates entities to initially undefined classes so that individuals are in some sense close to one another." In *Cluster Analysis*, Duran and Odell (1974) state that the determination of clusters should be such that, "Those individuals which are assigned to the same cluster are *similar*, yet individuals from different clusters are different (*not similar*)."

Cluster analysis is generally divided into nondeterministic pattern classification and deterministic pattern classification. The first approach usually fails when it is impossible to represent the highly joint and scattered data set by some known distribution functions. The deterministic approach cannot be implemented in a satisfactory manner whenever the data set consists of highly joint and scattered patterns. Both approaches tend to separate the concept of discrimination from the notion of clustering.

Discrimination techniques usually begin with either some data base separated into *a priori* known categories or *a priori* conceptual distinctions, and proceed to develop algorithms by which to separate incoming data into those *a priori* categories, whereas clustering algorithms use *a priori* selection of a *measure of similarity* in order to find an inherent structure in the data. The term "homogeneous" is applied in the sense that all elements in some category are similar (according to some predefined measure of similarity) to each other and dissimilar to elements that belong to other categories.

It is very interesting to visualize the different aims and goals that various users of recognition and classification techniques have used. Once it is realized that these goals are so different, it is easier to understand why we have such a variety of classification techniques and also to avoid somewhat fruitless arguments as to which techniques are the *most* useful in some global sense. The goals might vary from data exploration to data reduction, model fitting, hypothesis testing, data investigation to find useful hypotheses or a true typology, prediction based on categories of data, and many more. It must be clear to the users of clustering algorithms how and why they should consider quite consciously the interaction between technique and discipline context. The techniques, based on certain mathematical models, influence the selection and preprocessing of the data. The explicit definition of what the users mean intuitively by similarity between elements becomes embodied in their algorithms, which in essence lead to an interpretation of the structure of the data. In short, implicit and explicit judgments regarding data variables, scaling, selective criteria, and selection of elements will affect the outcome of a clustering method.

It should be clear that *explicitness is subjective*; just because Dr. Smith chose one method of classification does not mean that he has found Truth. He exemplified his reasoning why he has grouped elements into a given category, and therein lies the value of explicitness, but metaphysical questions regarding the *nature of reality* cannot be settled by making *a priori* methodological choices before analyzing a given data set (Anderberg, 1973). The problem of selecting an "appropriate" method for a particular data set will remain problematic even when results are obtained through clustering analysis, and it is our feeling that they will not be resolved through methodological decisions made out of context.

It is necessary for the users of classification techniques to regard several important questions that should provide them with a guide for developing a clustering method:

1   What are the goals of the classification?
2   How are my goals affected by my particular research problems and specific paradigms?
3   What are the basic assumptions as to the format and underlying structure of the data set?
4   Is there available a training sample or a relatively undifferentiated set of elements whose underlying structure has to be explored by classification on the basis of some similarity relation between the elements?
5   Feature vector analysis:
    (a) How can I best determine the elements' variables?
    (b) How shall I scale and count the variables?
    (c) Can individual variables or elements be combined into a single measure of similarity?
6   Can classes of classification techniques be identified? What are the monetary advantages of each class in connection with the resources of the user and the specific answers to questions 1–5.
7   How can the results be evaluated?

The resolution of these questions will enable the users to understand the relations between their classification problem and the relevant technique of clustering.

1.4 SIMILARITY MEASURES AND CLUSTERING

Let us now discuss the general ideas behind the question: How shall we devise some manner of grouping?

Figure 1.1, with its emphasis on the imprecision of the environment, shows a general route from the data collection to its processing. A more detailed diagram is presented in Figure 1.2.

The scientist selects an inexact environment and a set of tools to measure the data set. This data set is then preprocessed, normalized, and inverted into an associative set of descriptive numbers. Then the data is clustered or discriminated and modifications are fed back to improve data collection by selecting slightly different environments, improving instruments, or modifying the preprocessing by removing variations irrelevant to the scientist's clustering interest.

Throughout the clustering process, it is apparent that similarity measures can be constructed not only between pairs of elements, but also from an individual object to an entire category. Three primary forms of category description can be discussed:

1. Extracting descriptive states (average, etc.) by weighting the element in several ways. The most common one is the probability of an element belonging to that category.
2. Identification numbers of each element in a given category (used largely in single-linkage and graph-theoretic methods).
3. Functional forms (normal distribution with a standardized mean and covariance matrix).

The main problem in all these examples is boundary specification. Boundaries between categories can be specified implicitly by utilizing the

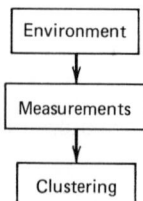

Figure 1.1 Environment and data processing.

Similarity Measures and Clustering

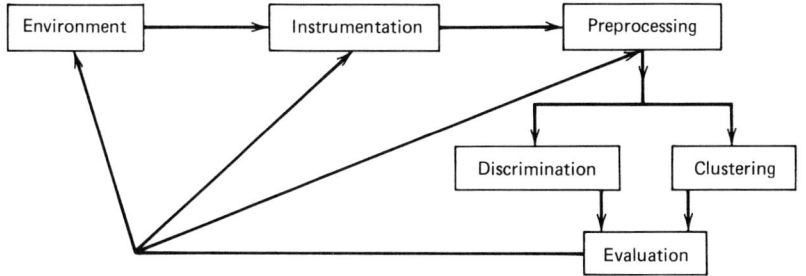

Figure 1.2 Context of clustering.

concept of "category position," or explicitly by parameterizing the boundaries themselves. Once we have resolved this problem, we can select the procedure for seeking optional partitions. However, these should be implemented with regard to the type of boundaries around the classification centers. These could be either soft boundaries (Figure 1.3a) or hard boundaries (Figure 1.3b).

If a description has hard boundaries, something either is or is not similar to a given concept. In soft boundaries we see a piece of information as varying in the degree of relevance to various concepts that are already internalized.

Even though a complete guiding philosophy of cluster analysis utilization is not available, some individual principles can be stated following Anderberg (1973):

1 Any given data set may admit many different but meaningful classifications. Each classification may pertain to a different aspect of the data. It is unnecessarily narrowing to seek a single "right" classification. Several different clustering procedures will be needed to discover multiple classifications. New insight and understanding might result from alternative classifications suggested through a cluster analysis and totally unexpected aspects of structure might be revealed in the process.

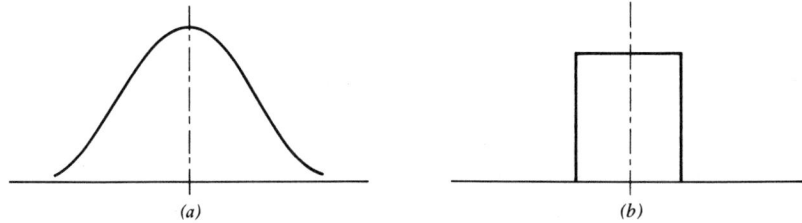

Figure 1.3 Classification boundaries. (a) Soft boundaries and (b) hard boundaries.

2   Cluster analysis is a device for suggesting hypotheses. The classification of data units or variables obtained from a cluster analysis procedure has no inherent validity. The analyst should not feel any pressure to embrace a particular classification, nor should he or she feel bound to the details of a classification that seems generally interesting. The worth of a particular classification and its underlying explanatory structure is to be justified by its consistency with known facts and without regard to the manner of its original generation.

3   A set of clusters is not itself a final result but only a possible outline. It follows then that there is little justification for using excessively detailed and expensive algorithms when results of the same general character can be achieved with low cost and intuitive procedures.

4   Cluster analysis methods involve a mixture of imposing a structure on the data and revealing that structure that actually exists in the data. The notion of finding "natural groups" tends to imply that the algorithm should passively conform like a wet tee shirt. Unfortunately, practical procedures involve fixed sequences of operations, which systematically ignore some aspects of structure while intensively dwelling on others. Such properties are often quite inadvertent and therefore are discovered only by observing their effects on data. To a considerable extent, a set of clusters reflects the degree to which the data set conforms to the structural forms embedded in the clustering algorithm.

5   In view of the preceding four points, it should come as no surprise that the results of a cluster analysis method rarely suggest a satisfactory structure for the total set of data. More commonly, one or more interesting clusters lead to inferences about part of the data. Along this line, certain clusters may be so natural and self-evident (once discovered) as to constitute "features" of the data likely to be revealed by almost any method. Rather than continue to recover the obvious, it is economical and relatively riskless to remove such features from the data set as they are found and concentrate attention on the more confused residue. Of course, adding or deleting variables alters the measurement space, and consequently such features may lose their distinctive character or alter their composition in response to such a change.

6   In the quest for clusters, two possibilities are often overlooked.
   (a)   The data may contain no clusters. When clustering variables, nearly complete independence or orthogonality would lead to such a result. When clustering data units, an absence of discriminating variables and a more or less uniform distribution of points in the measurement space would lead to a distinct lack of cohesion.

(b) The data may contain only one cluster. When clustering variables, nearly complete dependence or colinearity would lead to such a result. In the clustering of data units, the absence of discriminating variables combined with meaningful mutual association among all data units would give only one cluster.

Clearly, these two possibilities are opposite extremes with all other possibilities falling between them.

7   Prior knowledge about the population can be grossly misleading when applied to sample data, especially when the circumstances of data collection are not completely known. For example, suppose it is known that there are five groups in the population. If one group is rare or has been excluded systematically in data collection, there likely will be only four real groups in the sample. But forcing the clustering algorithm to produce five groups will create clusters that should be nonsense in this same framework of prior knowledge. The same result easily can arise when the chosen variables inadequately distinguish among the various classes or promote distinctions not relevant to the immediate purpose of the analysis.

In order to perform cluster analysis, it is necessary to select a criterion of similarity or dissimilarity. Similarity measure is an indication of how close two elements (or cluster descriptions) are to each other in a given set of features. Although features can be extremely dissimilar analytically, at the empirical level, the use of similarity measure assumes numerical comparability between features since it combines effects of individual features into a single number—the number indicating the similarity between two elements. This *a priori* assumption of numerical comparability allows relatively simple clustering processes that group elements by overall similarity, where a single measure of similarity can be used to determine the clustering.

When classification has been performed, the contribution of individual features can be determined by decomposing the sum of squared error into the contribution of each of the individual features. Thus we can first use the simplicity of single-similarity measures and then decompose each cluster into distributions of values for each feature, retaining distinctions between features while still being able to group elements as the basis of overall similarity. It is quite common to identify a number of different kinds of nonfuzzy measures of similarity following (Anderberg, 1973):

1   *Association*: Used primarily for information retrieval and biological taxonomy. Some forms of measures of association are also measures of correlation for binary variables.

2   *Correlation*: Primarily a function of the angle between a pair of object vectors, it can be normalized or left unnormalized. Normalized correlation depends only on the angle between the feature vectors. Normalized correlation is most useful as a measure of similarity when the magnitude of the vector does not affect the clustering. Unnormalized correlation weights the angle between two vectors by their magnitudes with respect to the present origin. Unnormalized correlation seems an unhappy compromise between normalized correlation and distance as a measure of similarity. In either case, a correlation is sensitive to the position of the origin, and values of correlation change if the position of the origin changes. Correlation is most useful as a measure of similarity when shape—the pattern of ratios of the various variables—is to be the prime determinant of similarity. For example, a person interested in the shape of human bodies might wish only to consider relative proportions of a group of people on whom he had collected measurements.

3 *Distances*: This measure, whose values run the reverse of similarity, may be determined in at least five ways:
    (a) *Absolute "City Block" Distance.* Less sensitive to a single variable having a large difference than are the following higher order measures of distance.
    (b) *Euclidean Distance.* Sensitive to the scaling of the variables that make up the feature vector but insensitive to the location of the origin. Geometric distances are used widely but can yield misleading results unless normalizations are performed properly.
    (c) *Mahalanobis Distance.* Weightings are data-derived in Mahalanobis distance by using a within-group covariance or correlation matrix. The within-group covariance matrix may change as the assignment of elements to classes changes. The choice of the within-group covariance matrix affects the value of Mahalanobis distance substantially. This matrix can be formed by considering the entire data set around the overall average, or by averaging together matrices around each class center (i.e., after subtracting the mean of the class), or by each class having a different matrix.
    (d) *General Weighted Distance.* Weightings can be derived either from an *a priori* evaluation of the importance of variables or from some other procedure, such as discriminant analysis.
    (e) *Minkowski Distance.* This measure of similarity includes the absolute and Euclidean forms of distance by setting the power of the expression $(x - y)$ appropriately. For example, if the power is 2, the Minkowski distance specializes to Euclidean distance.

4 *Probabilistic Measures*: Useful when it is appropriate to use the population statistics to modify weightings of variables. When the weighting is determined by the population statistics, however, these measures become quite sensitive to the choice of the population. The weighting depends on

Similarity Measures and Clustering

the value of the variable, as well as the particular variable used as in "weighted distance" above.

5 *Functional Measures of Similarity*: The value of the measure is a function of d, the distance from the other object or class.

Examples of these are given in Table 1.1.

We would like now to describe some classification methods in terms of the cluster description used, and with a rough description of the sorting procedure.

Table 1.1 Similarity Measures

Determination Method	Equations
Association	$S_{xy} = \dfrac{\eta_{xy}}{\eta_{xx} + \eta_{yy} - \eta_{xy}}$ where η_{ij} is the number of 1's that occur in x and y in the same variable
Correlation	$S_{xy} = \dfrac{x_1 y_1 + \cdots + x_\alpha y_\alpha}{\sqrt{(x_1^2 + \cdots + x_\alpha^2)(y_1^2 + \cdots + y_\alpha^2)}}$ where α is the number of variables
Distances	
Absolute	$D_{xy} = |x_1 - y_1| + \cdots + |x_\alpha - y_\alpha|$
Euclidean	$D_{xy} = \sqrt{(x_1 - y_1)^2 + \cdots + (x_\alpha - y_\alpha)^2}$
Weighted Euclidean	$D_{xy} = \sqrt{w_1(x_1 - y_1)^2 + \cdots + w_\alpha(x_\alpha - y_\alpha)^2}$ where w_i are variable weights
Mahalanobis	$D_{xy} = \sum_{i=1}^{\alpha} \sum_{j=1}^{\alpha} w_{ij}(x_i - y_i)(x_j - y_j)$ where the w_{ij} depend on the scatter matrix
Probabilistic (use $+1$ if $x_j = y_j$, use -1 if $x_j \neq y_j$)	$\sum_{j=1}^{\alpha} (\pm 1)\log\left[\dfrac{1}{\text{prob}_j(x_j, y_j)}\right]$
Functional	$S_{ij} = \dfrac{1}{(1 + D_{xy})}$

1.4.1 Clumping and Nearest Neighbor Techniques

In both clumping and nearest neighbor techniques, a class is described by a list of its elements, which are sorted on the basis of closeness to individual patterns. Conceptually, these are the simplest of the clustering techniques. Clumping techniques have been used widely in numerical taxonomy and, as commonly used, assume hierarchical data organization.

In clumping, distance can be measured to a cluster in terms of either the closest element in the cluster, the average element in the cluster, or the farthest element in the cluster. Clumping techniques become quite similar to partitioning techniques as the class description shifts from using the average of the members in the cluster. However, clumping procedures do not allow nonhierarchical groupings. Clumping procedures generally include:

1. Gathering the data and establishing a data matrix.
2. Calculating a similarity (or distance) matrix between all pairs of elements.
3. Joining the closest pair of patterns into a cluster.
4. Joining the next closest pair of elements. This can include the joining of a single element to the cluster created in a previous merger.
5. Continuing this process until all elements are together in a single cluster. To accomplish this requires a procedure for joining two clusters of patterns.

It should be noted that, in clumping, elements are never reshuffled and the hierarchical format is always maintained. Modifications of the simple procedure include removing "interior objects" from the class description, since interior elements do not affect discrimination between various categories. The nearest neighbor procedure, in essence, provides an empirical probability distribution, with the probability mass being concentrated at each of the objects.

1.4.2 Factorization

Factor analysis and discriminant analysis are based on a search for a method of obtaining new composite features by transformation of the axis of the measurement space. The methods under this class seek to find transformations that will maximize variations between categories and, at the same time, might take into account intraclass covariant structure. Thus by utilizing these methods, we are interested in separating, as much as possible, the average vectors of the various different categories by producing composite dimensions in which these means have maximum variation.

The methods are most appropriate where the data set (in clustering) or individual categories (in discrimination) are adequately characterized by the mean and covariance matrix. Principal components are particularly useful in obtaining a smaller number of composite variables from a large number of original variables.

1.4.3 Mixture Decomposition

In mixture decomposition (also in decision structures) the category of cluster is described by a probability distribution.

In most of these cases, the family of distributions is assumed and distributions of the devised form are fitted to the data set.

Decomposition techniques seem most appropriate where the degree of overlap among underlying clusters is assumed to be high, and where a paradigm provides sufficient guidance to support assumptions regarding the presumed form of the probability distribution of the data. Divisions found by decomposition are often similar to those found by partitioning when overlap between groups is small. Decomposition techniques are by far the most elegant mathematically, and yet often it is not clear whether the number of elements available to a user provides estimates of the distributional parameters sufficiently accurate to support the rigor of the techniques. An example of such structure is given in Figure 1.4.

1.4.4 Mode Analysis

Partitioning and mode seeking techniques describe a cluster or category in terms of an average position for each cluster or mode.

The distinguishing characteristic of these techniques is that they associate with every sample element a characteristic value proportional to the local concentration of elements or some other selected feature. This can be either a local maximum or any other designated mode structure.

One category can have several modes. Variation around the average can also be retained in the class description. The point of view for these techniques comes primarily from statistics, as opposed to probability theory. Partitioning can be divided into the following classes:

1 *Sequential Threshold Partitioning*: Here a cluster center is selected and all of the data subsets within some threshold distance of that center are placed in that cluster.

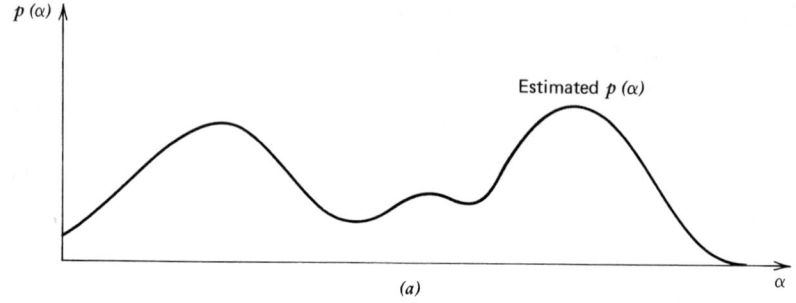

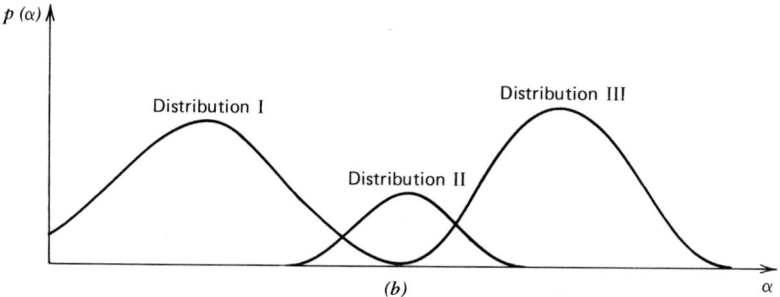

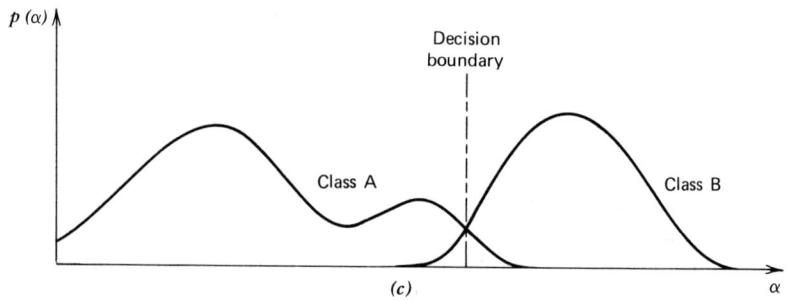

Figure 1.4 (*a*) Combined distribution, (*b*) decomposed distribution, mixture decomposition, and (*c*) decomposed distribution, decision structure.

2 *Parallel Threshold Partitioning*: In this method a number of cluster centers are chosen and a threshold value is set to determine the partitioning. Objects are introduced one at a time and associated with a cluster center. The location of that cluster center with which the data subset is associated is then modified, usually by updating a running average, as each object is added and the process is repeated, usually with a modified threshold.

3 *Parallel Partitioning*: As in parallel threshold partitioning, several cluster centers are chosen. The entire data set is then partitioned into disjoint subsets. Using the subsets, new cluster centers can be determined.

Mode analysis operates in a manner similar to partitioning, but it uses labeling information to influence the creation of new modes. Simply stated, a new mode for a category is created only when elements within that category are closer to a mode of a different category than they are to modes in their own category. Mode seeking techniques are most useful when a class has several modes, with objects from other classes lying in between the modes.

1.4.5 Adaptive Structures

Adaptive structures (also referred to as linear adaptive techniques or "corridor" methods) describe a set of methods where a category is described by characterizing the boundaries (soft or hard) between the categories.

These methods attempt to pass a plane through the data such that data are maximally distant from the plane while still having the plane cut somewhere through the middle of the data.

In the case of linear adaptive machines, the goal is to pass the plane between classes so that all elements in one class are separated from all elements in other classes. This approach is usually referred to as a learning machine.

1.4.6 Density Analysis

There are techniques in which the density of objects in a region around each object is used to indicate the relative weight of that element. Each object receives a value for its "characteristic function" that depends on the number of objects lying within a distance T of that object. Objects then are ordered by the value of their characteristic function, and clusters are built one object at a time *in an ordered sequence* around those objects having the maximum values. This approach can handle skewed distributions, since assignment to clusters is not primarily a symmetrical function of distance, but rather focuses on local object density and ordering of distances.

1.4.7 Graph Theoretic Models

Recent research [see Delattre and Hansen (1980) and Pavlidis (1977), for example] points out that data structure can be brought out by connecting close elements by some minimal spanning trees. It is pointed out that we can examine a minimal spanning tree and detect good links at which to cut a tree into various branches without having to assume a particular distributional form for the structure of the data.

1.4.8 Special Methods

There are several methods of clustering that are not as popular as the previously mentioned techniques but that are quite useful in specific cases. We list them in this section; the interested reader is invited to look into them in more detail by using the general reference list at the end of this chapter:

1. *Enumeration*: Evaluating all possible partitions of a data set into classes.
2. *Line Classification*: Fitting a line segment to a set of elements.
3. *Gravity Procedures*: Using gravitational force and its direction as a function of a distance between elements.
4. *Interactive Techniques*: Using interactive computer systems with graphic displays, or any other nonmachine system, to control clustering schemes.

1.5 APPLICATIONS

Pattern recognition and classification has evolved over the last decade from the use of computers for the recognition of handwritten characters to the recognition of sophisticated scenes. These include a wide range of applications within the scientific, industrial, medical, agricultural, governmental, and military communities, as mentioned before. We present here, for the sake of completeness, a more detailed list of applications:

1. *Scientific Applications*:
   (a) Astronomy: telescope resolution improvements and atmospheric degradation removal.
   (b) Geology—planetary exploration: crater counts, color analysis, robotics, topography, atmospheric measurements and analysis, landing site and related evaluations, and terrestrial geologic feature analysis and charting.
   (c) Geology—cartography and geodesy: mosaicing, surface model fitting, and maps (making and alteration).
   (d) Bubble chamber tracking and electron microscope crystallography.
   (e) Special image producing.
   (f) Satellite data analysis.
   (g) Sensing for life and data analysis on remote planets.

Applications

2. *Life and Behavioral Sciences*:
   (a) Anthropology.
   (b) Archeology.
   (c) Entomology.
   (d) Biology and botany: microbiology, ecology, and zoology.
   (e) Psychology: sociological aspects and criminological aspects.
   (f) Cybernetics.
   (g) Information management systems.
   (h) Education.
   (i) Communication.
3. *Industrial Applications*:
   (a) Character recognition.
   (b) Image controlled machines (process control).
   (c) Signature analysis.
   (d) Speech analysis.
   (e) Photographic recognition.
   (f) Mineral exploration (subsurface analysis).
   (g) Internal flow detection (X-ray and sonic).
   (h) Commercial photograph enhancement.
   (i) Filmed simulation.
   (j) Electronic toys design.
   (k) Automated cytology.
4. *Medical Applications*:
   (a) Microscopic examination and biomedical data: blood cell counting and blood tests, cancer cell identification and tests, neuron measurements, chromosome karyotyping, bone composition analysis, automated focusing and positioning, and brain-tissue studies.
   (b) Radioisotope examination.
   (c) X-ray examination and tomography: blood vessel thickness measurements, heart size measurements, breast cancer detection, intracranial blood vessel constriction detection, dental charting and analysis, bone structure analysis, pulmonary disease diagnosis, and skeletal structure analysis.
   (d) Electrocardiogram and vectorcardiogram analysis.
   (e) Electroencephalogram tracing and neurobiological signal processing.
   (f) Drug interaction.
   (g) Chromosome properties for genetic studies.
5. *Agricultural Applications*:
   (a) Crop analysis.
   (b) Soil evaluation.
   (c) Process control.
   (d) Earth-resource photography.

6 *Governmental Applications*:
   (a) Weather prediction: cloud tracking and water temperature measurements.
   (b) Public systems: traffic analysis and control, urban growth determination, smog detection and measurement, and air traffic radar data reduction.
   (c) Earth-resource data and remote sensing.
7 *Some Specific Military Applications*:
   (a) Aerial photography and remote sensing.
   (b) Sonar detection and classification.
   (c) Target identification.

BIBLIOGRAPHY

Anderberg, M. R. (1973). *Cluster Analysis for Applications*, Academic Press, New York.

Andrews, H. C. (1972). *Introduction to Mathematical Techniques in Pattern Recognition*, Wiley-Interscience, New York.

Bongard, M. (1970). *Pattern Recognition* (translated from Russian by T. Cheron; J. K. Hawkins, Ed.), Spartan Books, New York.

Braverman, D. (1964). Theories of pattern recognition, in *Advances in Communications Systems*, Vol. 1, (A. V. Balakrishnan, Ed.), Academic Press, New York.

Chen, C. H. (1973). *Statistical Pattern Recognition*. Hayden Book Company, Rochelle Park, NJ.

Chen, C. H. (Ed.) (1976). *Pattern Recognition and Artificial Intelligence*, Academic Press, New York.

Cormack, C. (1971). A review of classification, *J. Roy. Stat. Soc.*, Ser. A, **134**, 321–367.

Delattre, M., and Hansen, P. (1980). Bicriterion cluster analysis, *IEEE Trans. Pattern Anal. Mach. Intell.*, **PAMI-2** (4), July, 277–291.

Duda, R. O., and Hart, P. E. (1973). *Pattern Classification and Scene Analysis*, Wiley-Interscience, New York.

Duran, B. S., and Odell, P. L., (1974). *Cluster Analysis: A Survey*, Springer-Verlag, Heidelberg.

Fu, K. S. (1968). *Sequential Methods in Pattern Recognition and Machine Learning*, Academic Press, New York.

Fu, K. S. (Ed.) (1971). Pattern recognition and machine learning, in *Proc. Japan-US Semin. on the Learning Process in Control Syst.*, Nagoya, Japan, Aug., 1970; Plenum Press, New York.

Fu, K. S. (1974). *Systematic Methods in Pattern Recognition*, Academic Press, New York.

Fu, K. S. (Ed.) (1976a). *Digital Pattern Recognition*, Springer-Verlag, Heidelberg.

Fu, K. S. (Ed.) (1976b). *Applications of Syntactic Pattern Recognition*, Springer-Verlag, Heidelberg.

Fukunaga, K. (1972). *Introduction to Statistical Pattern Recognition*, Academic Press, New York.

Good, I. J. (1977). The botryology of botryology, in *Classification and Clustering* (J. Van Ryzin, Ed.), Academic Press, New York.

Bibliography

Grasselli, A. (Ed.) (1969). *Automatic Interpretation and Classification of Images*, Academic Press, New York.

Hartigan, J. A. (1975). *Clustering Algorithms*, Wiley, New York.

Ho, Y. C., and Agrawala, A. K. (1968). On pattern classification algorithms: Introduction and survey, *Proc. IEEE*, **56**, (12), Dec., 2101-2114.

Jordine, N., and Sibson, R. (1971). *Mathematical Taxonomy*, Wiley, London.

Kanal, L. N. (1968). Pattern recognition, in *Proc. IEEE Workshop on Pattern Recogn.*, Dorado, Puerto Rico; Thompson, Washington, D.C.

Klinger, A., Fu, K. S., and Kunii, T. (Eds.) (1976). *Data Structures, Computer Graphics and Pattern Recognition*, Academic Press, New York.

Lindsay, P. H., and Norman, D. A. (1972). *Human Information Processing*, Academic Press, New York.

Meisel, W. S. (1972). *Computer-Oriented Approaches to Pattern Recognition*, Academic Press, New York.

Mendel, J. M., and Fu, K. S. (Eds.) (1970). *Adaptive Learning and Pattern Recognition Systems: Theory and Applications*, Academic Press, New York.

Nagy, G. (1968). State of the art in pattern recognition, *Proc. IEEE*, **56**, May, 836-863.

Nilsson, N. J. (1965). *Learning Machines*, McGraw-Hill, New York.

Patrick, E. A. (1972). *Fundamentals of Pattern Recognition*, Prentice-Hall, Englewood Cliffs, NJ.

Pavlidis, T. (1977). *Structural Pattern Recognition*, Springer-Verlag, Heidelberg.

Rosenfeld, A. (1969). Picture processing by computer, *Comput. Surv.*, **1**, (3), 147-176.

Rosenfeld, A. (1976). *Digital Picture Analysis*, Springer-Verlag, Heidelberg.

Sebestyen, G. S. (1962). *Decision Making Processes in Pattern Recognition*, Macmillan, New York.

Sneath, P. H. A., and Sokal, R. R. (1974). *Numerical Taxonomy*, W. H. Freeman, San Francisco.

Sokal, R. R., and Sneath, P. H. A., (1963). *Principles of Numerical Taxonomy*, W. H. Freeman, San Francisco.

Späth, H. (1975). *Cluster-Analyse-Algorithmen*, R. Oldenbourg, Munich.

Tou, J. T., and Gonzalez, R. C. (1974). *Pattern Recognition Principles*, Addison-Wesley, Reading, PA.

Uhr, L. M. (1966). *Pattern Recognition Theory: Theory, Simulation, and Dynamic Models of Form Perception and Discovery*, Wiley, New York.

Ullman, J. R. (1973). *Pattern Recognition Techniques*, Crane Russak Co., New York.

Van Ryzin, J. (Ed.) (1977). *Classification and Clustering*, Academic Press, New York.

Vapnik, V. H., and Chezvonrnkis, A. Ya. (1974). *Theory of Pattern Recognition*, Science Press, Moscow.

Watanabe, S. (1969). *Methodologies of Pattern Recognition*, Academic Press, New York.

Watanabe, S. (Ed.) (1972). *Frontiers of Pattern Recognition*, Academic Press, New York.

Young, T. Y., and Calvert, T. W. (1974). *Classification, Estimation, and Pattern Recognition*, American Elsevier Publishing Company, New York.

CHAPTER TWO

The Algebra of Inexactness

2.1 INTRODUCTION

The "hard" sciences, such as engineering, chemistry, or physics, construct exact mathematical models of empirical phenomena, and then use these models to make predictions. Some aspects of the "real world" always escape such precise mathematical models and usually there is an elusive inexactness, a readjustment of structure, or an effect of observation over the original model that are so common in psychology, sociology, or any other of the "soft" sciences.

A central idea in the Platonic philosophy is that, in the real world, elements are perturbed by imperfection and thus, for example, there exists no element that is perfectly round. "Perfect notions" or "exact concepts" correspond to the sort of things envisaged in pure mathematics, while "inexact structures" are rampant in real life. It is our belief that inexact structures are rich enough in operations and properties to be of genuine use in constructing models for a wide variety of situations, and further that these mathematical properties will provide a practical guide for both philosophical and technical reasoning.

The purpose of this chapter is to introduce a unifying point of view to the notion of inexactness, based on the theory of fuzzy sets introduced by Zadeh in 1965 [2945]. The reason supporting the representation of inexact concepts by fuzzy sets has been given by Goguen [818]. Perhaps his most convincing argument was a representation theorem that says that any system satisfying certain axioms is equivalent to a system of fuzzy sets. Since the axioms are intuitively plausible for the system of all inexact concepts, the theorem allows us to conclude that inexact concepts can be represented by fuzzy sets. The representation theorem is a precise mathematical result in the theory of categories, so that a very precise sense is given to the concepts "system," "equivalent," and "represented."

Essentially, fuzziness is a type of imprecision that stems from a grouping of elements into classes that do not have sharply defined boundaries. Such classes—called fuzzy sets— arise, for example, whenever we describe ambiguity, vagueness, and ambivalence in mathematical models of empirical phenomena. Since certain aspects of reality always escape such models, the strictly binary (and even the ternary) approach to the treatment of physical phenomena is not always adequate to describe systems in the real world and the attributes of the system variables often emerge from an elusive fuzziness, a readjustment to context, or an effect of human imprecision. In many cases, however, even if the model is precise, fuzziness may be a concomitant of complexity. Systems of high cardinality are rampant in real life and their computer simulations require some kind of mathematical formulation to deal with the imprecise descriptions.

Ever since Zadeh introduced the idea of fuzzy set theory by utilizing the concept of membership grade, many researchers have been concerned with the properties and applications of fuzzy sets, as can be observed in the bibliography.

In what follows our attention is focused primarily on defining some of the basic notions within the conceptual framework of fuzzy set theory and exploring some of their elementary implications.

2.2 FUZZY SETS

The theory of fuzzy sets deals with a subset A of the universe of discourse X, where the transition between full membership and no membership is gradual rather than abrupt. The "fuzzy subset" has no well-defined boundaries where the universe of discourse (the universe) X covers a definite range of objects. Fuzzy classes of objects are often encountered in the real world. For instance, A may be the set of beautiful women in a town X, or A may be the set of long streets in town X. Traditionally, the grade of membership 1 is assigned to those objects that fully and completely belong to A, while 0 is assigned to objects that do not belong to A at all. The more an object x belongs to A, the closer to 1 is its grade of membership $\chi_A(x)$.

In abstract (or conventional, or nonfuzzy) set theory, the sets considered are defined as collections of objects having some very general property P; nothing special is assumed or considered about the nature of the individual objects. For example, we define a set X as the set of streets. Symbolically,

$$X = \{x \mid x \text{ is a street}\}.$$

Now what about the "class of long streets?" First of all, is it a set in the ordinary sense? Before we answer that, we may first ask: How "long" is a long street? Is a one-mile street a long street? If so, then is there any difference between a half-mile street and a one-mile long street? And so on. Frankly, we do not know how to answer these questions adequately from the information "long street" because the "class of long streets" does not constitute a set in the usual sense. In fact, most of the classes of objects encountered in the real physical world are of this fuzzy type; they do not have precisely defined criteria of membership. In such classes an object need not necessarily either belong or not belong to a class; there may be intermediate grades of membership. This is the concept of a fuzzy set, which is a "class" with a continuum of grades of membership.

Fuzzy set theory, introduced by Zadeh in 1965, is a generalization of abstract set theory. In other words, the former always includes the latter as a special case; definitions, theorems, proofs, and so on of fuzzy set theory always hold for nonfuzzy sets. Because of this generalization, fuzzy set theory has a wider scope of applicability than abstract set theory in solving problems that involve, to some degree, subjective evaluation.

Intuitively, a fuzzy set is a class that admits the possibility of partial membership in it. Let $X = \{x\}$ denote a space of objects. Then a fuzzy set A in X is a set of ordered pairs

$$A = \{(x, \chi_A(x))\}, \quad x \in X$$

where $\chi_A(x)$ is termed "the grade of membership of x in A." We assume for simplicity that $\chi_A(x)$ is a number in the interval [0, 1], with the grades 1 and 0 representing, respectively, full membership and nonmembership in a fuzzy set, as discussed before. We have assumed that an exact comparison is possible for the truths of any two inexact statements "$x \in A$" and "$y \in A$," and that the exact relation so obtained satisfies the minimal consistency requirements of transitivity and reflexivity; the ordering $\geq$ means "at least as true as" with $\leq$ denoting "not so true as."

The grades of membership reflect an "ordering" of the objects in the universe, induced by the predicate associated by A. It is interesting to note that the grade of membership value $\chi_A(x)$ of an object x in A can be interpreted as the degree of compatibility of the predicate associated with A and the object x. As we see later, it is also possible to interpret $\chi_A(x)$ as the degree of possibility that x is the value of a parameter fuzzily restricted by A.

In general, we distinguish three kinds of inexactness: generality, that a concept applies to a variety of situations; ambiguity, that it describes more than one distinguishable subconcept; and vagueness, that precise boundaries

Fuzzy Sets

are not defined. All three types of inexactness are represented by a fuzzy set: generality occurs when the universe is not just one point; ambiguity occurs when there is more than one local maximum of a membership function; and vagueness occurs when the function takes values other than just 0 and 1. Ambiguity and vagueness, therefore, depend upon there being some notion of nearness or continuity in the universe.

We now consider several examples of fuzzy sets.

EXAMPLE 2.2.1 In this example we consider the class of all real numbers that are much greater than 1. We can define this set as $A = \{x \mid x \text{ is a real number and } x \gg 1\}$. But it is not a well-defined set for the reasons mentioned before. This set may be defined subjectively by a membership function such as

$$\chi_A(x) = 0 \qquad \text{for } x \leq 1$$
$$\chi_A(x) = \frac{x-1}{x} \qquad \text{for } x > 1$$

The assignment of the membership function of a fuzzy set is subjective in nature and, in general, reflects the context in which the problem is viewed. Although the assignment of the membership function of a fuzzy set A is "subjective," it cannot be assigned arbitrarily. For example, it would be totally wrong to assign the membership function of Example 2.2.1 as

$$\chi_A(x) = \begin{cases} \dfrac{x-1}{x} & \text{for } x \leq 1 \\ 0 & \text{for } x > 1 \end{cases}$$

A function χ_A, such as

$$\chi_A(x) = \begin{cases} 0 & \text{for } x \leq 1 \\ e^{-(x-1)} & \text{for } x > 1, \end{cases}$$

which monotonically decreases as x increases, for $x > 1$ or

$$\chi_A(x) = \begin{cases} 0 & \text{for } x \leq 1 \\ 1 - e^{-1000(x-1)} & \text{for } x > 1, \end{cases}$$

which increases monotonically but is approximately equal to x for $x = 1.1$, should not be considered as they describe other classes of objects rather than the one required by Example 2.2.1. Functions such as these are called nonadmissible functions of the fuzzy set A. The function $\chi_A(x)$, such as defined in Example 2.2.1, and other functions, such as

1
$$\chi_A(x) = \begin{cases} 0 & \text{for } x \leq 1 \\ 1 - e^{-0.1(x-1)} & \text{for } x > 1 \end{cases}$$

2
$$\chi_A(x) = \begin{cases} 0 & \text{for } x \leq 1 \\ 1 - [\cosh(x-1)]^{-1} & \text{for } x > 1 \end{cases}$$

that satisfy the condition $0 \leq \chi_A(x) \leq 1$ for all $x \in X$ and that are consistent with the specification of the set, are called the admissible functions of A.

The *support* of A is the set of points in X at which $\chi_A(x)$ is positive. The *height* of A is the supremum of $\chi_A(x)$ over A. A *crossover point* of A is the point in X whose grade of membership in A is 0.5. A is *normal* if its height is 1; otherwise it is subnormal.

EXAMPLE 2.2.2 Let the universe be the interval [0, 120], with x interpreted as age. A fuzzy subset A of X labeled "old" may be defined by a grade of membership function such as

$$\chi_A(x) = \begin{cases} 0 & \text{for } 0 \leq x \leq 40 \\ \left[1 + \left(\dfrac{x-40}{5}\right)^{-2}\right]^{-1} & \text{for } 40 \leq x \leq 120. \end{cases}$$

In this example the support of old is the interval [40, 120], the height of old is effectively 1, and the crossover point of old is 45.

To simplify the representation of fuzzy sets, it is convenient to use the following notation.

A nonfuzzy finite set such as

$$X = \{x_1, x_2, \ldots, x_n\}$$

is expressed as

$$X = \sum_{j=1}^{n} x_j = x_1 + x_2 + \cdots + x_n$$

with the understanding that this is a representation of X as the union of its constituent singletons, with $+$ playing the role of the union rather than the arithmetic sum. Thus

$$x_j + x_k = x_k + x_j$$

and

$$x_j + x_j = x_j$$

for $j, k = 1, 2, \ldots, n$.

As a simple extension of this notation, a finite fuzzy set A on X is expressed as

$$A = \frac{\chi_A(x_1)}{x_1} + \cdots + \frac{\chi_A(x_n)}{x_n} = \sum_{j=1}^{n} \frac{\chi_A(x_j)}{x_j}$$

When X is not finite, we can use the notation

$$A = \int_X \frac{\chi_A(x)}{x}.$$

EXAMPLE 2.2.3 In the universe of discourse comprising $N = \{$positive integers$\}$, the fuzzy set A, labeled "integers approximately equal to 5," may be defined as

$$A = \frac{0.1}{2} + \frac{0.4}{3} + \frac{0.9}{4} + \frac{1.0}{5} + \frac{0.9}{6} + \frac{0.4}{7} + \frac{0.1}{8}.$$

EXAMPLE 2.2.4 In the universe of discourse comprising of $\mathbb{R} = \{$real numbers$\}$, the fuzzy set A, labeled "real numbers clustered around 5," may be defined by the grade of membership function

$$\chi_A(x) = \left\{1 + \left[\tfrac{1}{4}(x-5)\right]^2\right\}^{-1}$$

or as

$$A = \int_{\mathbb{R}} \frac{\left\{1 + \left[\tfrac{1}{4}(x-5)\right]^2\right\}^{-1}}{x}.$$

In many cases it is convenient to express the membership function of a fuzzy subset of the real line in terms of a standard function whose parameters may be adjusted to fit a specified membership function in an approximate fashion. Two such functions, the S-function and the π-function, are defined by

1
$$S(u; \alpha, \beta, \gamma) = 0 \qquad \text{for } u \leq \alpha$$

$$= 2\left(\frac{u-\alpha}{\gamma-\alpha}\right)^2 \qquad \text{for } \alpha \leq u \leq \beta$$

$$= 1 - 2\left(\frac{u-\gamma}{\gamma-\alpha}\right)^2 \qquad \text{for } \beta \leq u \leq \gamma$$

$$= 1 \qquad \text{for } u \geq \gamma$$

2
$$\pi(u; \beta, \gamma) = S\left(u; \gamma - \beta, \gamma - \frac{\beta}{2}, \gamma\right) \qquad \text{for } u \leq \gamma$$

$$= 1 - S\left(u; \gamma, \gamma + \frac{\beta}{2}, \gamma + \beta\right) \qquad \text{for } u \geq \gamma.$$

In $S(u; \alpha, \beta, \gamma)$, the parameter β, $\beta = (\alpha + \gamma)/2$, is the crossover point. In $\pi(u; \beta, \gamma)$, β is the bandwidth, that is, the separation between the crossover points of π, while γ is the point at which π is unity.

In some cases the assumption that χ_A is a mapping from X to $[0, 1]$ may be too restrictive, and it may be desirable to allow χ_A to take values in a lattice or, more particularly, in a Boolean algebra. For most purposes, however, it is sufficient to deal with the first two of the following hierarchy of fuzzy sets.

Definition 2.2.1

A fuzzy subset A of X is of *Type 1* if its membership function χ_A is a mapping from X to $[0, 1]$; and A is of Type k, $k = 2, 3, \ldots$, if χ_A is a mapping from X to the set of fuzzy subsets of Type $k - 1$. For simplicity it will always be understood that A is of Type 1 if it is not specified to be of a higher type.

EXAMPLE 2.2.5 Suppose that X is the set of all nonnegative integers and A is a fuzzy subset of X labeled "small integers." Then A is of Type 1 if the grade of membership of a generic element x in A is a number in the interval $[0, 1]$, for example,

$$\chi_{\text{small integers}}(x) = \left(1 + \left(\frac{x}{5}\right)^2\right)^{-1}, \qquad x = 0, 1, 2, \ldots.$$

Fuzzy Sets

On the other hand, A is of Type 2 if, for each x in X, $\chi_A(x)$ is a fuzzy subset of $[0, 1]$ of Type 1, for example, for $x = 10$,

$$\chi_{\text{small integers}}(10) = \text{low}$$

where low is a fuzzy subset of $[0, 1]$ whose membership function is defined by, say,

$$\chi_{\text{low}}(v) = 1 - S(v; 0, 0.25, 0.5), \quad v \in [0, 1],$$

which implies that

$$\text{low} = \int_0^1 \frac{1 - S(v; 0, 0.25, 0.5)}{v}.$$

If A is a fuzzy subset of X, then an α-level-set of A_α is a nonfuzzy set, denoted by A_α, which comprises all elements of X whose grade of membership in A is greater than or equal to α, namely,

$$A_\alpha = \{x \mid \chi_A(x) \geq \alpha\}.$$

A fuzzy set A may be decomposed into its level-sets through the resolution identity

$$A = \int_0^1 \alpha A_\alpha$$

or

$$A = \sum_\alpha \alpha A_\alpha$$

where αA_α is the product of a scalar α with the set A_α and $\int_0^1$ (or Σ_α) is the union of the A_α, with α ranging from 0 to 1.

The resolution identity may be viewed as the result of combining those elements in A that fall into the same level-set. More specifically, suppose that A is represented in the form

$$A = \frac{0.1}{2} + \frac{0.3}{4} + \frac{0.5}{7} + \frac{0.9}{8} + \frac{1}{9}.$$

Then A can be rewritten as

$$A = \frac{0.1}{2} + \frac{0.1}{4} + \frac{0.1}{7} + \frac{0.1}{8} + \frac{0.1}{9}$$

$$+ \frac{0.3}{4} + \frac{0.3}{7} + \frac{0.3}{8} + \frac{0.3}{9}$$

$$+ \frac{0.5}{7} + \frac{0.5}{8} + \frac{0.5}{9}$$

$$+ \frac{0.9}{8} + \frac{0.9}{9}$$

$$+ \frac{1}{9}$$

or

$$A = 0.1\left(\frac{1}{2} + \frac{1}{4} + \frac{1}{7} + \frac{1}{8} + \frac{1}{9}\right)$$

$$+ 0.3\left(\frac{1}{4} + \frac{1}{7} + \frac{1}{8} + \frac{1}{9}\right)$$

$$+ 0.5\left(\frac{1}{7} + \frac{1}{8} + \frac{1}{9}\right)$$

$$+ 0.9\left(\frac{1}{8} + \frac{1}{9}\right)$$

$$+ \frac{1}{9}$$

with the level-sets given by

$$A_{0.1} = 2 + 4 + 7 + 8 + 9$$

$$A_{0.3} = 4 + 7 + 8 + 9$$

$$A_{0.5} = 7 + 8 + 9$$

$$A_{0.9} = 8 + 9$$

$$A_1 = 9.$$

As can be seen easily, the resolution principle may provide a convenient way of generalizing various concepts associated with nonfuzzy sets to fuzzy sets. For example, let X be a linear vector space; then A is convex if and only if, for all $\lambda \in [0, 1]$ and all x_1, x_2 in X,

$$\chi_A(\lambda x_1 + (1 - \lambda)x_2) \geq \min(\chi_A(x_1), \chi_A(x_2))$$

In terms of the level-sets of A, A is convex if and only if the A_α are convex for all $\alpha \in [0, 1]$. Dually, A is concave if and only if

$$\chi_A(\lambda x_1 + (1 - \lambda)x_2) \leq \max(\chi_A(x_1), \chi_A(x_2)).$$

In many cases the fuzzy α-cuts can replace the classical grade-of-membership function as described above. For example, let A be a fuzzy set on X and A_α its α-cut. A_α can be written $\chi_A^{-1}([\alpha, 1])$; that is, the inverse image of the interval $[\alpha, 1]$. Let $\chi_{[\alpha, 1]}$ be the characteristic function of the interval $[\alpha, 1]$ in the universe $[0, 1]$.

$$\chi_{A_\alpha}(x) = \chi_{[\alpha, 1]}(\chi_A(x)), \quad \forall x \in X.$$

A fuzzy α-cut can be understood as the set of elements whose membership values are greater than "approximately α," that is, that belong to a fuzzy interval $(\tilde{\alpha}, 1]$, where $\chi_{(\tilde{\alpha}, 1]}$ is a continuous nondecreasing function from $[0, 1]$ to $[0, 1]$ and $\chi_{(\tilde{\alpha}, 1]}(1) = 1$. It is natural to extend $\chi_A(x)$ into

$$\chi_{A_{\tilde{\alpha}}}(x) = \chi_{(\tilde{\alpha}, 1]}(\chi_A(x)), \quad \forall x \in X$$

where $A_{\tilde{\alpha}}$ is the fuzzy α-cut of A. This can easily be derived from the extension principle; since $A_\alpha = \chi_A^{-1}([\alpha, 1])$, we also have $A_{\tilde{\alpha}} = \chi_A^{-1}((\tilde{\alpha}, 1])$. Hence we can extend χ_A^{-1}, viewed as a multivalued function, from $\mathcal{P}([0, 1])$ in X, and the extension principle can be generalized to deal with such functions. For example,

$$\chi_{A_{\tilde{\alpha}}}(x) = \sup_{\substack{\alpha \\ x \in \chi_A^{-1}([\alpha, 1])}} \chi^*_{(\tilde{\alpha}, 1]}([\alpha, 1])$$

where $\chi^*_{(\tilde{\alpha}, 1]}([\alpha, 1]) = \chi_{(\tilde{\alpha}, 1]}(\alpha)$. Note that $\{\alpha, x \in \chi_A^{-1}([\alpha, 1])\} = [0, \chi_A(x)]$; and since $\chi^*_{(\tilde{\alpha}, 1]}$ is nondecreasing and continuous, $\chi_{A_{\tilde{\alpha}}}(x) = \chi^*_{(\tilde{\alpha}, 1]}([\chi_A(x), 1])$, and thus $\chi_{A_{\tilde{\alpha}}}(x)$ is the greatest among the membership values of the sets $[\alpha, 1]$ whose images under χ_A^{-1} contain x.

2.2.1 Set-Theoretic Operations

Let A and B be fuzzy subsets of X. We can now discuss some basic operations performed on fuzzy sets.

1. Two fuzzy sets, A and B, are said to be *equal* ($A = B$) iff $\int_X \chi_A(x)/x = \int_X \chi_B(x)/x$ or $\forall x \in X$, $\chi_A(x) = \chi_B(x)$.
2. A is *contained* in B ($A \subseteq B$) iff $\int_X \chi_A(x)/x \leq \int_X \chi_B(x)/x$.
3. The *union* of fuzzy sets A and B ($A \cup B$) is defined by $A \cup B \triangleq \int_X (\chi_A(x) \vee \chi_B(x))/x$ where $\vee$ is the symbol for max.
4. The *intersection* of A and B is denoted by $A \cap B$ and is defined by $A \cap B \triangleq \int_X (\chi_A(x) \wedge \chi_B(x))/x$ where $\wedge$ is the symbol for min.

A justification of the choice of max and min was given by Bellman and Giertz [152]: max and min are the only operators f and g that meet the following requirements:

(a) The membership value of x in a compound fuzzy set depends on the membership value of x in the elementary fuzzy sets that form it, but not on anything else:

$$\forall x \in X, \quad \chi_{A \cup B}(x) = f(\chi_A(x), \chi_B(x))$$
$$\chi_{A \cap B}(x) = g(\chi_A(x), \chi_B(x))$$

(b) f and g are commutative, associative, and mutually distributive operators.

(c) f and g are continuous and nondecreasing with respect to each of their arguments. Intuitively, the membership of x in $A \cup B$ or $A \cap B$ cannot decrease when the membership of x in A or B increases. A small increase of $\chi_A(x)$ or $\chi_B(x)$ cannot induce a strong increase of $\chi_{A \cup B}(x)$ or $\chi_{A \cap B}(x)$.

(d) $f(u, u)$ and $g(u, u)$ are strictly increasing. If $\chi_A(x_1) = \chi_B(x_1) > \chi_A(x_2) = \chi_B(x_2)$, then the membership of x_1 in $A \cup B$ or $A \cap B$ is certainly strictly greater than that of x_2.

(e) Membership in $A \cap B$ requires more, and membership in $A \cup B$ less, than the membership in one of A or B:

$$\forall x \in X, \quad \chi_{A \cap B}(x) \leq \min(\chi_A(x), \chi_B(x))$$
$$\chi_{A \cup B}(x) \geq \max(\chi_A(x), \chi_B(x))$$

(f) Complete membership in A and in B implies complete membership in $A \cap B$. Complete lack of membership in A and in B implies complete lack of membership in $A \cup B$:

$$g(1,1) = 1, \qquad f(0,0) = 0$$

The above assumptions are consistent and sufficient to ensure the uniqueness of the choice of union and intersection operators.

Fung and Fu [736] also found max and min to be the only possible operators. They use a slightly different set of assumptions. They kept assumption (a) and added the following:

(b′) f and g are commutative, associative, and idempotent.
(c′) f and g are nondecreasing.
(g) f and g can be recursively extended to $m \geq 3$ arguments.
(h) $\forall x \in X, f(1, \chi_A(x)) = 1, g(0, \chi_A(x)) = 0$.

The interpretation of these axioms was given in the framework of group decision-making with a slightly more general valuation set.

5 The *complement* of A is denoted by $\bar{A}$ and is defined by

$$\bar{A} \triangleq \int_X \frac{1 - \chi_A(x)}{x}$$

Namely,

$$\forall x \in X, \qquad \chi_{\bar{A}}(x) = 1 - \chi_A(x)$$

The justification of this negation is more difficult than that of min and max. Natural conditions to impose on a complementation function h were proposed by Bellman and Giertz [152]:

(a) $\chi_{\bar{A}}(x)$ depends only on $\chi_A(x)$: $\chi_{\bar{A}}(x) = h(\chi_A(x))$.
(b) $h(0) = 1$ and $h(1) = 0$, to recover the usual complementation when A is an ordinary subset.
(c) h is continuous and strictly monotonically decreasing, since membership in $\bar{A}$ should become smaller when membership in A increases.
(d) h is involutive: $h(h(\chi_{\bar{A}}(x))) = \chi_A(x)$.

The above assumptions do not determine h uniquely, not even if we require in addition $h(\tfrac{1}{2}) = \tfrac{1}{2}$. However, $h(u) = 1 - u$ if the following requirement is introduced (Gaines [756]):

(e) $\forall x_1 \in X$, $\forall x_2 \in X$, if $\chi_A(x_1) + \chi_A(x_2) = 1$, then $\chi_{\bar{A}}(x_1) + \chi_{\bar{A}}(x_2) = 1$.

Instead of assumption (e), Bellman and Giertz have proposed the following very strong condition:

(f) $\forall x_1 \in X$, $\forall x_2 \in X$, $\chi_A(x_1) - \chi_A(x_2) = \chi_{\bar{A}}(x_2) - \chi_{\bar{A}}(x_1)$, which means that a certain change in the membership value in A should have the same effect on the membership in $\bar{A}$.

Assumltions (a), (b), and (f) entail $h(u) = 1 - u$.

However, there may be situations where assumptions (e) or (f) may not appear to be really necessary. Sugeno [2493] defines the λ-complement $\bar{A}^\lambda$ of A

$$\chi_{\bar{A}^\lambda}(x) = \frac{1 - \chi_A(x)}{1 + \lambda\chi_A(x))}, \quad \lambda \in (-1, +\infty)$$

λ-complementation satisfies assumptions (a)–(d).

6 The *product* of A and B is denoted by AB and is defined by

$$AB \triangleq \int_X \frac{\chi_A(x)\chi_B(x)}{x}$$

Thus A^α, where α is any positive number, should be interpreted as

$$A^\alpha \triangleq \int_X \frac{(\chi_A(x))^\alpha}{x}$$

Similarly, if α is any nonnegative real number such that $\alpha \sup_x \chi_A(x) \leq 1$, then

$$\alpha A \triangleq \int_X \frac{\alpha\chi_A(x)}{x}$$

As a special case the operation of *concentration* can be defined as

$$\operatorname{con}(A) \triangleq A^2$$

Fuzzy Sets

while that of *dilation* can be expressed by

$$\text{dil}(A) \triangleq A^{0.5}$$

7 The *bounded-sum* of A and B is denoted by $A \oplus B$ and is defined by $A \oplus B \triangleq \int_X 1 \wedge (\chi_A(x) + \chi_B(x))/x$ where $+$ is the arithmetic sum.

8 The *bounded-difference* of A and B is denoted by $A \ominus B$ and is defined by $A \ominus B \triangleq \int_X 0 \vee (\chi_A(x) - \chi_B(x))/x$ where $-$ is the arithmetic difference.

9 The *left-square* of A is denoted by ^2A and is defined by $^2A \triangleq \int_V \chi_A(x)/x^2$ where $V \triangleq \{x^2 \mid x \in X\}$. More generally,

$$^\alpha A \triangleq \int_V \frac{\chi_A(x)}{x^\alpha}$$

where $V \triangleq \{x^\alpha \mid x \in X\}$.

10 If $A_1,\ldots,A_k$ are fuzzy subsets of X, and $w_1,\ldots,w_k$ are nonnegative weights adding up to unity, then a *convex combination* of $A_1,\ldots,A_k$ is a fuzzy set A whose membership function is expressed by

$$\chi_A = w_1 \chi_{A_1} + \cdots + w_k \chi_{A_k} = \sum_{j=1}^{k} w_j \chi_{A_j}$$

where $+(\Sigma)$ denotes the arithmetic sum. The concept of a convex combination is useful in the representation of linguistic modifiers such as "essentially" and "typically," which modify the weights associated with the components of a fuzzy set.

11 If $A_1,\ldots,A_k$ are fuzzy subsets of $X_1,\ldots,X_k$, respectively, the Cartesian product of $A_1,\ldots,A_k$ is denoted by $A_1 \times \cdots \times A_k$ and is defined as a fuzzy subset of $X_1 \times \cdots \times X_k$ whose membership function is expressed by

$$\chi_{A_1 \times \cdots \times A_k}(x_1,\ldots,x_k) = \chi_{A_1}(x_1) \wedge \cdots \wedge \chi_{A_k}(x_k)$$

Equivalently,

$$A_1 \times \cdots \times A_k = \int_{X_1 \times \cdots \times X_k} \frac{\chi_{A_1}(x_1) \wedge \cdots \wedge \chi_{A_k}(x_k)}{x_1,\ldots,x_k}$$

EXAMPLE 2.2.6 Let $X = \{1, 2, 3, 4, 5, 6, 7\}$ and let

$$A = \frac{0.8}{3} + \frac{1}{5} + \frac{0.6}{6}$$

$$B = \frac{0.7}{3} + \frac{1}{4} + \frac{0.5}{6}.$$

Then

$$A \cup B = \frac{0.8}{3} + \frac{1}{4} + \frac{1}{5} + \frac{0.6}{6}$$

$$A \cap B = \frac{0.7}{3} + \frac{0.5}{6}$$

$$\bar{A} = \frac{1}{1} + \frac{1}{2} + \frac{0.2}{3} + \frac{1}{4} + \frac{0.4}{6} + \frac{1}{7}$$

$$AB = \frac{0.56}{3} + \frac{0.3}{6}$$

$$A^2 = \frac{0.64}{3} + \frac{1}{5} + \frac{0.36}{6}$$

$$0.5A = \frac{0.4}{3} + \frac{0.5}{5} + \frac{0.3}{6}$$

$$\text{con}(B) = \frac{0.49}{3} + \frac{1}{4} + \frac{0.25}{6}$$

$$\text{dil}(B) = \frac{0.84}{3} + \frac{1}{4} + \frac{0.7}{6}$$

$$A \oplus B = \frac{1}{3} + \frac{1}{4} + \frac{1}{5} + \frac{1}{6}$$

$$A \ominus B = \frac{0.1}{3} + \frac{1}{5} + \frac{0.1}{6}$$

$$^2A = \frac{0.8}{9} + \frac{1}{25} + \frac{0.6}{36}$$

$$^3A = \frac{0.8}{27} + \frac{1}{125} + \frac{0.6}{216}.$$

Fuzzy Sets

EXAMPLE 2.2.7 Let $X_1 = X_2 = \{2, 4, 6\}$ and let

$$A_1 = \frac{0.5}{2} + \frac{1}{4} + \frac{0.6}{6}$$

$$A_2 = \frac{1}{2} + \frac{0.6}{4}$$

Then

$$A_1 \times A_2 = \frac{0.5}{(2,2)} + \frac{1}{(4,2)} + \frac{0.6}{(6,2)} + \frac{0.5}{(2,4)} + \frac{0.6}{(4,4)} + \frac{0.6}{(6,4)}.$$

Following Dubois and Prade [574], suppose that the operations of union (U), intersection (I), and complement (C) are preserved in fuzzy sets. Namely,

1. $\forall a, b, c \in [0, 1]^8$: $U(a, I(b, c)) = I(U(a, b), U(a, c))$.
2. $I(a, U(b, c)) = U(I(a, b), I(a, c))$.
3. $U(a, a) = a$; $I(a, a) = a$.
4. $U(a, C(a)) = 1$; $I(a, C(a)) = 0$.

Now the following propositions hold:

Proposition 2.2.1 Assume the excluded-middle laws 4 hold for fuzzy sets; then union and intersection cannot be idempotent.

Proof Due to the strict decreasingness and continuity of C,

$$\exists p \in (0, 1): C(p) = p$$

Note that p cannot be 0 nor 1 since $C(0) = 1$ and $C(1) = 0$.

Now since the excluded-middle laws hold, $U(p, p) = 1$ and $I(p, p) = 0$, which is inconsistent with idempotency law 3 due to $p \notin \{0, 1\}$.   Q.E.D.

Proposition 2.2.2 Assume that the excluded-middle laws 4 hold for fuzzy sets and that both union and intersection satisfy:

5. $\forall a \in (0, 1), U(0, a) \neq 1$.
6. $\forall a \in (0, 1), I(1, a) \neq 0$.

Then $\cup$ and $\cap$ are no longer mutually distributive.

When we want to choose union and intersection to combine fuzzy sets, we have to give up either the excluded-middle laws or distributivity and idempotency.

Operators that satisfy excluded-middle laws are defined by

$$U(a, b) = \min(1, a + b)$$

$$I(a, b) = \max(0, a + b - 1)$$

$$C(a) = 1 - a.$$

$\tilde{P}(X)$, the set of subsets of X, is now a nondistributive complemented structure. Yager [2894] has shown that $\max(0, \cdot + \cdot - 1)$ was the only mapping generating an intersection operator that was associative, and that:

7   $\forall k: a - k \in [0, 1]; b + k \in [0, 1]; \forall a, b \in [0, 1]; I(a, b) = I(a - k, b + k)$.

This is so when condition 7 is interpreted as a linear compensation effect between membership values.

Another interesting operator is an averaging operator; an example of an averaging operator is the arithmetic mean

$$M(a, b) = \frac{a + b}{2}.$$

But others are worth considering, such as

$$M(a, b) = \frac{\min(a, b)}{1 - |a - b|}$$

and

$$M(a, b) = \frac{\max(a, b)}{1 + |a - b|}.$$

Another interesting combination of sets is the symmetrical difference, in which we only keep elements that belong to only one of two sets.

Consider $A \triangle B = B \triangle A$ where $\triangle$ denotes the symmetrical differences between fuzzy sets. The symmetrical differences can be generated by using mapping D from $[0, 1]^2$ to $[0, 1]$ such that:

8   $D(0, 1) = D(1, 0) = 1; D(0, 0) = D(1, 1) = 0.$
9   $\forall a, b \in [0, 1], D(a, b) = D(1 - a, 1 - b).$

A rather general form of symmetrical differences is $D(a, b) = f(g(a, b), g(1 - a, 1 - b))$, where f and g are any continuous mappings from $[0, 1]^2$ and $[0, 1]$, f being commutative and such that $f(g(0, 0), g(1, 1)) = 0$; $f(g(0, 1), g(1, 0)) = 1$. Examples of symmetrical differences of fuzzy sets are discussed by Dubois and Prade [574], and include:

1. $g(a, b) = \min(1 - a, b)$ and $f(a, b) = \max(a, b)$; hence $D(a, b) = \max(\min(1 - p, q), \min(p, 1 - q))$. This is consistent with the usual definition of $A \triangle B = (\bar{A} \cap B) \cup (A \cap \bar{B})$ translating $\cap$ and $\cup$ into "min" and "max." Note that $A \triangle B = (A \cup B) \cap (\bar{A} \cup \bar{B})$ also holds with these fuzzy set operators. Hence $\triangle$ is associative.

2. $g(a, b) = 1 - a + b$ and $f(a, b) = |a - b|/2$; hence $D(a, b) = |a - b|$, which is consistent with $A \triangle B = (\bar{A} \cap B) \cup (A \cap \bar{B})$ translating $\cap$ and $\cup$ into $\max(0, \cdot + \cdot - 1)$ and $\min(1, \cdot + \cdot)$. Here, $\triangle$ is no longer associative.

2.2.2 The Extension Principle

One of the basic ideas of fuzzy set theory, which provides a general extension of nonfuzzy mathematical concepts to fuzzy environments, is the extension principle.

The extension principle is a basic identity that allows the domain of the definition of a mapping or a relation to be extended from points in X to fuzzy subsets of X. More specifically, suppose that f is a mapping from X to Y and A is a fuzzy subset of X expressed as

$$A = \frac{\chi_1}{x_1} + \cdots + \frac{\chi_n}{x_n}.$$

Then the extension principle asserts that

$$f(A) = f\left(\frac{\chi_1}{x_1} + \cdots + \frac{\chi_n}{x_n}\right) \equiv \frac{\chi_1}{f(x_1)} + \cdots + \frac{\chi_n}{f(x_n)}.$$

Thus the image of A under f can be deduced from the knowledge of the images of $x_1, \ldots, x_n$ under f.

If we have an n-ary function f, which is a mapping from the Cartesian product $X_1 \times X_2 \times \cdots \times X_n$ to a universe Y such that $y = f(x_1, \ldots, x_n)$, and $A_1, \ldots, A_n$, which are n fuzzy sets in $X_1, \ldots, X_n$, respectively, which are characterized by a set of membership functions $\{\chi_{A_i}(x_i)\}_{i=1}^n$, then the extension principle allows us to induce from n fuzzy sets $\{A_i\}_{i=1}^n$ a fuzzy set

F on Y such that

$$\chi_F(y) = \sup_{\substack{x_1,\ldots,x_n \\ y=f(x_1,\ldots,x_n)}} \min[\chi_{A_1}(x_1),\ldots,\chi_{A_n}(x_n)]$$

$$\chi_F(y) = 0 \quad \text{if } f^{-1}(y) = \phi$$

EXAMPLE 2.2.8 Let $X = 1 + 2 + \cdots + 7$ and "small" be a fuzzy subset on X defined by "small" $= 1/1 + 1/2 + 0.8/3 + 0.5/4$. If we take f as the operation of squaring, then

$$f(\text{small}) = {}^2\text{small} = (\text{small}^2) = \frac{1}{1} + \frac{1}{4} + \frac{0.8}{9} + \frac{0.5}{16}.$$

EXAMPLE 2.2.9 Let $X_1 = X_2 = 1 + 2 + \cdots + 7$ and let

$$A_1 = \text{approximately } 2 = \frac{0.6}{1} + \frac{1}{2} + \frac{0.8}{3}$$

$$A_2 = \text{approximately } 4 = \frac{0.8}{3} + \frac{1}{4} + \frac{0.7}{5}.$$

Let $*$ be a binary operation defined on $X_1 \times X_2$ with values in Y, and let A_1 and A_2 be fuzzy subsets on X_1 and X_2, respectively, such that

$$A_1 = \chi_{A_{11}} x_{11} + \cdots + \chi_{A_{1n}} x_{1n}$$

and

$$A_2 = \chi_{A_{21}} x_{21} + \cdots + \chi_{A_{2n}} x_{2n}$$

The operation $*$ may be extended to fuzzy subsets of X_1 and X_2 by defining

$$A_1 * A_2 = \left(\sum_i \chi_{A_{1i}} x_{1i}\right) * \left(\sum_j \chi_{A_{2j}} x_{2j}\right)$$

$$= \sum_{i,j} (\chi_{A_{1i}} \wedge \chi_{A_{2j}})(x_{1i} * x_{2j}).$$

Fuzzy Sets

In our case we take ∗ as the arithmetic product of A_1 and A_2 and we get

$$A_1 * A_2 = (\text{approximately } 2 \times \text{approximately } 4)$$

$$= \left(\frac{0.6}{1} + \frac{1}{2} + \frac{0.7}{3}\right) \times \left(\frac{0.8}{3} + \frac{1}{4} + \frac{0.7}{5}\right)$$

$$= \frac{0.6}{3} + \frac{0.6}{4} + \frac{0.6}{5} + \frac{0.8}{6} + \frac{1}{8} + \frac{0.7}{9}$$

$$+ \frac{0.7}{10} + \frac{0.7}{12} + \frac{0.7}{15}.$$

If the support of A is a continuum, that is

$$A = \int_X \frac{\chi_A(x)}{x},$$

then the statement of the extension principle assumes the following form:

$$f(A) \triangleq f\left(\int_X \frac{\chi_A(x)}{x}\right) \triangleq \int_Y \frac{\chi_A(x)}{f(x)}$$

with the understanding that $f(x)$ is a point in Y and $\chi_A(x)$ is its grade of membership in $f(A)$, which is a fuzzy subset of Y.

In some applications it is convenient to use a modified form of the extension principle by decomposing A into its constituent level-sets rather than its fuzzy singletons. Thus on writing

$$A = \int_0^1 \alpha A_\alpha$$

where A_α is an α-level-set of A, the statement of the extension principle takes the form

$$f(A) = f\left(\int_0^1 \alpha A_\alpha\right) \equiv \int_0^1 \alpha f(A_\alpha)$$

when the support of A is a continuum, and

$$f(A) = f\left(\sum_\alpha \alpha A_\alpha\right) = \sum_\alpha \alpha f(A_\alpha)$$

when either the support of A is a countable set or the distinct level-sets of A form a countable collection.

Clearly, with the principle of extension, we can fuzzify any domain of mathematical reasoning using set theory. It should be clear however, that another way of fuzzifying a structure is just to replace nonfuzzy sets by fuzzy sets or by the family of their α-cuts in the framework of the structure.

EXAMPLE 2.2.10 (DUBOIS AND PRADE [574]): Let X be a metric space equipped with the pseudometric d, that is:

1. d is a mapping from X^2 to $\mathbb{R}^+$.
2. $d(x, x) = 0, \forall x$.
3. $d(x_1, x_2) = d(x_2, x_1), \forall x_1, \forall x_2$.
4. $d(x_1, x_3) \leq d(x_1, x_2) + d(x_2, x_3), \forall x_1, \forall x_2, \forall x_3$.

A *fuzzy distance* $\tilde{d}$ between fuzzy sets A and B on X is defined by

$$\forall \delta \in \mathbb{R}^+, \quad \chi_{\tilde{d}(A, B)}(\delta) = \sup_{\delta = d(u, v)} \left[\min(\chi_A(u), \chi_B(v)) \right]$$

$\tilde{d}(A, B)$ models a distance between fuzzy "spots." When A and B are connected subsets of X, $\tilde{d}(A, B)$ is an ordinary interval whose extremities are, respectively, the shortest and greatest distances between a point of A and a point of B. $\tilde{d}$ is a mapping from $[\tilde{\mathcal{P}}(X)]^2$ to the set of fuzzy sets on $\mathbb{R}^+$ (i.e., positive real fuzzy sets). $\tilde{d}(A, A)$ can be interpreted as the fuzzy diameter of A and $\chi_{\tilde{d}(A, A)}(0) = \text{hgt}(A)$. It is clear that we have $\tilde{d}(A, B) = \tilde{d}(B, A)$.

The question of knowing whether some triangular inequality still holds for $\tilde{d}$ is less straightforward. Let $A_\alpha, B_\alpha, C_\alpha$ be the α-cuts of three fuzzy sets on X. Let us denote by $u, v,$ and w, respectively, any element of $A_\alpha, B_\alpha,$ and C_α. The following inequalities hold:

$$\sup_{u, w} d(u, w) = d(u^*, w^*) \leq d(u^*, v) + d(v, w^*) \leq \sup_v (d(u^*, v) + d(v, w^*))$$

$$\sup_v (d(u^*, v) + d(v, w^*)) \leq \sup_{u, v, w} (d(u, v) + d(v, w))$$

$$\sup_{u, v, w} (d(u, v) + d(v, w)) \leq \sup_{u, v} d(u, v) + \sup_{v, w} d(v, w)$$

$$\inf_{u, v} d(u, v) + \inf_{v, w} d(v, w) \leq \inf_{u, v, w} (d(u, v) + d(v, w)).$$

The sides of the two last inequalities correspond to two different fuzzifications:

$$\chi_{\tilde{d}(A,B,C)}(\delta) = \sup_{\substack{u,v,w \\ \delta = d(u,v)+d(v,w)}} \left[\min(\chi_A(u), \chi_B(v), \chi_C(w))\right]$$

$$\chi_{\tilde{\Delta}(A,B,C)}(\Delta) = \sup_{\substack{\delta, \delta' \\ \Delta = \delta + \delta'}} \left\{ \min\left[\sup_{\substack{u,v \\ \delta = d(u,v)}} \min(\chi_A(u), \chi_B(v)), \sup_{\substack{v,w \\ \delta' = d(v,w)}} \min(\chi_B(v), \chi_C(w)) \right] \right\}.$$

In $\chi_{\tilde{d}(A,B,C)}$ we consider all the paths between A and C with a detour in B, while in $\chi_{\tilde{\Delta}(A,B,C)}$ the arrival point in B is no longer constrained to be the departure point. Hence we have, if A_α, B_α, and C_α are connected and without holes, $\forall \alpha$, $\tilde{d}(A,B,C) \subseteq \tilde{\Delta}(A,B,C)$.

Let us denote by $\tilde{d}(A,B,C)_\alpha$ and $\tilde{\Delta}(A,B,C)_\alpha$ the α-cuts of $\tilde{d}(A,B,C)$ and $\tilde{\Delta}(A,B,C)$, respectively. It is easy to show that $\forall \alpha \in (0,1]$,

$$\inf(\tilde{d}(A,C)_\alpha) \leq \inf(\tilde{d}(A,B,C)_\alpha)$$

$$\sup(\tilde{d}(A,C)_\alpha) \leq \sup(\tilde{d}(A,B,C)_\alpha)$$

$$\sup(\tilde{d}(A,C)_\alpha) \leq \sup(\tilde{\Delta}(A,B,C)_\alpha),$$

provided that these bounds are attained for some elements of X. Nothing can be said when comparing $\inf(\tilde{d}(A,C)_\alpha)$ and $\inf(\tilde{\Delta}(A,B,C)_\alpha)$. The main reason is that, when A, B, and C are ordinary sets, the triangular inequality does not hold for the minimal distances between the subsets. The two first inequalities can be interpreted as a triangular inequality for fuzzy distances. $\tilde{d}$ may also be viewed as a fuzzy measure of dissimilarity between fuzzy sets.

2.3 INEXACT ALGEBRA AND UNCERTAINTY

We may view fuzzy logic as a special kind of many-valued logic. In fuzzy logic the truth value of a formula, instead of assuming two values (0 and 1), can assume any value in the interval [0, 1] and is used to indicate the degree of truth represented by the formula. For example, let $P(x)$ represent "x is a large number compared with unity"; then the truth value of $P(10^5)$ and

$P(10^{-5})$ are certainly 1 and 0, respectively. As for $P(125)$, the truth value of it may be some value between 0 and 1, say 0.5.

We assume that well-formed formulas are defined to be exactly the same as those in two-valued logic. Letting $T(S)$ denote the truth value of a formula S, the evaluation procedure for a formula in fuzzy logic can be described as follows:

1. $T(S) = T(A)$ if $S = A$ and A is a ground atomic formula.
2. $T(S) = 1 - T(R)$ if $S = \overline{R}$.
3. $T(S) = \min[T(S_1), T(S_2)]$ if $S = S_1 \wedge S_2$.
4. $T(S) = \max[T(S_1), T(S_2)]$ if $S = S_1 \vee S_2$.
5. $T(S) = \inf[T(B(x)) \times D]$ if $S = (x) B$ and D is the domain of x.
6. $T(S) = \sup[T(B(x)) \times D]$ if $S = (Ex) B$ and D is the domain of x.

Note that, if D is a finite set, then conditions 5 and 6 become

5'. $T(S) = T(B(x_1) \cdots B(a_n))$ if $S = (x) B$ and x assume values $a_1, \ldots, a_n$.
6'. $T(S) = T(B(a_1) + \cdots + B(a_n))$ if $S = (Ex) B$ and x assume values $a_1, \ldots, a_n$.

The reader should note that two-valued logic is a special case of fuzzy logic; all the rules stated above are applicable in two-valued logic.

EXAMPLE 2.3.1 Consider $S = (P \vee Q) \wedge (\overline{R})$. Assume $T(P) = 0.5$, $T(Q) = 0.7$, and $T(R) = 0.8$. Then

$$T(S) = \min\{\max[T(P), T(Q)], 1 - T(R)\}$$

$$= \min[\max(0.5, 0.7), 1 - 0.8]$$

$$= \min(0.7, 0.2)$$

$$= 0.2.$$

Let $\mathcal{P}(X)$ be the set of ordinary subsets of X. $\mathcal{P}(X)$ is a Boolean lattice for $\cup$ and $\cap$, namely, L is a lattice if it is a partially ordered set such that

$$\forall a \in L, \forall b \in L, \quad \begin{cases} \exists! c \in L, c = \inf(a, b) \\ \exists! d \in L, d = \sup(a, b) \end{cases}$$

where inf and sup are, respectively, the greatest lower bound and the least upper bound.

L is complemented iff

$$\exists \mathbf{0} \in X, \exists \mathbf{1} \in X, \forall a \in L, \exists \bar{a} \in L, \quad \inf(a, \bar{a}) = \mathbf{0} \quad \text{and}$$

$$\sup(a, \bar{a}) = \mathbf{1}$$

and

$$\bar{a} \neq \mathbf{0} \quad \text{if } a \neq \mathbf{1}, \quad \bar{a} \neq \mathbf{1} \quad \text{if } a \neq \mathbf{0}$$

0 and **1** are, respectively, the least and the greatest element of L ($\forall a \in L$, $\inf(a, \mathbf{0}) = \mathbf{0}$, $\sup(a, \mathbf{1}) = \mathbf{1}$). A lattice with a **0** and a **1** is a complete lattice. L is distributive iff sup and inf are mutually distributive.

Let $\tilde{\mathcal{P}}(X)$ be the set of fuzzy subsets of X. Its structure can be induced from that of the real interval [0, 1]. [0, 1] is a pseudocomplemented distributive lattice where max and min play the role of sup and inf, respectively. The pseudocomplementation is complementation to 1, and $\tilde{\mathcal{P}}(X)$, the set of mappings from X to [0, 1], is a pseudocomplemented distributive lattice. More particularly, we have the following properties for $\cup$, $\cap$, and $\bar{\ }$:

1. *Commutativity*: $A \cup B = B \cup A$; $A \cap B = B \cap A$.
2. *Associativity*: $A \cup (B \cup C) = (A \cup B) \cup C$, $A \cap (B \cap C) = (A \cap B) \cap C$.
3. *Idempotency*: $A \cup A = A$, $A \cap A = A$.
4. *Distributivity*: $A \cup (B \cap C) = (A \cup B) \cap (A \cup C)$, $A \cap (B \cup C) = (A \cap B) \cup (A \cap C)$.
5. $A \cap \emptyset = \emptyset$, $A \cup X = X$
6. *Identity*: $A \cup \emptyset = A$, $A \cap X = A$.
7. *Absorption*: $A \cup (A \cap B) = A$, $A \cap (A \cup B) = A$.
8. *De Morgan's Laws*: $\overline{(A \cap B)} = \bar{A} \cup \bar{B}$, $\overline{(A \cup B)} = \bar{A} \cap \bar{B}$.
9. *Involution*: $\bar{\bar{A}} = A$.
10. *Equivalence Formula*: $(\bar{A} \cup B) \cap (A \cup \bar{B}) = (\bar{A} \cap \bar{B}) \cup (A \cap B)$.
11. *Symmetrical Difference Formula*: $(\bar{A} \cap B) \cup (A \cap \bar{B}) = (\bar{A} \cup \bar{B}) \cap (A \cup B)$.

The only law of ordinary fuzzy set theory that is no longer true is the excluded-middle law:

$$A \cap \bar{A} = \emptyset, \quad A \cup \bar{A} \neq X.$$

However,

$$\forall A, \forall x \in X, \quad \begin{cases} \min[\chi_A(x), \chi_{\bar{A}}(x)] \leq 0.5 \\ \max[\chi_A(x), \chi_{\bar{A}}(x)] \geq 0.5, \end{cases}$$

namely, A and $\bar{A}$ might overlap and $A \cup \bar{A}$ do not cover X exactly.

To check whether a formula A is valid or inconsistent in fuzzy logic, the simplest approach involves expanding A into conjunctive and disjunctive forms. Before defining these normal forms, we give some definitions.

A literal is a variable x_i or $\bar{x}_i$, the complement of x_i.
A clause is a disjunction of one or more than one literal.
A phrase is a conjunction of one or more than one literal.
A formula A is said to be in conjunctive normal form if $A = C_1 \wedge C_2 \wedge \cdots \wedge C_m$, $m \geq 1$ and every C_i, $1 \leq i \leq m$, is a clause.
A formula A is said to be in disjunctive normal form if $A = P_1 \vee P_2 \vee \cdots \vee P_m$, $m \geq 1$ and every P_i, $1 \leq i \leq m$, is a phrase.

In two-valued logic it can be shown that every formula can be expressed in conjunctive and disjunctive normal forms, owing to the existence of the distributive laws and De Morgan's laws. Since both of the laws mentioned above hold in fuzzy logic and since there is no syntactical difference between formulas in fuzzy logic and formulas in two-valued logic, we can easily see that formulas in fuzzy logic can also be expressed in conjunctive and disjunctive normal form.

Lemma 2.3.1 Let C be a clause. If C contains a complementary pair of literals, then C is fuzzily valid.

Proof Let $C = L_1 \vee L_2 \vee \cdots \vee L_n$. Assume L_i and L_j form such a complementary pair. Then $T(L_i) = 1 - T(L_j)$. For every possible assignment, either $T(L_i)$ or $T(L_j)$ will be greater than or equal to 0.5. Therefore, $\max[T(L_i), T(L_j)] \geq 0.5$ for all possible assignments. Since

$$T(C) = \max[T(L_1), T(L_2), \ldots, T(L_n)],$$

then

$$T(C) = \max[T(L_1), T(L_2), \ldots, T(L_i), \ldots, T(L_j), \ldots, T(L_n)]$$

$$\geq \max[T(L_i), T(L_j)] \geq 0.5.$$

Thus C is fuzzily valid. Q.E.D.

Lemma 2.3.2 Let C be a clause. If C is fuzzily valid, then C contains a complementary pair of literals.

Proof Consider the assignment in which every literal of C is assigned a truth value smaller than 0.5. $T(C)$ will be smaller than 0.5 under this assignment, so C is not fuzzily valid. This is contradictory to the assumption that C is fuzzily valid. Q.E.D.

Combining Lemmas 2.3.1 and 2.3.2, we have the following theorem.

Theorem 2.3.1 Let C be a clause. C is fuzzily valid iff C contains a complementary pair of literals.

Similarly, we can prove the following theorem concerning inconsistency in fuzzy logic.

Theorem 2.3.2 Let P be a phrase. P is fuzzily inconsistent iff P contains a complementary pair of literals.

Theorems 2.3.1 and 2.3.2 can be utilized to check the consistency of formulas in fuzzy logic. Suppose we want to see whether a formula A is valid. We can expand A into a conjunctive normal form:

$$A = C_1 \wedge C_2 \wedge \cdots \wedge C_n.$$

Then A is fuzzily valid iff every C_i is fuzzily valid. However, the fuzzy validity of a clause can be established through Theorem 2.3.1. Similarly, in case we want to check the fuzzy inconsistency of formula A, we can expand A into disjunctive normal form:

$$A = P_1 \vee P_2 \vee \cdots \vee P_n.$$

Then A is fuzzily inconsistent iff every phrase P_i is fuzzily inconsistent and the fuzzy inconsistency of a phrase can be established through Theorem 2.3.2.

It is quite clear that among the infinite number of distinct assignments of grade membership to the variables, there are a finite number of binary assignments (binary assignments of 0 or 1 to every variable).

It can be shown that the set of fuzzy functions compatible with a given Boolean function $F_B(x)$ is a sublattice of the lattice of fuzzy functions. It can also be shown that the "soft" algebra is a bounded, distributive, and symmetric lattice. Hence it is clear that a finite fuzzy lattice is isomorphic to a ring of sets. In the case of Boolean lattices, it is easy to prove that a finite lattice is Boolean if and only if it is isomorphic to the Boolean lattice of all subsets of a finite set. In any lattice we have the standard definition that an element ξ is an atom if $\xi \succ\!\!- 0$ (ξ covers 0) and a dual atom if $\xi -\!\!\prec 1$ (1 covers ξ), and thus all the elements that are immediate successors of the lower bound 0 are the atoms of the lattice.

Regarding Boolean algebra, it is obvious that the above definition of an atom and the classical definition that a nonzero element ρ in a Boolean algebra B is an atom iff $\rho x = \rho$ or $\rho x = 0$ for every x in B are identical.

Because the n-variable switching functions form a Boolean algebra of order 2^{2^n}, the algebra has 2^n atoms that are just the minterms. In fuzzy algebra, however, we prefer the original definition of an atom, which implies that the atom is the minimal nonzero element of the lattice. Thus an element ρ is an atom of the fuzzy lattice L iff $\rho \neq 0$ and for every x, if $x \leq \rho$, then $x = \rho$ or $x = 0$.

In order for an element to be the minimal element of L, it must be the conjunction of all fuzzy variables and their complements, and thus ensure the minimal grade membership required. Hence it is clear that the fuzzy lattice has only a single atom, which has the form of $\Pi_j x_j \bar{x}_j$. The fuzzy algebra over n variables $(x_1, x_2, \ldots, x_n)$ has as its atom the element $x_1 \bar{x}_1 x_2 \bar{x}_2 \ldots x_j \bar{x}_j \ldots x_n \bar{x}_n$. This notion is a fuzzy analog of a one-point set. A fuzzy algebra Z is said to be *atomic* if, for every nonzero element y in Z, there exists the atom $\Pi_j x_j \bar{x}_j \leq y$. In a similar way it is clear that, in a fuzzy algebra over n variables, there exists a set of 2^n dual atoms that consists of the fuzzy variables and their complements.

The measure of uncertainty or the degree of fuzziness of a fuzzy set is assumed to express, on a global level, the difficulty of deciding the belongingness of elements to the fuzzy set.

Let us consider a set I and a lattice L; any map from I to L is called an L-fuzzy set. Let us denote by $L(I)$ the class of all maps from I to L. It is possible to induce a lattice structure to $L(I)$ by the binary operations $\vee$ and $\wedge$ associating to any pair of elements f and g of $L(I)$ the elements $f \vee g$ and $f \wedge g$ of $L(I)$, defined point by point as

$$(f \vee g)(x) \equiv \text{l.u.b.} \{f(x), g(x)\}$$

$$(f \wedge g)(x) \equiv \text{g.l.b.} \{f(x), g(x)\}$$

where l.u.b. and g.l.b. denote, respectively, the least upper bound and the greatest lower bound of $f(x)$ and $g(x)$ in the lattice L.

For $L = [0, 1]$ we have

$$(f \vee g)(x) = \max\{f(x), g(x)\}$$

$$(f \wedge g)(x) = \min\{f(x), g(x)\}.$$

We try to introduce, for every element, or "fuzzy set" $f \in L(I)$, a measure of the degree of its "fuzziness." We require of this quantity, which we denote by $d(f)$, that is must depend only on the values assumed by f on I and satisfy at least the following properties.

1. $d(f)$ must be 0 if and only if f takes on I the values 0 or 1.
2. $d(f)$ must assume the maximum value if and only if f assumes always the value $\frac{1}{2}$.
3. $d(f)$ must be greater or equal to $d(f^*)$ where f^* is any "sharpened" version of f, that is any fuzzy set such that $f^*(x) \geq f(x)$ if $f(x) \geq \frac{1}{2}$ and $f^*(x) \leq f(x)$ if $f(x) \leq \frac{1}{2}$.

Let I be a finite set; this assumption and some others that we make in the following simplify the mathematical formalism but may be suitably weakened in future generalizations. We note, however, that the finiteness of I corresponds to a large class of actual situations.

By defining $H(f)$, in a way similar to Shannon entropy, as

$$H(f) = -K \sum_{i=1}^{N} f(x_i) \ln f(x_i)$$

where N is the number of elements in I and K is a positive constant, we get

$$H(f \vee g) = -K \sum_{i=1}^{N} \max[f(x_i), g(x_i)] \ln \max[f(x_i), g(x_i)]$$

$$H(f \wedge g) = -K \sum_{i=1}^{N} \min[f(x_i), g(x_i)] \ln \min[f(x_i), g(x_i)].$$

Now let $d(f) = H(f) + H(\bar{f})$ be defined as the entropy of the fuzzy set f_1 where

$$\bar{f}(x) = 1 - f(x)$$

satisfies the following properties:

$$\bar{\bar{f}} = f \quad \text{(involution law)}$$

$$\overline{f \vee g} = \bar{f} \wedge \bar{g} \quad \text{(De Morgan's laws)}$$

$$\overline{f \wedge g} = \bar{f} \vee \bar{g}$$

We explicitly note that $\bar{f}$, usually called the complement of f, is not the algebraic complement of f with respect to the lattice operations.

Clearly, $d(f) = d(\bar{f})$; moreover, $d(f)$ can be written using Shannon's function

$$S(x) = -x \ln x - (1-x)\ln(1-x)$$

as

$$d(f) = K \sum_{h=1}^{N} S(f(x_h))$$

and $d(f)$ satisfies properties 1 and 2. Property 3 is also satisfied. In fact if f^* is a sharpened version of f, we have by definition:

1. $0 \leq f^*(x) \leq f(x) \leq \frac{1}{2}$, for $0 \leq f(x) \leq \frac{1}{2}$.
2. $1 \geq f^*(x) \geq f(x) \geq \frac{1}{2}$, for $\frac{1}{2} \leq f(x) \leq 1$.

By the well-known property of Shannon's function $S(x)$, monotonically increasing in the interval $[0, \frac{1}{2}]$ and monotonically decreasing in $[\frac{1}{2}, 1]$ with a maximum at $x = \frac{1}{2}$, we immediately obtain from statements 1 and 2 above that, for any value of $f(x)$,

$$S(f^*(x)) \leq S(f(x)), \quad x \in I$$

and

$$d(f^*) \leq d(f).$$

If we assume that $K = 1/N$, we obtain the functional

$$v(f) = \frac{1}{N} \sum_{h=1}^{N} S(f(x_h)),$$

which we call the "normalized entropy." This name is appropriate because, taking the logarithm in base 2, we have

$$0 \leq v(f) \leq 1 \quad \text{for all } f \in L(I).$$

The functional $d(f)$ has been assumed to give a measure of the fuzziness of f; this quantity may also be considered as measuring an amount of information even if its meaning is different from the standard one of Shannon's information theory.

An interesting case is when the fuzzy set f is random, that is, when f is a map

$$f: \Omega \times I \to [0, 1]$$

such that, for any fixed x, $f(\xi, x)$ is a random variable with respect to a given probability space (Ω, F, p) where Ω is the nonempty set of sample points, F a σ-field of subsets of Ω, and p a probability measure. For any fixed ξ, $f(\xi, x)$ is a fuzzy set. Let us consider the case when Ω has only a finite number M of elements $\xi_1, \ldots, \xi_M$, which may occur with probabilities $p(\xi_1), \ldots, p(\xi_M)$; we may introduce an average fuzzy set $\langle f \rangle$ as

$$\langle f(x) \rangle \equiv \sum_{i=1}^{M} f(x, \xi_i) p(\xi_i).$$

In such a case, the entropy of the fuzzy set is itself a random variable; in fact, if the event ξ_i happens, we have the fuzzy set $f(\xi_i, x)$ whose entropy is $d_i(f)$. In this case it is meaningful to consider the average entropy given by

$$\sum_{i=1}^{M} p(\xi_i) d_i = \sum_{i=1}^{M} \sum_{j=1}^{N} p(\xi_i) S(f(\xi_i, x_j)).$$

Other proposed measures of fuzziness that satisfy the required properties 1–3 are:

1 $d(A) = F[\sum_{i=1}^{|X|} c_i f_i(\chi_A(x_i))]$ for finite X, where $c_i \in \mathbb{R}^+$, $\forall i$; f_i is a real-valued function satisfying (a) $f_i(0) = f_i(1) = 0$, (b) $f_i(x) = f_i(1 - x)$, $\forall x \in [0, 1]$, and (c) f_i is strictly increasing on $[0, \frac{1}{2}]$; F is a positive increasing function. For linear F the following holds:

$$d(A_1) + d(A_2) = d(A_1 \cup A_2) + d(A_1 \cap A_2).$$

2 $d(A) = \sum_{i=1}^{|M|} |\chi_A(x_i) - \chi_{A_{0.5}}(x_i)|$ where $A_{0.5}$ is the 0.5-cut of A.

Clearly, measure 2 is a special case of measure 1 when F is the identity and $\forall i$, $c_i = 1$, $f_i(x) = x$ when $x \in [0, 0.5]$.

It is interesting to note that the entropy is a special case of measure 1 when $F(x) = kx$, $k > 0$ and $\forall i$, $c_i = 1$ and $f_i(x) = -x \ln(x) - (1 - x)\ln(1 - x)$.

3 $d(A) = [1/p(X)] \int_X F[\chi_A(x) \, dp(x)]$ where $F(y) = F(1 - y)$, $y \in [0, 1]$; $F(0) = F(1) = 0$; and F is strictly increasing in $[0, 0.5]$.

Related measures of fuzziness are discussed in the next section, introducing the theory of possibility.

2.4 TOWARD THE THEORY OF POSSIBILITY

This section is devoted to the subject of possibility developed by Zadeh [2994] in 1978.

Let Y be a variable taking values in X; then a *possibility distribution*, Π_Y, associated with Y, may be viewed as a fuzzy constraint on the values that may be assigned to Y. Such a distribution is characterized by a *possibility distribution function* $\pi_Y: X \to [0, 1]$, which associates with each $x \in X$ the "degree of ease" or the possibility that Y may take x as a value.

In some cases the constraint on the values of Y is physical in origin; in many cases, however, the possibility distribution that is associated with a variable is epistemic rather than physical. A basic assumption in fuzzy logic is that such epistemic possibility distributions are induced by propositions expressed in a natural language. In more concrete terms, this assumption may be stated as follows.

Possibility Postulate If F is a fuzzy subset of X characterized by its membership function $\chi_F: X \to [0, 1]$, then the proposition "Y is F" induces a possibility distribution Π_Y that is equal to F. Equivalently, "Y is F" translates into the *possibility assignment equation* $\Pi_Y = F$, that is,

$$Y \text{ is } F \to \Pi_Y = F,$$

which signifies that the proposition "Y is F" has the effect of constraining the values that may be assumed by Y, with the possibility distribution Π_Y identified with F.

An important aspect of the concept of a possibility distribution is that it is nonstatistical in nature. As a consequence, if P_Y is a probability distribution associated with Y, then the only connection between Π_Y and P_Y is that impossibility (i.e., zero possibility) implies improbability but not vice versa. Thus Π_Y cannot be inferred from P_Y, nor can P_Y be inferred from Π_Y.

As in the case of probabilities, we can define joint and conditional possibilities. Thus if Y_1 and Y_2 are variables taking values in X_1 and X_2, respectively, then we can define the joint and conditional possibility distributions through their respective distribution functions:

$$\pi_{(Y_1, Y_2)}(x_1, x_2) = \text{poss}\{Y_1 = x_1, Y_2 = x_2\}, \quad x_1 \in X_1, x_2 \in X_2$$

and

$$\pi_{(Y_1|Y_2)}(x_1|x_2) = \text{poss}\{Y_1 = x_1 | Y_2 = x_2\}$$

where the last equation represents the conditional distribution function of Y_1 given Y_2.

If we know the distribution function of Y_1 and the conditional distribution function of Y_2 given Y_1, then we can construct the joint distribution function of Y_1 and Y_2 by forming the conjunction

$$\pi_{(Y_1, Y_2)}(x_1, x_2) = \pi_{Y_1}(x_1) \wedge \pi_{(Y_2|Y_1)}(x_2|x_1).$$

Several forms of the marginal possibility distribution function of Y_1 are:

1. $\pi_{Y_1}(x_1) = \sup_{x_2 \in X_2} \pi_{(Y_1, Y_2)}(x_1, x_2)$.
2. $\pi_{Y_1}(x_1) = \sup_{x_2 \in X_2} \pi_{(Y_1|Y_2)}(x_1|x_2)$.
3. $\pi_{Y_1}(x_1) = \pi_{(Y_1|Y_2)}(x_1, \tilde{x}_2(x_1))$.

Here $\tilde{x}_2(x_1)$ is the value of x_2 (for a given x_1) at which

$$\pi_{(Y_1|Y_2)}(x_1, x_2) = 1$$

if $\tilde{x}_2(x_1)$ is defined for every $x_1 \in X_1$.

Intuitively, case 1 represents the possibility of assigning a value to Y_1 as perceived by an observer (Y_1, Y_2) observing the joint possibility distribution $\Pi_{(Y_1, Y_2)}$. Case 2 represents the perception of an observer $(Y_1 | Y_2)$ observing only the conditional possibility distribution $\Pi_{(Y_1|Y_2)}$ and is unconcerned with or unaware of $\Pi_{(Y_2|Y_1)}$. Similarly, case 3 expresses the perception of an observer who assumes that x_2 is assigned a value (if it exists) that makes $\pi_{(Y_1|Y_2)}(x_1, x_2)$ equal to unity.

In relating π_{Y_1} to $\pi_{(Y_1|Y_2)}$ through the supremum operator, we are tacitly invoking the principle of maximal restriction, which asserts that, in the absence of complete information about Π_{Y_1}, we should equate Π_{Y_1} to the maximal possibility distribution that is consistent with the partial information about Π_{Y_1}.

Let π be a possibility distribution induced by a fuzzy set F in X. Let G be a nonfuzzy set of X. The possibility that x belongs to G is $\Pi(G)$ where

$$\Pi(G) = \sup_{x \in G} \chi_F(x) = \sup_{x \in G} \pi(G).$$

EXAMPLE 2.4.1 If p is a proposition of the form "$p \triangleq X$ is F," which translates into the possibility assignment equation

$$\Pi_{A(X)} = F$$

where F is a fuzzy subset of U and $A(X)$ is an implied attribute of X taking values in U, then the information conveyed by p, $I(p)$, may be identified with the possibility distribution, $\Pi_{A(X)}$, of the fuzzy variable $A(X)$. Thus the connection between $I(p)$, $\Pi_{A(X)}$, $R(A(X))$, and F is expressed by

$$I(p) \triangleq \Pi_{A(X)}$$

where

$$\Pi_{A(X)} = R(A(X)) = F.$$

The proposition "$p \triangleq$ Ted is young" translates into the possibility assignment equation

$$\Pi_{\text{age(Ted)}} = \text{young},$$

where χ_{young} is given. Then

$$I(\text{Ted is young}) = \Pi_{\text{age(Ted)}}$$

in which the possibility distribution function of age(Ted) is given by

$$\pi_{\text{age(Ted)}}(u) = 1 - S(u; 20, 30, 40), \quad u \in [0, 100].$$

From the definition of $I(p)$ it follows that, if $P \triangleq X$ is F and $q \triangleq X$ is G, then p is at least as informative as q, expressed as $I(p) \geq I(q)$, if $F \subset G$. Thus we have a partial ordering of the $I(p)$ defined by

$$F \subset G \Rightarrow I(X \text{ is } F) \geq I(X \text{ is } G),$$

which implies that the more restrictive a possibility distribution is, the more informative is the proposition with which it is associated.

The following is a list of translation rules for fuzzy propositions:

1 *Modifier Rule*: If

$$Y \text{ is } F \rightarrow \Pi_Y = F,$$

Toward the Theory of Possibility

then

$$Y \text{ is } mF \rightarrow \Pi_Y = F^+$$

where m is a modifier such as *not*, *very*, *more or less*, and so on, and F^+ is a modification of F induced by m. More specifically: If $m = $ not, then $F^+ = \bar{F} = $ complement of F, that is,

$$\chi_{F^+}(x) = 1 - \chi_F(x), \quad x \in X.$$

If $m = $ very, then $F^+ = F^2$, that is,

$$\chi_{F^+}(x) = \chi_F^2(x), \quad x \in X.$$

If $m = $ more or less, then $F^+ = \sqrt{F}$, that is,

$$\chi_{F^+}(x) = \sqrt{\chi_F(x)}, \quad x \in X.$$

2 *Conjunctive, Disjunctive, and Implicational Rules*: If

$$Y \text{ is } F \rightarrow \Pi_Y = F \quad \text{and} \quad Z \text{ is } G \rightarrow \Pi_Z = G$$

where F and G are fuzzy subsets of X_1 and X_2, respectively, then:
  (a) Y is F and Z is $G \rightarrow \Pi_{(Y,Z)} = F \times G$ where $\chi_{F \times G}(x_1, x_2) \stackrel{\Delta}{=} \chi_F(x_1) \wedge \chi_G(x_2)$.
  (b) Y is F or Z is $G \rightarrow \Pi_{(Y,Z)} = \bar{F} \cup \bar{G}$ where $\bar{F} \stackrel{\Delta}{=} F \times X_2$, $\bar{G} \stackrel{\Delta}{=} X_1 \times G$, and $\chi_{\bar{F} \cup \bar{G}}(x_1, x_2) = \chi_F(x_1) \vee \chi_G(x_2)$.
  (c) If Y is F, then Z is $G \rightarrow \Pi_{(Z|Y)} = \bar{F}' \oplus \bar{G}$ where $\Pi_{(Z|Y)}$ denotes the conditional possibility distribution of Z given Y, and the bounded sum $\oplus$ is defined by $\chi_{\bar{F}' \oplus \bar{G}}(x_1, x_2) = 1 \wedge (1 - \chi_F(x_1) + \chi_G(x_2))$.

3 *Quantification Rule*: If $X = \{x_1, \ldots, x_n\}$, Q is a quantifier such as *many*, *few*, *several*, *all*, *some*, *most*, and so on, and

$$Y \text{ is } F \rightarrow \Pi_Y = F,$$

then the proposition "QY are F" translates into

$$\Pi_{\text{count}(F)} = Q$$

where count(F) denotes the number of elements of X that are in F, namely,

$$\text{count}(F) = \sum_{i=1}^{n} \chi_i$$

4 *Truth Qualification Rule*: Let τ be a fuzzy truth-value, for example, *very true*, *quite true*, *more or less true*. Such a truth-value may be regarded as a fuzzy subset of the unit interval that is characterized by a membership function $\chi_\tau : [0, 1] \to [0, 1]$.

A truth-qualified proposition can be expressed as "Y is F is τ." The translation rule for such propositions can be given by

$$Y \text{ is } F \text{ is } \tau \to \Pi_Y = F^+$$

where

$$\chi_{F^+}(x) = \chi_\tau(\chi_F(x)).$$

In his approach to approximate reasoning, Zadeh [2978] uses various rules of inference. These rules, when applied, yield some very interesting conclusions drawn from imprecise propositions. Some of the rules of inference are:

1 *Projection Rule*: Consider a fuzzy proposition whose translation is expressed as

$$p \to \Pi_{(Y_1, \ldots, Y_n)} = F$$

and let $Y_{(s)}$ denote a subvariable of the variable $Y \triangleq (Y_1, \ldots, Y_n)$. Furthermore, let $\Pi_{Y_{(s)}}$ denote the marginal possibility distribution of $Y_{(s)}$, that is,

$$\Pi_{Y_{(s)}} = \text{proj}_{X_{(s)}} F$$

where X_i, $i = 1, \ldots, n$, is the universe of discourse associated with Y_i:

$$X_{(s)} = X_{i_1} \times \cdots \times X_{i_k}$$

and the projection of F on $X_{(s)}$ is defined by the possibility distribution function

$$\pi_{Y_s}(x_{i_1}, \ldots, x_{i_k}) = \sup_{x_{j_1}, \ldots, x_{j_m}} \chi_F(x_1, \ldots, x_n)$$

where $\bar{s} \triangleq (j_1,\ldots,j_m)$ is the index subsequence that is complementary to s, and χ_F is the membership function of F.

Now let q be a retranslation of the possibility assignment equation

$$\Pi_{Y_{(s)}} = \text{proj}_{X_{(s)}} F.$$

Then the projection rule asserts that q may be inferred from p, for example,

$$p \to \Pi_{(Y_1,\ldots,Y_n)} = F$$

$$\downarrow$$

$$q \leftarrow \Pi_{Y_{(s)}} = \text{proj}_{X_{(s)}} F.$$

2 *Conjunction Rule*: Consider a proposition p, which is an assertion concerning the possible values of, say, two variables X and Y that take values in U and V, respectively. Similarly, let q be an assertion concerning the possible values of the variables Y and Z, taking values in V and W. With these assumptions, the translations of p and q may be expressed as

$$p \to \Pi^p_{(X,Y)} = F$$

$$q \to \Pi^q_{(Y,Z)} = G.$$

Let $\bar{F}$ and $\bar{G}$ be, respectively, the cylindrical extensions of F and G in $U \times V \times W$. Thus

$$\bar{F} = F \times W$$

and

$$\bar{G} = U \times G.$$

Using the conjunction rule, one can infer from p and q a proposition that is defined by the following scheme:

$$r \to \Pi^p_{(X,Y)} = F$$

$$\underline{q \to \Pi^q_{(Y,Z)} = G}$$

$$r \leftarrow \Pi_{(X,Y,Z)} = \bar{F} \cap \bar{G}.$$

On combining the projection and conjunction rules, Zadeh obtains the *compositional rule of inference*, which includes the classical *modus ponens* as a special case.

More specifically, on applying the projection rule, we obtain the following inference scheme:

$$p \to \Pi^p_{(X,Y)} = F$$

$$q \to \Pi^q_{(Y,Z)} = G$$

$$\overline{r \leftarrow \Pi^r_{(X,Z)} = F \circ G}$$

where the composition of F and G is defined by

$$\chi_{F \circ G}(u, w) = \sup_v (\chi_F(u, v) \wedge \chi_G(v, w)).$$

In particular, if p is a proposition of the form "X is F" and q is a proposition of the form "if X is G, then Y is H," then we get

$$p \to \Pi_X = F$$

$$\underline{q \to \Pi_{(Y|X)} = \overline{G}' \oplus \overline{H}}$$

$$r \leftarrow \Pi_{(Y)} = F \circ (\overline{G}' \oplus \overline{H}).$$

EXAMPLE 2.4.2 Premises:

$$p \stackrel{\triangle}{=} \text{item } x \text{ is near pattern } D$$

$$q \stackrel{\triangle}{=} \text{item } y \text{ is near pattern } C.$$

Question: What is the distance between items x and y?

Let (x_1, x_2) and (y_1, y_2) be the coordinates of items x and y, respectively. Let $\Pi_{(x_1, x_2)}$ and $\Pi_{(y_1, y_2)}$ be the possibility distributions induced by p and q, namely, derived from the definition of the binary relation near.

Now, let d be the distance between items x and y, where

$$d = \left[(x_1 - y_1)^2 + (x_2 - y_2)^2\right]^{1/2}.$$

Using the extension principle, the possibility distribution function of d is given by

$$\Pi_d(z) = \sup_{u_1, v_1, u_2, v_2} \left[\Pi_{(x_1, x_2)}(u_1, v_1) \wedge \Pi_{(y_1, y_2)}(u_2, v_2) \right]$$

where $z = [(u_1 - u_2)^2 + (v_1 - v_2)^2]^{1/2}$ and the supremum is taken over all possible values of x_1, x_2, y_1, and y_2 under the definition of z.

This is basically a construction of a conceptual framework for inference from propositions whose meaning is not sharply defined. Through the use of fuzzy logic, the answer to a query is usually expressed in the form of a possibility distribution of one or more variables. In contrast to the conventional techniques of inference, the standards of precision in fuzzy logic are generally not high, but through the use of linguistic variables and linguistic approximation, these standards can be adjusted to fit the imprecision and unreliability of the information that is resident in the data.

Although in principle there is no connection between probabilities and possibilities, in practice the knowledge of possibilities conveys some information about the probabilities but not vice versa. Certainly, if an event is impossible, then it is also improbable. However, it is not true that an event that is possible is also probable. This rather weak connection between the two may be stated more precisely in the form of the *possibility/probability consistency principle*, namely: If Y is a variable that takes the values $y_1, \ldots, y_n$ with probabilities $p_1, \ldots, p_n$ and possibilities $\chi_1, \ldots, \chi_n$, respectively, then the *degree of consistency* of the probabilities $p_1, \ldots, p_n$ with the possibilities $\chi_1, \ldots, \chi_n$ is given by

$$\rho = \chi_1 p_1 + \chi_2 p_2 + \cdots + \chi_n p_n.$$

Intuitively, this means that, in order to be consistent with χ's, high probabilities should not be assigned to those values of Y that are associated with low degrees of possibility.

2.5 FUZZY RELATIONS AND ESTIMATED SIMILARITIES

If X is the Cartesian product of n universes of discourse $X_1, \ldots, X_n$, then an n-ary *fuzzy relation*, R, in X is a fuzzy subset of X. R may be expressed as the union of its consistent fuzzy singletons $\chi_R(x_1, \ldots, x_n)/(x_1, \ldots, x_n)$, that is,

$$R = \int_{X_1 \times \cdots \times X_n} \frac{\chi_R(x_1, \ldots, x_n)}{(x_1, \ldots, x_n)}$$

where χ_R is the membership function of R.

Common examples of (binary) fuzzy relations are: *much greater than*, *resembles*, *is relevant to*, and *is close to*. For example, if $X_1 = X_2 = (-\infty, \infty)$, the relation *is close to* may be defined by

$$\text{is close to} \triangleq \int_{X_1 \times X_2} \frac{e^{-\alpha|x_1 - x_2|}}{(x_1, x_2)}$$

where α is a scale factor. Similarly, if $X_1 = X_2 = 1 + 2 + 3 + 4$, then the relation *much greater than* may be defined by the relation matrix

R	1	2	3	4
1	0	0.4	0.8	1
2	0	0	0	0.8
3	0	0	0	0.4
4	0	0	0	0

in which the (i, j)th element is the value of $\chi_R(x_1, x_2)$ for the ith value of x_1 and the jth value of x_2.

The fuzzy relation "x is much greater than y" in N may be defined subjectively by a membership function such as

$$\chi(x, y) = \begin{cases} 0 & \text{if } x - y \leq 0 \\ \left[1 + 10(x - y)^{-1}\right]^{-1} & \text{if } x - y > 0. \end{cases}$$

EXAMPLE 2.5.1 Let $x = (\alpha_1, \alpha_2, \ldots, \alpha_n)$ and $y = (\beta_1, \beta_2, \ldots, \beta_n)$ be two points in the n-dimensional Euclidean space R^n. The fuzzy relation "y is in the neighborhood of x" is a fuzzy set in R^n, which may be defined subjectively by a membership function such as

$$\chi(x, y) = \frac{1}{\exp\|x - y\|}$$

where $\|x - y\| = [(\alpha_1 - \beta_1)^2 + (\alpha_2 - \beta_2)^2 + \cdots + (\alpha_n - \beta_n)^2]^{1/2}$.

Definition 2.5.1

The *composition* of two fuzzy relations A and B, denoted by $B \circ A$, is defined as a fuzzy relation in X whose membership function is related to those of A and B by

$$\chi_{B \circ A}(x, y) = \sup_{v} \left[\min[\chi_A(x, v), \chi_B(v, y)]\right], \qquad v, x, y \in X$$

Fuzzy Relations and Estimated Similarities

Definition 2.5.2 Let X and Y be two spaces of objects, and let h be a mapping from X to Y. Let B be a fuzzy set in Y with membership function $\chi_B(y)$. The fuzzy set A in X induced by the inverse mapping h^{-1} is defined by

$$A = \{(x, \chi_A(x)) \mid x = h^{-1}(y) \text{ and } \chi_A(x) = \chi_B(y), y \in B\}$$

Now consider the converse problem. Suppose A is a given fuzzy set in X and h is a mapping from X to Y. What is the membership function for fuzzy set B in Y that is induced by this mapping? To answer this question, we consider the following two cases separately.

Case A

The mapping h is one to one. The fuzzy set B in Y induced by the mapping h is defined as

$$B = \{(y, \chi_B(y)) \mid y = h(x) \text{ and } \chi_B(y) = \chi_A(x), x \in A\}.$$

Note that since h is one to one, for any $y \in Y$ where y is an image of some x in A, that is $y = h(x)$, there does not exist x' other than x such that $y = h(x')$. Hence $\chi_B(y)$ is uniquely defined by $\chi_A(x)$.

Case B

The mapping is many to one. In this case, the following ambiguity arises. Suppose x_1 and x_2 are two distinct points with different grades of membership in A and are mapped to the same point y in Y. Then what grade of membership in B should be assigned to y? To resolve this ambiguity, we agree to assign the larger of the two grades of membership to y. If there are n distinct points $x_1, x_2, \ldots, x_n$ in A with different grades of membership that are mapped to the same point y in Y, the grade of membership in Y will be the largest of these grades of membership.

To combine the above two cases, we give the following general definition.

Definition 2.5.3

Let X and Y be two spaces of objects, and let h be a mapping from X to Y. Let A be a fuzzy set in X with membership function $\chi_A(x)$. The fuzzy set B in Y induced by the mapping h is defined by

$$B = \left\{(y, \chi_B(y)) \mid y = h(x) \text{ and } \chi_B(y) = \max_{x \in T^{-1}(y)} \chi_A(x), x \in A\right\}$$

where $T^{-1}(y)$ is the set of points in X that are mapped into y by h.

If R is a relation from U to V (or, equivalently, a relation in $U \times V$) and S is a relation from V to W, then the *composition* of R and S is a fuzzy relation from U to W denoted by $R \circ S$ and defined by

$$R \circ S = \int_{U \times W} \frac{\sup_{v} \chi_R(u,v) \wedge \chi_S(v,w)}{(u,w)}.$$

If U, V, and W are finite sets, then the relation matrix for $R \circ S$ is the max-min product of the relation matrices for R and S.

If R is an n-ary fuzzy relation in $X_1 \times \cdots \times X_n$, then its *projection* (shadow) on $X_{i_1} \times \cdots X_{i_k}$ is a k-ary fuzzy relation R_q in X, which is defined by

$$R_q \stackrel{\Delta}{=} \operatorname{proj} R \text{ on } X_{i_1} \times \cdots \times X_{i_k} \stackrel{\Delta}{=} P_q R$$

$$\stackrel{\Delta}{=} \int_{X_{i_1} \times \cdots \times X_{i_k}} \frac{V_{x_{(\bar{q})}} \chi_R(x_1, \ldots, x_n)}{(x_{i_1}, \ldots, x_{i_k})}$$

where q is the index sequence $(i_1, \ldots, i_k)$; $x_{(q)} \stackrel{\Delta}{=} (x_{i_1}, \ldots, x_{i_k})$; $\bar{q}$ is the complement of q; and $V_{x_{(\bar{q})}}$ is the supremum of $\chi_R(x_1, \ldots, x_n)$ over the x's that are in $x_{(\bar{q})}$.

The domain of a fuzzy relation R is denoted by dom R and is a fuzzy set defined by

$$\chi_{\operatorname{dom} R}(x) = \sup_{y \in Y} [\chi_R(x,y)], \quad x \in X.$$

Similarly, the range of R is denoted by ran R and is defined by

$$\chi_{\operatorname{ran} R}(y) = \sup_{x \in X} [\chi_R(x,y)], \quad y \in Y.$$

The height of R is denoted by $h(R)$ and is defined by

$$h(R) = \sup_{x} \left\{ \sup_{y} [\chi_R(x,y)] \right\}.$$

A fuzzy relation is subnormal if $h(R) < 1$ and normal if $h(R) = 1$.

The support of R is denoted by $S(R)$ and is defined to be the exact subset of $X \times Y$ over which $\chi_R(x,y) > 0$.

Specifically, a *similarity region* S in X is a fuzzy relation that is:

1 *Reflexive*:

$$\chi_s(x,y) = 1, \quad \text{iff } x \equiv y.$$

Fuzzy Relations and Estimated Similarities

2 *Symmetric*:
$$\chi_s(x, y) = \chi_s(y, x), \quad \forall x, y \in X.$$

3 *Transitive*.
$$\chi_s(x, z) \geq \max_y \{\min[\chi_s(x, y), \chi_s(y, z)]\}$$

Similarly, a dissimilarity relation D can be defined as the complement of S with

$$\chi_d(x, y) = 1 - \chi_s(x, y), \quad x, y \in X$$

If $\chi_d(x, y)$ is interpreted as a distance function $d(x, y)$, then transitivity implies that

$$1 - d(x, z) \geq \max_y \{\min[1 - d(x, y), 1 - d(y, z)]\}$$

and since

$$\min[1 - d(x, y), 1 - d(y, z)] = 1 - \max[d(x, y, d(y, z)],$$

we can conclude that

$$d(x, z) \leq \max[d(x, y), d(y, z)], \quad \forall x, y, z \in X,$$

which implies the triangle inequality.

A *fuzzy restriction* is a fuzzy relation that acts as an elastic constraint on the values that may be assigned to a variable. More specifically, if Y is a variable that takes values in a universe of discourse X, then a fuzzy restriction $R(Y)$ on the values that may be assigned to Y is a fuzzy relation in X such that the assignment of a value y to Y requires a stretch of the restriction expressed by

$$\text{degree of stretch} = 1 - \chi_{R(Y)}(y)$$

where $\chi_{R(Y)}(y)$ is the grade of membership of y in $R(Y)$, namely,

$$g = u : \chi_{R(Y)}(y)$$

where g denotes a generic value of Y and $\chi_{R(Y)}(y)$ is the "degree of ease" with which y may be assigned to Y.

As a simple illustration, suppose that $X = 0 + 1 + 2 + \cdots$ and that Y is a variable labeled "small integer." Assume that the fuzzy set *small integer* is defined by

$$\text{small integer} = \frac{1}{0} + \frac{1}{1} + \frac{0.8}{2} + \frac{0.6}{3} + \frac{0.4}{4} + \frac{0.2}{5}$$

Then if g is a generic value of the variable "small integer" and we assign the value 2 to this variable, we have

$$g = 2 : 0.8$$

which implies that the fuzzy restriction labeled *small integer* must be stretched to the degree 0.2 to allow the assignment of the value 2 to the variable "small integer."

More generally, if $Y = (Y_1, \ldots, Y_n)$ is an *n*-ary variable taking values in the Cartesian product space

$$X = X_1 \times \cdots \times X_n,$$

then an *n*-ary fuzzy relation $R(Y_1, \ldots, Y_n)$ in X is a *fuzzy restriction* if it acts as an elastic constraint on the values that may be assigned to Y. An *n*-ary variable that is associated with a fuzzy restriction on the values that may be assigned to it is said to be an *n-ary fuzzy variable*.

Until now we have focused our attention upon fuzzy relations in the sense of fuzzy sets on a Cartesian product of universes. Other kinds of relations involving fuzziness include, for example, *tableaus* of fuzzy sets whose columns refer to the universes and the rows contain $(k + 1)$-tuples of labels of fuzzy sets.

Other extensions are interval-valued fuzzy relations discussed by Ponsard [2026] and fuzzy relations between fuzzy sets studied by C. L. Chang [342] and Sanchez [2276].

2.6 FUZZY STATISTICS

Ordinarily, imprecision and indeterminacy are considered to be statistical, random characteristics and are taken into account by the methods of probability theory. In real situations a frequent source of imprecision is not only the presence of random variables, but the impossibility, in principle, of operating with exact data as a result of the complexity of the system, imprecision of the constraints and objectives. At the same time, classes of objects that do not have clear boundaries appear in the problems; the imprecision of such classes is expressed in the possibility that an element does not only belong or not belong to a certain class but intermediate grades of membership are also possible.

Intuitively, a similarity is felt between the concepts of fuzziness and probability. The problems in which they are used are similar or coincide. These are problems in which indeterminacy is encountered due to random factors, inexact knowledge, or the theoretical impossibility or lack of necessity of obtaining exact solutions. The similarity is also underscored by the fact that the intervals of variation of the membership grade of fuzzy sets $\chi \in [0, 1]$ coincide. However, between the concepts of fuzziness and probability there are also essential differences.

Probability is an *objective* characteristic; the conclusions of probability theory can, in general, be *tested by experience*.

The membership grade is *subjective*, although it is natural to assign a lower membership grade to an event that, considered from the aspect of probability, would have a lower probability of occurrence. The fact that the assignment of a membership function of a fuzzy set is "nonstatistical" does not mean that we cannot use probability distribution functions in assigning membership functions. As a matter of fact, a careful examination of the variables of fuzzy sets reveals that they may be classified into two types: statistical and nonstatistical. The variable "magnitude of x" is an example of the former type. However, if we consider the "class of tall men," the "height of a man" can be considered to be a statistical variable. In this case, for instance, if a man is under five feet tall, we would not call him a tall man by "everybody's" standard, and we would assign to him a low grade of membership in the class of tall men. Similarly, if a man is over 7 feet, he certainly deserves a high grade of membership in the class of tall men.

The motivation for the development of fuzzy statistics is its philosophical and conceptual relation to subjective probability. In the subjectivistic view probability represents the *degree of belief* that a given person has in a given event on the basis of given evidence. This view (called also personalistic or judgmental probability) can best be described by an early article by James Bernoulli,[†] who defines probability as a degree of confidence in a proposition of whose truth we cannot be certain. His "degree of confidence" is identified with the probability of an event and depends on the knowledge that the individual has at his disposal. Thus it varies from individual to individual and can be best described as the art of guessing (*Ars conjectandi*).

The difficulty in applying subjective probability stems from the vagueness associated with judgments made via subjective analysis. The postulates of subjective probability cannot be applied, as is widely recognized, to many

---

[†] Bernoulli, J. (1713). *Ars Conjectandi*, Basel, Switzerland.

interesting and useful theories of modern science that are inexact. Subjective probability can be regarded as a personal way of treating objective views in that it is concerned with individual judgments. These judgments, however, are not additive, since human behavior often contradicts the assumption of subjective probability, that an individual is using additive measures in his or her criteria for evaluation.

We remove the restrictive device known as additivity and formulate the basic structure of a fuzzy statistic to be applied as our analytic model of investigation of imprecise data.

In the classical approach a *probability system* is a triple (Ω', S, P), where Ω' is an arbitrary set (the sample space that includes all possible outcomes), S is a set of events, and P is a real-valued function defined for each $A \in S$ such that:

1. $0 \leq P(A) \leq 1, \forall A \in S$.
2. $P(\Omega') = 1$.
3. If $A_1, A_2, \ldots$ is any sequence of pairwise disjoints sets in S, then

$$P\left(\bigcup_n A_n\right) = \sum_n P(A_n).$$

Definition 2.6.1

Let B be a Borel field (σ-algebra) of subsets of the real line Ω. A set function $\mu(\cdot)$ defined on B is called a *fuzzy measure* if it has the following properties:

1. $\mu(\Phi) = 0$ (Φ is the empty set).
2. $\mu(\Omega) = 1$.
3. If $\alpha, \beta \in B$ with $\alpha \subset \beta$, then $\mu(\alpha) \leq \mu(\beta)$.
4. If $\{\alpha_j \mid 1 \leq j < \infty\}$ is a monotone sequence, then

$$\lim_{j \to \infty} [\mu(\alpha_j)] = \mu\left[\lim_{j \to \infty}(\alpha_j)\right].$$

Clearly, $\Phi, \Omega \in B$; also, if $\alpha_j \in B$ and $\{\alpha_j \mid 1 \leq j < \infty\}$ is a monotonic sequence, then $\lim_{j \to \infty}(\alpha_j) \in B$. In the above definition 1 and 2 mean that the fuzzy measure is bounded and nonnegative, 3 means monotonicity (in a way similar to that in which finite additive measures are used in probability), and 4 means continuity. It should be noted that, if Ω is a finite set, then the continuity requirement can be deleted.

(Ω, B, μ) is called a fuzzy measure space. $\mu(\cdot)$ is the fuzzy measure of (Ω, B).

Fuzzy Statistics

The fuzzy measure μ is defined on subsets of the real line. Clearly, $\mu[\chi_A \geq T]$ is a nonincreasing, real-valued function of T when χ_A is the membership function of set A. Throughout our discussion we use ξ_T to represent $\{x \mid \chi_A(x) \geq T\}$ and $\mu(\xi_T)$ to represent this function assuming that our set A is well specified.

Let $\chi_A: \Omega \to [0, 1]$ and $\xi_T = \{x \mid \chi_A(x) \geq T\}$. The function χ_A is called a B-measurable function if $\xi_T \in B$, $\forall T \in [0, 1]$. Definition 2.6.2 defines the fuzzy expected value of χ_A when $\chi_A \in [0, 1]$. Extension of this definition when $\chi_A \in [a, b]$, $a < b < \infty$, is presented later.

Definition 2.6.2

Let χ_A be a B-measurable function such that $\chi_A \in [0, 1]$. The *fuzzy expected value* (FEV) of χ_A over a set A, with respect to the measure $\mu(\cdot)$, is defined as

$$\sup_{T \in [0, 1]} \{\min[T, \mu(\xi_T)]\}$$

where $\xi_T = \{x \mid \chi_A(x) \geq T\}$.

Now, $\mu\{x \mid \chi_A(x) \geq T\} = f_A(T)$ is a function of the threshold T. The actual calculation of FEV(χ_A) then consists of finding the intersection of the curves $T = f_A(T)$.

The intersection of the two curves will be at a value $T = H$, so that FEV(χ_A) = $H \in [0, 1]$. Figure 2.1 illustrates the above remarks.

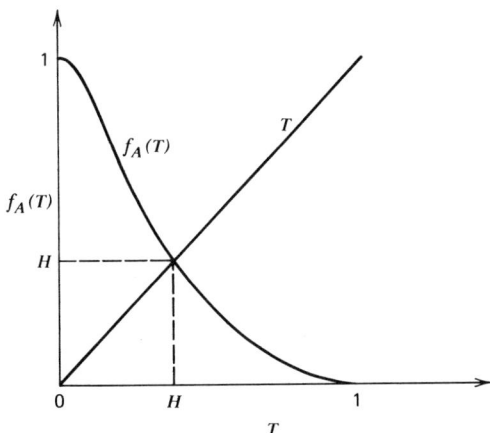

Figure 2.1 The evaluation of FEV(χ_A).

It should be noted that in Definition 2.6.2 we define the FEV only where χ_A is limted to the interval $[0, 1]$. However, if $\chi_A \in [0, 1]$ is replaced by $\eta = a\chi_A + b$, where a and b are real-valued quantities, Definition 2.6.2 does not hold. In this case we can extend the notion of the FEV by defining

$$\text{FEB}(\eta) = \text{FEV}(a\chi_A + b) \triangleq \sup_{T^* \in [b, a+b]} \{\min[T^*, \mu^*(\xi_{T^*})]\}$$

where T^* and μ^* are defined as $aT + b$ and $a\mu + b$, respectively, and $\xi_{T^*} = \{x \mid a\chi_A(x) + b \geq T^*\}$. It should be noted that, when we are discussing the $\text{FEV}(\chi_A)$, $\chi_A \in [0, 1]$, we are using the fuzzy measure μ in the evaluation of the FEV, whereas in the case of evaluating $\text{FEV}(\eta)$, $\eta = a\chi_A + b$, we are using a *function* of the fuzzy measure, μ^*, which, in general, is not a fuzzy measure. However, when $b \equiv 0$ and $a \equiv 1$, namely $\eta \equiv \chi_A \in [0, 1]$, it is clear that μ^* is a fuzzy measure, $\mu^* = \mu$, and thus the use of a function of a fuzzy measure in evaluating the FEV is consistent with the definition of the FEV.

Proposition 2.6.1 Let a and b be constants and let $\chi_A \to [0, 1]$. Then

$$\text{FEV}(a\chi_A + b) = b + a \cdot \text{FEV}(\chi_A).$$

Proof **Case 1** $a \geq 0$.

$$\text{FEV}(a\chi_A + b) = \sup_{T^* \in [b, a+b]} \{\min[T^*, \mu^*(\xi_{T^*})]\}$$

where $T^* = aT + b$, $\mu^* = a\mu + b$, and

$$\xi_T^* = \{x \mid a\chi_A(x) + b \geq T^*\}$$
$$= \{x \mid a\chi_A(x) + b \geq aT + b\} = \{x \mid \chi_A(x) \geq T\} = \xi_T$$

Thus

$$\text{FEV}(a\chi_A + b) = \sup_{T \in [0, 1]} \{\min[aT + b, a\mu(\xi_T) + b]\}$$

$$= b + \sup_{T \in [0, 1]} \{\min[aT, a\mu(\xi_T)]\}$$

$$= b + a \sup_{T \in [0, 1]} \{\min[T, \mu(\xi_T)]\} = b + a\,\text{FEV}(\chi_A).$$

Case 2 $a < 0$.

$$\text{FEV}(a\chi_A + b) = \sup_{T^* \in [a+b, b]} \{\min[T^*, \mu^*(\xi_{T^*})]\}$$

Fuzzy Statistics

where $T^* = aT + b$, $\mu^* = a\mu + b$, and

$$\begin{aligned}\xi_{T^*} &= \{x \mid a\chi_A(x) + b \leq T^*\} \\ &= \{x \mid a\chi_A(x) + b \leq aT + b\} \\ &= \{x \mid a\chi_A(x) \leq aT\} \\ &= \{x \mid \chi_A(x) \geq T\} = \xi_T.\end{aligned}$$

Thus

$$\begin{aligned}\text{FEV}(a\chi_A + b) &= \sup_{T \in [0,1]} \left\{ \min[aT + b, a\mu(\xi_T) + b] \right\} \\ &= b + \sup_{T \in [0,1]} \left\{ \min[aT, a\mu(\xi_T)] \right\} \\ &= b + a \inf_{T \in [0,1]} \left\{ \max[T\mu(\xi_T)] \right\} \\ &= b + a\,\text{FEV}(\chi_A). \quad\quad \text{Q.E.D.}\end{aligned}$$

It should be noted again that, when dealing with the FEV(η) where $\eta \notin [0, 1]$, we are not using a fuzzy measure in our evaluation but rather a function of the fuzzy measure, μ^*, which transforms μ under the same transformation that χ and T undergo to η and T^*, respectively.

EXAMPLE 2.6.1 Using the base variable "tall," consider a population X that consists of z people and y people who have been assigned grades of membership χ_1 and χ_2, respectively, in the set of tall people, A. We require that $0 \leq \chi_1 < \chi_2 < 1$. The fuzzy measure $\mu(\xi_T)$ is such that $T \leq \chi_1$, $\mu(\xi_T) = 1$, while if $T > \chi_2$, $\mu(\xi_T) = 0$. For $\chi_1 < T \leq \chi_2$, $\mu(\xi_T) = y/(z + y)$. We have therefore defined the measure as the proportion of the population whose grade of membership is larger than or equal to T over the entire population Ω. Thus

$$\mu(\xi_T) = \begin{cases} 0 & \text{if } \chi_2 < T \leq 1 \\ \dfrac{y}{z+y} & \text{if } \chi_1 < T \leq \chi_2 \\ 1 & \text{if } 0 \leq T \leq \chi_1. \end{cases}$$

The graph for the determination of the FEV(χ_A) for this population is shown in Figure 2.2.

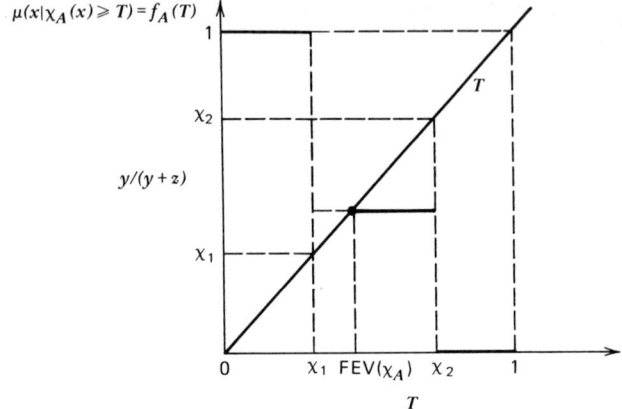

Figure 2.2 Graphical evaluation of FEV in the case $\chi_1 \leq y/(y+z) \leq \chi_2$.

Figure 2.2 is not the only possible value of FEV(χ_A). We can also easily distinguish two other cases:

1 FEV(χ_A) = χ_1 if $\chi_1 \geq y/(y+z)$.
2 FEV(χ_A) = χ_2 if $\chi_2 \leq y/(y+z)$.

We expect that the calculation of the FEV(χ_A) for the discrete case will involve finding the intersection of the line $T = T$ with a function $\mu(x \mid \chi_A(x) \geq T) = f_A(T)$ as shown in Figure 2.3.

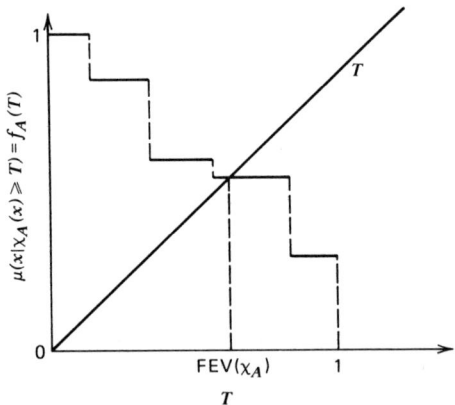

Figure 2.3 Graphical calculation of FEV(χ_A).

We now illustrate and obtain the relations between the FEV and the standard probabilistic expectation on one hand, and between the FEV and measures of central tendency, representing typical values of data sets, on the other.

Definition 2.6.3

A value $T \in [0, 1]$ is a *subtypical* value of a function ξ if $T \leq \mu(\xi_T)$, *supertypical* if $T \geq \mu(\xi_T)$, and typical if $T = \mu(\xi_T)$. Note that a typical value is both subtypical and supertypical.

Theorem 2.6.1 If a number T_1 is a subtypical value of two functions, ξ and η, and if for every $T \geq T_1$, $\xi_T = \eta_T$, then $\text{FEV}(\xi) = \text{FEV}(\eta)$.

Proof For $T < T_1$ we have

$$T < T_1 \leq \mu(\xi_{T_1}) \leq \mu(\xi_T)$$

and

$$T \leq T_1 \leq \mu(\eta_{T_1}) \leq \mu(\eta_T).$$

Thus

$$\min[T, \mu(\xi_T)] = T$$

and

$$\min[T, \mu(\eta_T)] = T$$

so that

$$\min[T, \mu(\xi_T)] = \min[T, \mu(\eta_T)].$$

For $T \geq T_1$ it is given that $\xi_T = \eta_T$ and therefore

$$\mu(\xi_T) = \mu(\eta_T)$$

and

$$\min[T, \mu(\xi_T)] = \min[T, \mu(\eta_T)].$$

Hence

$$\sup_{T \in [0,1]} \{\min[T, \mu(\xi_T)]\} = \sup_{T \in [0,1]} \{\min[T, \mu(\eta_T)]\}$$

and the result follows. Q.E.D.

Theorem 2.6.1 implies that values of a function that are below a subtypical value may be ignored in calculating the FEV. Similarly, we show in Theorem 2.6.3, which is the dual of Theorem 2.6.1, that values of a function that are above a supertypical value may be ignored in calculating the FEV.

The B-measurable function χ is called the compatibility function. In the case of linguistic variables, the numerical variables whose values constitute what may be called the base variable describe a fuzzy restriction on the meaning of the linguistic variable. This fuzzy restriction (which is clearly subjective) on the values of the base variable is characterized by the compatibility function χ, which associates with each value of the base variable a number in the interval [0, 1], representing its compatibility with the fuzzy restriction.

If the composition established in Definition 2.6.2 under the operation "max-min" seems like a pessimistic evaluation of the FEV, does a similar composition using the operation "min-max" act as an optimistic model?

The answer to the above question lies in the following theorem.

Theorem 2.6.2 Let $\chi: \Omega \to [0, 1]$ and let

$$\xi_T = \{x \mid \chi(x) \geq T\} \in B.$$

Then

$$\sup_{T \in [0,1]} \{\min[T, \mu(\xi_T)]\} = \inf_{T \in [0,1]} \{\max[T, \mu(\xi_T)]\},$$

where μ is a fuzzy measure.

Proof Let $\sup_{T \in [0,1]}\{\min[T, \mu(\xi_T)]\} = H$. Then

$$H = \max\left\{\sup_{T \in [0, H]} \{\min[T, \mu(\xi_T)]\}, \sup_{T \in (H, 1]} \{\min[T, \mu(\xi_T)]\}\right\}.$$

Since $\mu(\xi_T)$ increases as T decreases, we have

$$\sup_{T \in [0, H]} \{\min[T, \mu(\xi_T)]\} \leq H$$

and

$$\sup_{T \in (H, 1]} \{\min[T, \mu(\xi_T)]\} \leq \sup_{T \in [0, H]} \{\min[T, \mu(\xi_T)]\}.$$

Fuzzy Statistics

Thus

$$\sup_{T \in [0, H]} \{\min[T, \mu(\xi_T)]\} = H$$

and

$$\mu(\xi_H) \geq H, \mu(\xi_H^+) = \mu\{x \mid \chi(x) > H\} \leq H.$$

Assume

$$\inf_{T \in [0, 1]} \{\max[T, \mu(\xi_T)]\} = R$$

such that $R \neq H$. Then

$$R = \min\left\{ \inf_{T \in [0, R]} \{\max[T, \mu(\xi_T)]\}, \inf_{T \in (R, 1]} \{\max[T, \mu(\xi_T)]\} \right\},$$

since

$$\inf_{T \in (R, 1]} \{\max[T, \mu(\xi_T)]\} \geq R.$$

Clearly,

$$\inf_{T \in [0, R]} \{\max[T, \mu(\xi_T)]\} \geq \inf_{\overline{T} \in (R, 1]} \{\max[T, \mu(\xi_T)]\}$$

and thus we define

$$\inf_{T \in (R, 1]} \{\max[T, \mu(\xi_T)]\} = R.$$

Since the turning point between $\mu(\xi_T)$ and T is at H where

$$\mu(\xi_H) \geq H \quad \text{and} \quad \mu(\xi_H^+) \leq H,$$

it is quite obvious that R and H must be identical. Namely,

$$\sup_{T \in [0, 1]} \{\min[T, \mu(\xi_T)]\} = \inf_{T \in [0, 1]} \{\max[T, \mu(\xi_T)]\} = H.$$

Q.E.D.

Theorem 2.6.2 can also be derived via a set of interesting lemmas. We feel that the following is not just a repetition of a proof, but that the lemmas give more insight to the properties of the FEV.

Lemma 2.6.1 For every $T_1, T_2 \in [0, 1]$,

$$\min[T_1, \mu(\xi_T)] \leq \max[T_2, \mu(\xi_{T_2})].$$

Proof **Case 1** $T_1 \leq T_2$. In this case

$$\min[T_1, \mu(\xi_{T_1})] \leq T_1 \leq T_2 \leq \max[T_2, \mu(\xi_{T_2})].$$

Case 2 $T_2 < T_1$. In this case

$$\mu(\xi_{T_1}) \leq \mu(\xi_{T_2}).$$

Hence

$$\min[T_1, \mu(\xi_{T_1})] \leq \mu(\xi_{T_1}) \leq \mu(\xi_{T_2}) \leq \max[T_2, \mu(\xi_{T_2})] \quad \text{Q.E.D.}$$

from which the lemma follows.

Lemma 2.6.2

$$\sup_{T \in [0,1]} \{\min[T, \mu(\xi_T)]\} \leq \inf_{T \in [0,1]} \{\max[T, \mu(\xi_T)]\}.$$

Proof By Lemma 2.6.1 every member of $\{\min[T, \mu(\xi_T)], T \in [0, 1]\}$ is less than every member of $\{\max[T, \mu(\xi_T)], T \in [0, 1]\}$. The lemma follows.

Lemma 2.6.3 There is no real number s such that

$$\sup_{T \in [0,1]} \{\min[T, \mu(\xi_T)]\} < s < \inf_{T \in [0,1]} \{\max[T, \mu(\xi_T)]\}.$$

Proof Suppose such an s exists. We distinguish two cases: Case 1—$\mu(\xi_s) \leq s$; Case 2—$\mu(\xi_s) > s$. Each case leads to a contradiction.

Case 1

$$s < \inf_{T \in [0,1]} \{\max[T, \mu(\xi_T)]\} \leq \max[s, \mu(\xi_s)] = s.$$

Fuzzy Statistics

Case 2

$$s = \min[s, \mu(\xi_s)] \leq \sup_{T \in [0,1]} \{\min[T, \mu(\xi_T)]\} < s.$$

Since both cases lead to a contradiction, the lemma follows. Q.E.D.

Lemma 2.6.4 It is not true that

$$\sup_{T \in [0,1]} \min[T, \mu(\xi_T)]\} < \inf_{T \in [0,1]} \{\max[T, \mu(\xi_T)]\}.$$

Proof If this inequality held, there would be a real number between the left- and right-hand members, contradicting Lemma 2.6.3. Q.E.D.

Hence, Theorem 2.6.2 follows from Lemmas 2.6.2 and 2.6.4.

Theorem 2.6.3 If a number T_2 is a supertypical value of two functions, ξ and η, and if, for every $T \leq T_2$, $\xi_T = \eta_T$, then FEV(η) = FEV(ξ).

Proof For $T > T_2$ we have

$$T > T_2 \geq \mu(\xi_{T_2}) \geq \mu(\xi_T)$$

and

$$T > T_2 \geq \mu(\eta_{T_2}) \geq \mu(\eta_T).$$

Thus

$$\max[T, \mu(\xi_T)] = \max[T, \mu(\eta_T)].$$

For $T \leq T_2$ it is given that $\xi_T = \eta_T$ and therefore

$$\mu(\xi_T) = \mu(\eta_T)$$

and

$$\max[T, \mu(\xi_T)] = \max[T, \mu(\eta_T)].$$

Hence

$$\inf_{T \in [0,1]} \{\max[T, \mu(\xi_T)]\} = \inf_{T \in [0,1]} \{\max[T, \mu(\eta_T)]\}$$

and by use of Theorem 2.6.2 the result follows. Q.E.D.

An average is a value that is typical of a set of data. Since such typical values tend to lie centrally within a set of data arranged according to magnitude, averages are also known as *measures of central tendency*. Several types of averages can be defined, the most common being the arithmetic mean, the median, the mode, the geometric mean, and the harmonic mean.

Before discussing the relations of the FEV to the mean and to the median, we evaluate the use of the concept by constructing a combinatorial scheme to generalize Example 2.6.1.

Assume a finite set of data points where there are $n + 1$ distinct levels of compatibility such that

$$0 \leq \chi_1 < \chi_2 < \cdots < \chi_{n+1} \leq 1$$

which implies n distinct levels of fuzzy measure $\mu(\xi_T)$, excluding 0 and 1.

Clearly, the two sets representing

$$\{\chi_i\}_{i=1}^{n+1} \quad \text{and} \quad \{\mu_j(\xi_T)\}_{j=1}^{n}$$

are in increasing order, and we trivially exclude any permutation in each of the sets. Thus the set representing the union of these two sets has $(2n + 1)$ elements; in order to find all possible arrangements of these elements we can view the problem of finding the number of arrangements as follows.

Find the number of n-arrangements of $(2n + 1)$ objects, of which exactly $(n + 1)$ are alike of one kind and exactly n are alike of the second kind.

To find the solution, we let γ be the required number of arrangements. Let the $(2n + 1)$ objects be labeled $x_1, \ldots, x_{2n+1}$ with the $(n + 1)$ alike of one kind labeled $x_1, \ldots, x_{n+1}$, and the n alike of the second kind labeled $x_{n+2}, \ldots, x_{2n+1}$. Then there will be $(2n + 1)!$ n-arrangements of these (now distinct) labeled objects, such that we can set

$$(2n + 1)! = \gamma n! (n + 1)!$$

since, for each of the γ required arrangements, the first $(n + 1)$ alike, being labeled, can be interchanged in $(n + 1)!$ ways and the n alike of the second kind, being labeled, can be interchanged in $n!$ ways. Thus $\gamma = [(2n + 1)!]/[n!(n + 1)!]$.

We now present our claim.

Theorem 2.6.4 The median of the set of $(2n + 1)$ numbers, representing $\{\chi_i\}_{i=1}^{n+1}$ and $\{\mu_j(\xi_T)\}_{j=1}^{n}$ as described above, where $0 \leq \chi_1 < \chi_2 < \cdots < \chi_{n+1} \leq 1$ for a finite n, arranged in order of magnitude (i.e., in an array left to right), is the FEV.

Fuzzy Statistics　　　　　　　　　　　　　　　　　　　　　　　　　　77

Proof In order to find the FEV we have to compare χ_i's, $i > 1$, with the proper $\mu_i(\xi_T)$, take the minimum of these two, and then the maximum over all minimum values. Clearly, $\min(\chi_1, 1) = \chi_1$. It should be noted that χ_1 cannot appear to the right of the middle of the array since it has to be followed by at least all the members of the sequence $\{\chi_k\}_{k=2}^{n+1}$. It is important to note that the values of the two sequences are independent but not their place in the array. This is due to the fact that, if $\chi_j \geq \chi_i$, then $\mu_j(\xi_T) \leq \mu_i(\xi_T)$. Because of that and because the array is arranged in an increasing order, the first part of the array, including n numbers [$(n + 1)$ left to right and excluding χ_1] is compared with the second part of the array, including the last n numbers (right to left). Since we take the minimum of every comparison, the results must lie in the first $(n + 1)$ numbers of the array. Out of these $(n + 1)$ numbers we are interested in the maximum, which is obviously the number in the $(n + 1)$th place since the array is increasing in order. This number is the *median* of the array.　　Q.E.D.

EXAMPLE 2.6.2　Let X be our tested population such that

x people are of age α (compatible value χ_1)
y people are of age β (compatible value χ_2)
z people are of age γ (compatible value χ_3)

where $x + y + z = X$, $1 \geq \chi_3 > \chi_2 > \chi_1 \geq 0$.

Since $n = 2$, there are 10 different arrangements as given below; the FEV's can easily be verified graphically and thus are omitted here.

1　$\chi_1, \chi_2, \chi_3, \dfrac{z}{X}, \dfrac{z+y}{X}$;　　FEV $= \chi_3$.

2　$\dfrac{z}{X}, \dfrac{z+y}{X}, \chi_1, \chi_2, \chi_3$;　　FEV $= \chi_1$.

3　$\chi_1, \dfrac{z}{X}, \chi_2, \chi_3, \dfrac{z+y}{X}$;　⎫

4　$\chi_1, \dfrac{z}{X}, \chi_2, \dfrac{z+y}{X}, \chi_3$;　⎬　FEV $= \chi_2$.

5　$\dfrac{z}{X}, \chi_1, \chi_2, \dfrac{z+y}{X}, \chi_3$;

6　$\dfrac{z}{X}, \chi_1, \chi_2, \chi_3, \dfrac{z+y}{X}$;　⎭

7　$\chi_1, \chi_2, \dfrac{z}{X}, \dfrac{z+y}{X}, \chi_3$;　⎫　FEV $= \dfrac{z}{X}$.

8　$\chi_1, \chi_2, \dfrac{z}{X}, \chi_3, \dfrac{z+y}{X}$;　⎭

9　$\chi_1, \dfrac{z}{X}, \dfrac{z+y}{X}, \chi_2, \chi_3$;　⎫　FEV $= \dfrac{z+y}{X}$.

10　$\dfrac{z}{X}, \chi_1, \dfrac{z+y}{X}, \chi_2, \chi_3$;　⎭

EXAMPLE 2.6.3 Using the base variable "hourly wages," let us assume a given population and a given subjective compatibility curve such that

1 person is making $3.00 \rightarrow \chi_1 = 0.40$

3 persons are making $4.00 \rightarrow \chi_2 = 0.50$

4 persons are making $4.20 \rightarrow \chi_3 = 0.55$

2 persons are making $4.50 \rightarrow \chi_4 = 0.60$

2 persons are making $10.00 \rightarrow \chi_5 = 1.00$.

Arranging the sequences $\{\chi_i\}_{i=1}^5$ and $\{\mu_j(\xi_T)\}_{j=1}^4$, where $\mu(\xi_T) = \mu\{w \mid \chi(w) \geq T\} = |\xi_T|/12$, in an array, we get the values

$$\left\{\frac{1}{6}, \frac{1}{3}, \frac{2}{5}, \frac{1}{2}, \frac{11}{20}, \frac{3}{5}, \frac{2}{3}, \frac{11}{12}, 1\right\}$$

Both the FEV and the median are 11/20. The probabilistic expected value (mean) is 0.61 (using the same compatibility curve on [0, 1] interval).

In this example the FEV gives a better indication of the average hourly wage than the mean, by not being affected by the extreme value of $10.00.

It should be noted that a different population distribution might give a FEV that is different from the median.

EXAMPLE 2.6.4 The hourly wages of five people are $2.20, $2.50, $2.70, $3.50, and $10.00; using the same compatibility data of Example 2.6.3, these wages give the following sequence of χ_i's, $1 \leq i \leq 5$:

$$0.25, 0.3, 0.35, 0.45, 1$$

Namely, the median is 0.35 ($2.70), the FEV is 0.4 ($3.00), and the mean is 0.47 ($3.80).

Since the median of a data set is a quantile that, like the expected value, acts as a *center* for a given distribution, we claim that the *first* $\mu_j(\xi_T)$ in the array (from left to right) that is greater than or equal to $\frac{1}{2}$ indicates the respective χ_j as the median of the data set. If no such $\mu_j(\xi_T)$ exists, then χ_1 is the median. We denote the above specified $\mu_j(\xi_T)$ and χ_j as μ_{median} and χ_{median}, respectively. Clearly, if $\mu_{\text{median}} \geq$ FEV then $\chi_{\text{median}} \leq$ FEV and if $\mu_{\text{median}} <$ FEV then $\chi_{\text{median}} >$ FEV; since $\mu_{\text{median}} \geq \frac{1}{2}$, the mean as obtained is closer to the FEV than to the median. Namely,

$$|\text{mean} - \text{FEV}| \leq |\text{mean} - \text{median}|$$

or, using the empirical result for unimodal frequency curves that are moderately skewed (asymmetrical), we have

$$|\text{mean} - \text{FEV}| \leq \tfrac{1}{3} |\text{mean} - \text{mode}|.$$

Similar results can be derived for grouped data. It should be pointed out that *different* compatibility curves with the same fuzzy measure and the same frequency distribution might yield a different FEV. This is not so when the standard median is used, since the FEV is also a function of the subjective evaluation.

Now let the domain of a function χ be the union of a finite number of subspaces $K = \{s_1, s_2, \ldots, s_n\}$ such that $\chi: K \to [0, 1]$. If we consider a fuzzy measure space $(K, 2^K, \mu)$, we can write the FEV of χ as

$$\text{FEV}(\chi) = \max_{K' \in 2^K} \left\{ \min\left[\min_{s \in K'} \chi(s), \mu(K') \right] \right\}.$$

We assume that $\chi(s_i) \leq \chi(s_{i+1})$ for $1 \leq i \leq n - 1$. If this is not the case then rearrangement of $\chi(s_i)$ is necessary. Then the following holds.

Theorem 2.6.5 The FEV of χ in $(K, 2^K, \mu)$ can be written as

$$\max_i \left\{ \min[\chi(s_i), \mu(K_i)] \right\}$$

where $K_i = \{s_i, s_{i+1}, \ldots, s_n\}$, $1 \leq i \leq n$.

Proof Let $\chi(s_i) = \min_{s \in K'} \chi(s)$. Then

$$\max_{K' \in 2^K} \left\{ \min\left[\min_{s \in K'} \chi(s), \mu(K') \right] \right\} \leq \max_i \left\{ \min[\chi(s_i), \mu(K_i)] \right\},$$

and, since $\{K_i \mid 1 \leq i \leq n < 2^K\}$, the reverse inequality holds too. The equality follows. Q.E.D.

Even though the power set 2^K has 2^n members, Theorem 2.6.5 calls for a monotone sequence of subsets of K such that $K_1 > K_2 > \cdots > K_n$ so that we can find the FEV by calculating $\min[\chi(s_i), \mu(K_i)]$ at n points at most.

Theorem 2.6.6 The FEV of χ in $(K, 2^K, \mu)$ can be written as

$$\min[\chi(s_j), \mu(K_j)]$$

iff

$$\chi(s_{j-1}) \leq \mu(K_j) \leq \chi(s_j)$$

or

$$\mu(K_j) > \chi(s_j) \geq \mu(K_{j+1}).$$

Proof (a) *Necessity*: Clearly, since $\mu(K_i)$ is a monotonic decreasing function and $\chi(s_i)$ monotonic increasing, the function

$$\min[\chi(s_i), \mu(K_i)]$$

has a single peak for some i. Thus it is necessary that

$$\min[\chi(s_{j-1}), \mu(K_{j-1})] \leq \min[\chi(s_j), \mu(K_j)] \geq \min[\chi(s_{j+1}), \mu(K_{j+1})]$$

(b) *Sufficiency*: Assume $\chi(s_j) \geq \mu(K_j)$. Then we have

$$\chi(s_{j-1}) \leq \mu(K_i) \leq \chi(s_j)$$

from the above inequality. If we assume

$$\chi(s_j) < \mu(K_j)$$

then

$$\mu(K_j) > \chi(s_j) > \mu(K_{j+1}).$$

Hence the proof is complete.   Q.E.D.

Thus there is no need to evaluate $\min[\chi(s_j), \mu(K_j)]$ for all i but only at point j, fulfilling the requirements of Theorem 2.6.6. Moreover, since χ is a known function, there is a need to evaluate $\mu(K_i)$ for three different points only.

The relation between the FEV, when $\mu(\cdot)$ is taken to be additive (probability measure), and the probabilistic expected value can be established via the following theorem.

Theorem 2.6.7 Let (Ω, B, p) be a probability space and let $\chi: \Omega \to [0, 1]$ be a B-measurable function. Then

$$|\Delta| = \left| \left\{ \int_\Omega \chi(x) \, dp - \sup_{T \in [0, 1]} \{\min[T, p(\xi_T)]\} \right\} \right| \leq \tfrac{1}{4}$$

where $\xi_T = \{x \mid \chi(x) \geq T\}$.

Fuzzy Statistics

In what follows we construct a combinational scheme to obtain the numerical value of the FEV for a finite set of data.

Proof Let $\hat{\chi}(x) \triangleq \sup[\chi(x)]$ and $\underset{\wedge}{\chi}(x) \triangleq \inf[\chi(x)]$. Clearly, $\Omega - \xi_T = \{x \mid \chi(x) < T\}$ and therefore

$$\int_\Omega \chi(x)\,dp = \int_{\Omega-\xi_T} \chi(x)\,dp + \int_{\xi_T} \chi(x)\,dp$$

$$\leq \int_{\Omega-\xi_T} T\,dp + \int_{\xi_T} \chi(x)\,dp$$

$$\leq T \cdot p(\Omega - \xi_T) + \hat{\chi}(x) \cdot p(\xi_T)$$

$$= T \cdot [1 - p(\xi_T)] + \hat{\chi}(x) p(\xi_T)$$

$$= T + p(\xi_T) \cdot [\hat{\chi}(x) - T].$$

Similarly,

$$\int_\Omega \chi(x)\,dp \geq \int_{\xi_T} T\,dp + \int_{\Omega-\xi_T} \chi(x)\,dp$$

$$\geq T \cdot p(\xi_T) + \underset{\wedge}{\chi}(x) \cdot p(\Omega - \xi_T)$$

$$= T \cdot p(\xi_T) + \underset{\wedge}{\chi}(x) \cdot [1 - p(\xi_T)]$$

$$= \underset{\wedge}{\chi}(x) + p(\xi_T) \cdot [T - \underset{\wedge}{\chi}(x)].$$

Define the quantity H by

$$H \triangleq \sup_{T \in [0,1]} \{\min[T, p(\xi_T)]\}.$$

Then

$$H = \max\left\{ \sup_{T \in [0, H]} \{\min[T, p(\xi_T)]\},\ \sup_{T \in (H, 1]} \{\min[T, p(\xi_T)]\} \right\};$$

clearly,

$$\sup_{T \in [0, H]} \{\min[T, p(\xi_T)]\} \leq H$$

since $p(\xi_T)$ increases as T decreases, and therefore it is trivial that

$$\sup_{T\in(H,1]} \{\min[T, p(\xi_T)]\} \leq \sup_{T\in[0, H]} \{\min[T, p(\xi_T)]\}$$

and thus

$$\sup_{T\in[0, H]} \{\min[T, p(\xi_T)]\} = H,$$

and

$$p(\xi_H) \geq H.$$

Similarly,

$$p(\xi_{H^+}) \leq H$$

since

$$\xi_{H^+} = \{x \mid \chi(x) > H\}$$

and

$$\sup_{T\in(H,1]} \{\min[T, p(\xi_T)]\} \leq \sup_{T\in[0, H]} \{\min[T, p(\xi_T)]\}.$$

Hence

$$\lim_{T\to H^+} \left| \int_\Omega \chi(x)\, dp - \sup_T \{\min[T, p(\xi_T)]\} \right|$$

$$\leq H^+ + p(\xi_{H^+}) \cdot [\hat{\chi}(x) - H^+] - H$$

$$\leq p(\xi_{H^+}) \cdot [\hat{\chi}(x) - H^+]$$

and

$$\lim_{T\to H^-} \left| \int_\Omega \chi(x)\, dp - \sup_T \{\min[T, p(\xi_T)]\} \right|$$

$$\geq \underline{\chi}(x) + p(\xi_{H^-}) \cdot [H^- - \underline{\chi}(x)] - H$$

$$\geq [1 - p(\xi_{H^-})] \cdot [\underline{\chi}(x) - H^-].$$

Fuzzy Statistics

Hence

$$[1 - p(\xi_{H^-})] \cdot [\underset{\wedge}{\chi}(x) - H^-] \leq \Delta \leq p(\xi_{H^+}) \cdot [\hat{\chi}(x) - H^+].$$

Clearly,

$$\hat{\chi}(x) = 1 \quad \text{and} \quad \underset{\wedge}{\chi}(x) = 0.$$

Thus

$$H^- \cdot [p(\xi_{H^-}) - 1] \leq \Delta \leq (1 - H^+) \cdot p(\xi_{H^+})$$

and hence in the limit

$$|\Delta| \leq H(1 - H).$$

Thus

$$\frac{d[H(1-H)]}{dH} = 1 - 2H = 0$$

implies $H = \frac{1}{2}$, and $|\Delta| \leq \frac{1}{4}$. Q.E.D.

We now investigate some other interesting properties of the FEV.

Let $\chi_1 \in [0, 1]$ and $\chi_2 \in [0, 1]$; clearly, if χ_1 and χ_2 are B-measurable, then $\max(\chi_1, \chi_2)$, $\min(\chi_1, \chi_2)$, and $1 - \chi_1$ are also B-measurable. Thus it is trivial to show that the following hold, if computed under the same measure μ:

1. Let $K \in [0, 1]$, then FEV $K = K$.
2. If $\chi_1 \leq \chi_2$, then FEV $\chi_1 \leq$ FEV χ_2.
3. FEV$\{\min(\chi_1, \chi_2)\} \leq \min\{$FEV χ_1, FEV $\chi_2\}$.
4. FEV$\{\max(\chi_1, \chi_2)\} \geq \max\{$FEV χ_1, FEV $\chi_2\}$.
5. If $A \subset B$, then FEV χ over set A is smaller than or equal to FEV χ over set B.
6. FEV χ over set C, where $C = A \cup B$, is larger than or equal to the maximum of the FEV χ over set A and FEV χ over set B.
7. FEV χ over set D, where $D = A \cap B$, is smaller than or equal to the minimum of the FEV χ over set A and the FEV χ over set B.
8. Let $a \in A$ and $x \in X$ such that

$$\chi: X \times A \to [0, 1]$$

and let $\chi(x, a)$ be a B-measurable function of x for an arbitrary a. Then

(a) $\text{FEV}\left\{\sup_{a \in A} \chi(x, a)\right\} \geq \sup_{a \in A} \{\text{FEV}\,\chi(x, a)\}$

(b) $\text{FEV}\left\{\inf_{a \in A} \chi(a, x)\right\} \leq \inf_{a \in A} \{\text{FEV}\,\chi(x, a)\}.$

9. Let $K \in [0, 1]$. Then $\text{FEV}\{\min(K, \chi)\} = \min\{K, \text{FEV}\,\chi\}$.
10. Let $K \in [0, 1]$. Then $\text{FEV}\{\max(K, \chi)\} = \max\{K, \text{FEV}\,\chi\}$.

Just to illustrate how these properties could be proven, we prove the last one.

Proof of Proposition 10 By definition

$$\text{FEV}\{\max(K, \chi)\} = \sup_{T \in [0, 1]} \left\{\min\left[T, \mu(\xi_{T_1})\right]\right\}$$

where $\xi_{T_1} = \{x \mid \max(K, \chi) \geq T\}$. Thus

$\text{FEV}\{\max(K, \chi)\}$

$$= \max\left\{\sup_{T \in [0, K]} \left\{\min\left[T, \mu(\xi_{T_1})\right]\right\}, \sup_{T \in (K, 1]} \left\{\min\left[T, \mu(\xi_{T_1})\right]\right\}\right\}.$$

Two cases must be checked.

Case 1 Let

$$\sup_{T \in [0, K]} \left\{\min\left[T, \mu(\xi_{T_1})\right]\right\} \geq \sup_{T \in (K, 1]} \left\{\min\left[T, \mu(\xi_{T_2})\right]\right\};$$

then

$$\text{FEV}\{\max(K, \chi)\} = \sup_{T \in [0, K]} \left\{\min\left[T, \mu(\xi_{T_1})\right]\right\} = \text{FEV}\,K$$

since $\xi_{T_1} = \chi$ for $T \in [0, K]$. Hence, by proposition 4,

$$\text{FEV}\{\max(K, \chi)\} = \max\{K, \text{FEV}\,\chi\}.$$

Case 2 Let

$$\sup_{T \in [0, K]} \left\{\min\left[T, \mu(\xi_{T_1})\right]\right\} < \sup_{T \in (K, 1]} \left\{\min\left[T, \mu(\xi_{T_1})\right]\right\};$$

then

$$\text{FEV}\{\max(K, \chi)\} = \sup_{T \in (K, 1]} \{\min[T, \mu(\xi_{T_1})]\}.$$

It is clear that

$$\text{FEV}\,\chi = \max\left\{\sup_{T \in [0, K]} \{\min[T, \mu(\xi_T)]\}, \sup_{T \in (K, 1]} \{\min[T, \mu(\xi_T)]\}\right\}$$

where $\xi_T = \{x \mid \chi \geq T\}$, and since

$$\sup_{T \in [0, K]} \{\min[T, \mu(\xi_T)]\} \leq K \quad \text{and} \quad \xi_T = \xi_{T_1}$$

for $T \in (K, 1]$, we can obtain the result that

$$\text{FEV}\,\chi = \text{FEV}\{\max[K, \chi]\}$$

and thus

$$\text{FEV}\{\max(K, \chi)\} = \max\{K, \text{FEV}\,\chi\}. \qquad \text{Q.E.D.}$$

Another approach to the subject can be taken via the concept of the probability of fuzzy events. In the last part of this section, we briefly discuss this approach, first introduced by Zadeh [2948].

In ordinary probability theory, given a random variable X in one dimension, we define

$$P\{X = x\} = f(x),$$

where $f(x)$ is the probability density of the random variable X. Then we may define the (cumulative) distribution function $P(x)$, which gives $P\{X < x\}$, as

$$P(x) = \int_{-\infty}^{x} f(x')\, dx',$$

for $x' \in [-\infty, \infty]$. Then with $dp(x) \equiv f(x)\, dx$, $P(x)$ can be written as

$$P(x) = \int_{-\infty}^{x} dp(x').$$

It is held that $\lim_{x \to \infty} P(x) \equiv \int_{-\infty}^{\infty} dp(x) = 1$, that is, that the distribution function is normalizable. For a fuzzy random variable X_A, we write

$$P\{X_A = x\} \equiv \chi_A(x) f(X),$$

thereby associating with each x a grade of membership in the set A. Then we define a quantity $P(A; x)$ as

$$P(A; x) \equiv P\{X_A \leq x\} = \int_{-\infty}^{x} \chi_A(x')\, dp(x').$$

Corresponding to the normalization condition of ordinary probability theory, we write that $\lim_{x \to \infty} P(A, x) \equiv P(A)$, where

$$P(A) = \int_{-\infty}^{\infty} \chi_A(x)\, dp(x).$$

The last equation is Zadeh's definition of a fuzzy event A in a one-dimensional space. We now seek to formalize the above argument to multidimensional spaces by referring directly to Zadeh's paper [2948].

A probability space is assumed to be a triplet $(\mathbb{R}^n, B, P)$, where B is the σ-field of Borel sets in $\mathbb{R}^n$ and P is a probability measure over $\mathbb{R}^n$. A point in $\mathbb{R}^n$ is denoted by $\mathbf{x}$.

Let a set $A \in B$. Then on defining a characteristic function of or grade of membership in the set A by $\chi_A(\mathbf{x}): \mathbb{R}^n \to [0, 1]$, we define the probability of A as

$$P(A) = \int_{\mathbb{R}^n} \chi_A(\mathbf{x})\, dp(\mathbf{x}),$$

where

$$\int_{\mathbb{R}^n} dp(\mathbf{x}) = 1,$$

or

$$P(A) = \langle \chi_A \rangle = E\{\chi_A\}$$

where we use the notation $\langle \ \rangle$ to denote the expected value of a function, and where $E\{\chi_A\}$ is the expected value of χ_A.

In a multidimensional case,

$$P\{(X_{A_1} \leq x_1) \cap (X_{A_2} \leq x_2) \cap \cdots \cap (X_{A_n} \leq x_n)\}$$

$$\equiv \int_{-\infty}^{x_1} \cdots \int_{-\infty}^{x_n} \chi_A(\mathbf{x})\, dp(\mathbf{x}),$$

Fuzzy Statistics

for $x_i \in [-\infty, \infty]$ $\forall i$ in $\mathbb{R}^n$, and where

$$\chi_A(\mathbf{x}) = \chi_A(x_1,\ldots,x_n) \quad \text{and} \quad dp(\mathbf{x}) = f(x_1,\ldots,x_n) \cdot dx_1\ldots dx_n.$$

Clearly, if $A \subset C$, then

$$P(A) \leq P(C).$$

We also have

$$P(A \cup C) = P(A) + P(C) - P(A \cap C),$$

where $P(A \cap C)$ is the probability of intersection of A and C, for which

$$\chi_{A \cap C} = \min[\chi_A(x), \chi_C(x)] \quad \forall x \in X.$$

Let A and C be two fuzzy events in the space $(\mathbb{R}^n, B, P)$. A and C are said to be independent if

$$P(AC) = P(A)P(C).$$

The mean of a fuzzy event A relative to a probability measure P may be defined as

$$P(A)m_P(A) = \int_{\mathbb{R}^n} x\chi_A(x)\,dp(x).$$

EXAMPLE 2.6.5 Consider a population that consists of members of the set of aged people defined by

$$A = \{x \mid x > x_m\}$$

where x is defined as chronological age in years, and x_m is some (subjectively) chosen threshold age. Let $\chi_A(x)$, the membership function in the set of aged people, be given by

$$\chi_A(x) = 1 - e^{-x/x_m},$$

from which it follows that $\chi_{\bar{A}}(x) = e^{-x/x_m}$. Assume that the density of chronological ages in the population is given by

$$dp(x) = \frac{1}{x_0^2} x e^{-x/x_0}\,dx$$

so that for $x \in [0, \infty]$, $\int_0^\infty dp(x) = 1$, and

$$P(A) = \frac{1}{x_0^2} \int_0^\infty (1 - e^{-x/x_m}) x e^{-x/x_0} \, dx.$$

Evaluation of the integral yields

$$P(A) = 1 - \left(1 + \frac{x_0}{x_m}\right)^{-2}.$$

As $x_0/x_m \to 0$, $P(A) \to 0$, while as $x_0/x_m \to \infty$, $P(A) \to 1$.

The probability $P(A)$ of being found or classified as a member of the set A in this example depends on the ratio of x_0/x_m where x_0 is a fixed number of years, and is the most probable age in the distribution, while x_m, a subjectively chosen threshold age, will vary among observers. We reach the rather interesting conclusion that the probability of a fuzzy event may not be uniquely defined. As a first extension of Zadeh's work, we claim that the number

$$\langle P_{x_0}(A) \rangle = \int_\mathbb{R} P(A; x_0, x_m) \, dp(x_m),$$

the expected value of $P(A; x_0, x_m)$, over the distribution of the answers to the question "What do you think a threshold age, x_m, for old age is?" gives additional precision to the notion of aged, where $\mathbb{R}$ is the domain over which the x_m are defined.

Let two probability distributions be independent and given as $dp(x_1) = e^{-x_1} dx_1$ and $dp(x_2) = x_2 e^{-x_2} dx_2$. Since both $dp(x_1)$ and $dp(x_2)$ must be positive and normalizable, both x_1 and $x_2 \in [0, \infty]$. We ask: Are x_1 and x_2 similar? To answer this question, we must define a set A that includes the notion of similarity. We write $A = \{x_1, x_2 \,|\, |x_1 - x_2| \approx 1\}$. A suitable grade of membership in such a set is given by

$$\chi_A(x_1, x_2) = e^{-|x_1 - x_2|}$$

If $x_1 = x_2$, $\chi_A(x_1, x_2) = 1$, in which case x_1 and x_2 are identical. If $x_1 \approx x_2$, then they are similar. We now ask for the occurrence of such a set. We have

$$P(A) = \int_0^\infty \int_0^\infty e^{-|x_1 - x_2|} e^{-(x_1 + x_2)} x_2 \, dx_1 \, dx_2 = \tfrac{3}{8}.$$

Consider now, rather than the chronological age, "biological" age defined by

$$x_B = \alpha t,$$

when the quantity α, an aging rate, is distributed according to a distribution $dp(\alpha)$ such that $\int_{\mathbf{R}} dp(\alpha) = 1$.

Let membership in the set of aged people at time t be denoted by

$$A(t) = \{\alpha \mid \alpha > \alpha_m; t\}.$$

Set

$$\chi_{A(t)}(\alpha) \equiv \chi_A(\alpha, t) = 1 - e^{-\alpha t/\alpha_m t_m}$$

and let, for $\alpha \in [0, \infty]$,

$$dP(\alpha) = \frac{1}{\alpha_0^2} \alpha e^{-\alpha/\alpha_0} \, d\alpha.$$

Then

$$P(A(t)) \equiv P_A(t) = \frac{1}{\alpha_0^2} \int_0^\infty (1 - e^{-\alpha t/\alpha_m t_m}) \alpha e^{-\alpha/\alpha_0} \, d\alpha.$$

We find that

$$P_A(t) = 1 - (1 + \lambda t)^{-2}$$

where $\lambda = \alpha_0/\alpha_m t_m$. For positive, finite α_0, α_m, and t_m,

$$\lim_{t \to 0} P_A(t) \to 0$$

and

$$\lim_{t \to \infty} P_A(t) \to 1.$$

For any finite $t > 0$,

$$\lim_{\lambda \to 0} P_A(t) \to 0$$

and

$$\lim_{\lambda \to \infty} P_A(t) \to 1.$$

It should be noted that $\chi_A(\alpha, t)$ satisfies the differential equation

$$\frac{\partial \chi_A}{\partial t} + \frac{\alpha}{\alpha_m t_m} \chi_A = \frac{\alpha}{\alpha_m t_m}.$$

Corresponding to the relation given in this equation, there is a differential equation satisfied by $P_A(t)$. It is

$$\frac{dP_A(t)}{dt} + \left(\frac{2\lambda}{1+\lambda t}\right)P_A(t) = \frac{2\lambda}{1+\lambda t}.$$

In general, let

$$P_A(t) = \int_\mathbb{R} \chi_A(x) p(x,t)\, dx,$$

and form

$$\frac{dP_A(t)}{dt} = \int_\mathbb{R} \chi_A(x) \frac{\partial p}{\partial t}\, dx.$$

Assume that a normalizable $p(x,t)$ satisfies the differential equation

$$\frac{\partial p}{\partial t} + \lambda(x) p = q(x,t).$$

Thus

$$\frac{dP_A(t)}{dt} = \int_\mathbb{R} \chi_A(x)\{q(x,t) - \lambda(x) p(x,t)\}\, dx$$

or

$$\frac{dP_A(t)}{dt} = \langle q(t) \rangle_z - \int_\mathbb{R} \chi_A(x) \lambda(x) p(x,t)\, dx$$

where

$$\langle q(t) \rangle_z = \int_\mathbb{R} \chi_A(x) q(x,t)\, dx$$

and

$$\langle q(t) \rangle_z \equiv \langle \chi_A q(t) \rangle.$$

Details of applying both this technique and the FEV to imprecise data and to the analysis of fuzzy differential equations representing imprecise or incomplete models of dynamical systems can be found in Kandel [1220] and Kandel and Byatt [1231].

CHAPTER THREE

Pattern Classification and Fuzzy Sets

3.1 INTRODUCTION

The early development of fuzzy sets drew much of its inspiration from a study in the early sixties (Bellman et al. [153]) related to the subject of pattern classification. The use of fuzzy sets in pattern recognition and classification may throw some light on the general problem of decision-making and fuzzy processes in general. Although a great amount of literature has been published dealing with fuzzy techniques in pattern recognition, cluster analysis, and related topics, a unified approach is not yet available.

In a very fundamental way, the intimate relation between the theory of fuzzy sets and the theory of pattern recognition and classification rests on the fact that most real-world classes are fuzzy in nature. Thus, given an object p and a cluster C, the basic question in most problems related to cluster analysis is not whether p is or is not an element in C, but the degree to which p belongs to C [grade of membership of p in C, $\chi_C(p)$]. The problem is that most practical problems in pattern analysis do not lend themselves to a precise formulation, and thus less precise techniques might have some solution to the intrinsic imprecision that in most cases, is incorporated in the recognition and classification problems.

The objectives of the present chapter are to outline a conceptual framework for pattern recognition and cluster analysis based on the theory of fuzzy sets, and to discuss some significant contributions in the field.

3.2 PATTERN RECOGNITION AND CLUSTER ANALYSIS IN A FUZZY-SET-THEORETIC FRAMEWORK

In principle, a recognition algorithm, when applied to an object p, yields the grade of membership of p in a class C, $\chi_C(p)$. For instance, let $p =$ Tallahassee, and $C =$ class of large cities; then a recognition algorithm applied to Tallahassee should yield a numerical answer (0.4, for example) to the query "What is the degree to which Tallahassee is a large city?" As a simple illustration, we may consider p to be a string of primitives that can be derived from a formal grammar. In this case a recognition algorithm may consist of a parsing procedure.

The grade of membership of an object p in a class C may also be regarded as the degree of similarity between p and a *typical* (or ideal) object representing class C. Such a prototype (ideal prototype), or a template, is sometimes thought of as being the average vector in the pattern space, and this is essentially a template-similarity procedure. When the explicit description of the recognition algorithm is known, this algorithm is said to be *transparent*; if such a description is not available, it is said to be *opaque* [2981]. For example, the user of a computer may not know the specific algorithm used in the machine to perform matrix inversion. Similarly, human beings may not be able to articulate the algorithm that they use to assign a grade of membership to a certain scene, classifying it in the fuzzy set of "impressive scenes." Human perception usually uses such opaque algorithms to recognize and classify objects.

Within the framework of the theory of fuzzy sets, the problem of pattern classification may be viewed—in its essential form—as that of conversion of an opaque recognition algorithm into a transparent recognition algorithm. More specifically, let X be a universe of objects and let R_{op} be an opaque recognition algorithm that defines a fuzzy subset C of X. Then pattern recognition may be viewed as the process of converting an opaque recognition algorithm R_{op} into a transparent recognition algorithm R_{tr}.

We assume for simplicity that only one fuzzy subset of X is defined by R_{op}. More generally, there may be a number of such subsets, say $C_1, \ldots, C_n$, with R_{op} yielding the grade of membership of p in each of these subsets.

As an illustration of this formulation, consider the following typical problem. Suppose that X is the universe of handwritten letters and that when a letter, p, is presented to a person, that person can specify the grade of membership, $\chi_C(p)$, of p in the fuzzy set C, of handwritten a's. Namely,

$$\chi_C(p) = R_{op}(p), \quad \text{for } p \text{ in } X.$$

Usually, a person is presented with a finite set of sample letters $p_1, \ldots, p_r$, so that the result of application of R_{op} to $p_1, \ldots, p_r$ is a set of ordered pairs

$(p_1, \chi_C(p_1)), \ldots, (p_r, \chi_C(p_r))$, which, in the notation of fuzzy sets, may be expressed as the linear form

$$S_C = \frac{\chi_C(p_1)}{p_1} + \cdots + \frac{\chi_C(p_r)}{p_r}$$

where S_C stands for a fuzzy set of samples from C, and a term of the form $\chi_C(p_i)/p_i$, $i = 1, \ldots, r$, signifies that $\chi_C(p_i)$ is the grade of membership of p_i in C.

If, based on the knowledge of S_C, we could convert the opaque recognition algorithm R_{op} into a transparent recognition algorithm R_{tr}, then given any p we could deduce $\chi_C(p)$ by applying R_{tr} to p. Equivalently, we may view this as the process of interpolation of the membership function of C from the knowledge of the values that it takes at the points $p_1, \ldots, p_r$.

It is interesting to note that originally this is the way in which the problem of pattern classification was introduced by Bellman, Kalaba, and Zadeh [153], using a fuzzy-set-theoretic framework. However, the later formulation by Zadeh [2981], based on the conversion of R_{op} to R_{tr}, appears to be more natural.

A desired property in pattern classification is that the recognition process be fully automated in the sense that it may be performed by a computer rather than by a human. This requires that the transparent recognition algorithm R_{tr} act on $f(p)$, a mathematical object resulting by associating a measurement procedure f with each object p, rather than on p itself, since an object must be well-defined in order to be capable of manipulation by the computer.

Let A be a fuzzy subset of X that is defined by an opaque recognition algorithm R_{op} in the sense that

$$\chi_C(p) = R_{op}(p), \quad p \in X.$$

Denote by R_{tr} a transparent recognition algorithm that, acting on the mathematical object $f(p)$, yields $\chi_C(p)$. Then the problem of *automatic* (or *machine*) *pattern recognition* may be expressed in symbols as that of determining f and R_{tr} such that

$$\chi_C(p) = R_{op}(p)$$

$$R_{tr}(f(p)) = R_{op}(p), \quad p \in X.$$

Thus automatic pattern recognition involves two distinct kinds of problems: (1) conversion of the object p into a mathematical object, $f(p)$; and (2) conversion of the opaque recognition algorithm R_{op} that acts on p's into a transparent recognition algorithm that acts on $f(p)$'s. Of these, problem (1) is by far the more difficult since it is closely related to the problem of feature selection—a problem that falls into the least well-defined and least well-developed area in pattern recognition.

It is important to observe that, from a practical point of view, it is desirable that (1) $f(p)$ be defined by a small number of attributes, and (2) that the measurement of these attributes be relatively simple. With these added considerations, the problem of pattern classification may be reformulated in the following terms.

Given an opaque recognition algorithm R_{op} that defines a fuzzy subset of objects p in X:

Problem 1 Feature Extraction: Select a small set of measurement procedures f_1 and/or a set of primitives in order to turn p into a mathematical object **x** (vector in a pattern space and/or formal structure).

Problem 2 Define a transparent algorithm R_{tr} that from $f(p)$ yields the grade of membership of p in a class C, $\chi_C(p)$.

It is clear that in the above formulation fuzziness may be present at several levels of the recognition or classification processes. First, the notion of an object does not admit a precise definition, and hence the functions $\{f_i\}_{i=1}^n$ cannot be regarded as precise functions in the classical mathematical sense. Second, since the derived equality

$$R_{tr}[f(p)] = R_{op}(p), \quad p \in X$$

cannot be realized precisely, the classification problem is quite inexact. Furthermore, we get imprecision from the difficulty and ambiguity of assessing the goodness of a transparent recognition algorithm that may be offered as a solution to a given problem.

The main thrust of the above comments is that the problem of pattern classification is intrinsically incapable of precise mathematical formulation. For this reason the conceptual structure of the theory of fuzzy sets may well provide a more natural setting for the formulation and approximate solution of problems in pattern classification than the more traditional approaches based on classical set theory, probability theory, and two-valued logic.

3.3 SYNTACTIC AND SEMANTIC TECHNIQUES

Syntactic pattern recognition makes use of the idea that certain pattern classes that have a hierarchical structure can be described by a formal grammar, known as the pattern grammar. In Section 3.3.1 a brief summary of fuzzy formal languages and finite fuzzy automata is given, based on the material in Lee and Zadeh [1489], Kandel and Lee [1237], and Dubois and Prade [574].

3.3.1 Fuzzy Languages

Let V_T be a finite set called an alphabet. We denote by V_T^* the set of finite strings constructed by concatenation of elements of V_T, including the null string Λ. V_T^* is a free monoid over V_T. A language is a subset of V_T^*, and a *fuzzy formal language* is a fuzzy set R on V_T^*, that is,

$$R = \sum_{x \in V_T^*} \frac{\chi_R(x)}{x}$$

with χ_R a function from V_T^* to $[0, 1]$. $\chi_R(x)$ is the degree of membership of x in R.

Union and intersection of fuzzy languages can be defined as usual:

$$R_1 \cup R_2 : \chi_{R_1 \cup R_2}(x) = \max(\chi_{R_1}(x), \chi_{R_2}(x)), \quad \forall x \in V_T^*$$

$$R_1 \cap R_2 : \chi_{R_1 \cap R_2}(x) = \min(\chi_{R_1}(x), \chi_{R_2}(x)), \quad \forall x \in V_T^*$$

And the complement $\bar{R}$ of R has membership function $(1 - \chi_R)$.

A specific operation between languages is concatenation: any string x in V_T^* is the concatenation of a prefix string u and a suffix string v: $x = uv$. According to the extension principle, the concatenation $R_1 R_2$ of two fuzzy languages R_1 and R_2 is defined by

$$\chi_{R_1 R_2}(x) = \sup_{x = uv} \min(\chi_{R_1}(u), \chi_{R_2}(v)).$$

The concatenation of fuzzy languages is associative. Denoting by R^t the concatenation of R t times, the Kleene closure of R is $\hat{R} = \{\Lambda\} \cup R \cup R^2 \cup R^3 \cup \cdots \cup R^t \cup \cdots$. Note that $\forall x \in V_T^*$, if $x = a_1 a_2 \ldots a_k$, $a_i \in V_T$, $i = 1, k$, then

$$\chi_{\hat{R}}(x) = \sup_{i=1,k} \left[\sup_{\substack{x = u_1 u_2 \ldots u_i \\ \forall j, u_j \in V_T^*}} \left[\min_{j=1,i} \chi_R(u_j) \right] \right]$$

k is the length of x, denoted $l(x)$.

The following property is due to Negoita and Ralescu [1835]: $R = \hat{R}$ iff $\chi_R(\Lambda) = 1$ and $\chi_R(uv) \geq \min(\chi_R(u), \chi_R(v))$, $\forall u, v \in V_T^*$.

A fuzzy grammar is a qaudruple $G = (V_N, V_T, P, s)$ where: V_T is a set of terminals or alphabet; V_N is a set of nonterminals ($V_N \cap V_T = \phi$), that is, labels of certain fuzzy sets on V_T^* called fuzzy syntactic categories; P is a finite set of rules called productions; and $s \in V_N$ is the initial symbol, that is, the label of the syntactic category "string." The elements of P are expressions of the form $\alpha \xrightarrow{\rho} \beta$; $\rho \in [0, 1]$ where α and β are strings in $(V_T \cup V_N)^*$. ρ is the grade of membership of β given α. The symbol * always indicates a free monoid X^* over the set X. ρ also expresses a degree of properness of the rule $\alpha \rightarrow \beta$.

Let $\alpha_1, \ldots, \alpha_m$ be strings in $(V_T \cup V_N)^*$, and let $\alpha_1 \xrightarrow{\rho_2} \alpha_2, \ldots, \alpha_{m-1} \xrightarrow{\rho_m} \alpha_m$ be productions. Then α_m is said to be derivable from α_1 in G, more briefly $\alpha_1 \Rightarrow_G \alpha_m$. The expression $\alpha_1 \xrightarrow{\rho_2} \alpha_2, \ldots, \xrightarrow{\rho_m} \alpha_m$ is referred to as a derivation chain from α_1 to α_m.

A fuzzy grammar G generates a fuzzy language $R(G)$ in the following manner. A string x of V_T^* is said to be in $R(G)$ iff x is derivable from s. The grade of membership $\chi_G(x)$ of x in $R(G)$ is

$$\chi_G(x) = \sup \min(\chi(s \rightarrow \alpha_1), \chi(\alpha_1 \rightarrow \alpha_2), \ldots, \chi(\alpha_m \rightarrow x)) > 0$$

where $\chi(\alpha_i \rightarrow \alpha_{i+1})$ is the nonnull ρ_{i+1} such that

$$\left(\alpha_i \xrightarrow{\rho_{i+1}} \alpha_{i+1}\right) \in P, \qquad \forall i = 0, m$$

if $\alpha_0 = s$ and $\alpha_{m+1} = x$.

The supremum is taken over all derivation chains from s to x. The degree or properness of a derivation chain is that of its least proper link, and $\chi_G(x)$ is calculated on the "best" chain. Two fuzzy grammars G_1 and G_2 are said to be equivalent iff $\forall x \in V_T^*$, $\chi_{G_1}(x) = \chi_{G_2}(x)$.

Paralleling the standard classification of ordinary grammars, we distinguish four types of fuzzy grammars:

Type 0 grammar. The allowed productions are of the general form $\alpha \xrightarrow{\rho} \beta$, $\rho > 0$, $\alpha, \beta \in (V_T \cup V_N)^*$.

Type 1 grammar (context-sensitive). The productions are of the form $\alpha_1 A \alpha_2 \xrightarrow{\rho} \alpha_1 \beta \alpha_2$, $\rho > 0$, $\alpha_1, \alpha_2, \beta \in (V_T \cup V_N)^*$, $A \in V_N$, $\beta \neq \Lambda$. $s \xrightarrow{1} \Lambda$ is also allowed.

Type 2 grammar (context-free). The allowable productions are now $A \xrightarrow{\rho} \beta$, $\rho > 0$, $A \in V_N$, $\beta \in (V_N \cup V_T)^*$, $\beta \neq \Lambda$, and $s \xrightarrow{1} \Lambda$.

Syntactic and Semantic Techniques

Type 3 grammar (regular). The allowable productions are $A \xrightarrow{\rho} aB$ or $A \xrightarrow{\rho} a$, $\rho > 0$, where $a \in V_T$; $A, B \in V_N$, and $s \xrightarrow{1} \Lambda$.

A grammar is said to be recursive iff there is an algorithm that computes $\chi_G(x)$. Lee and Zadeh [1489] showed that a fuzzy context-sensitive grammar was recursive.

Let $R(G)$ be a fuzzy language and let G be a grammar that generates $R(G)$. Several nonfuzzy languages can be generated from $R(G)$. For instance,

$$R(G, \lambda) = \{x \in V_T^* \mid \chi_G(x) > \lambda\}$$

$$R(G, \geq, \lambda) = \{x \in V_T^* \mid \chi_G(x) \geq \lambda\}$$

$$R(G, =, \lambda) = \{x \in V_T^* \mid \chi_G(x) = \lambda\}$$

$$R(G, \lambda_1, \lambda_2) = \{x \in V_T^* \mid \lambda_1 < \chi_G(x) \leq \lambda_2\},$$

where $\lambda, \lambda_1,$ and λ_2 are thresholds that belong to $[0, 1]$. These languages are called *cut-point* languages. Note that, since P is finite, the image of V_T^* through χ_G, $\chi_G(V_T^*) \subseteq [0, 1]$, is also finite because we use only max and min operators to valuate strings.

Mizumoto et al. [1729] have proven the following properties:

1. $\forall \lambda, \forall i = 0, 3$, if G is a fuzzy grammar of type i, then $R(G, \lambda)$ is of type i.
2. $\forall \lambda$, for $i = 0$ and 2, if G is a fuzzy grammar of type i, then $R(G, \lambda_1, \lambda_2)$ and $L(G, =, \lambda)$ may not be of type i. For $i = 3$, $R(G, \lambda_1, \lambda_2)$ and $R(G, =, \lambda)$ are of type 3. For $i = 1$, the result is unknown.
3. $\forall \lambda$, $i = 0, 3$, if G is a fuzzy grammar of type i, then $R(G, \geq, \lambda)$ is of type i.

An important development in fuzzy formal languages is the notion of fuzzy syntax directed translations developed by Thomason [2572].

A translation T of a language R_1 in V_T^* into a language R_2 in W_T^*, where V_T and W_T are alphabets, is a fuzzy relation on $V_T^* \times W_T^*$ such that $\text{dom } T = R_1$ and $\text{ran } T = R_2$, where $\text{dom } T$ and $\text{ran } T$ are, respectively, the domain and the range of T. $\chi_T(x, y)$ is the grade of properness of translating x by y, $x \in \text{supp } R_1, y \in \text{supp } R_2$.

An efficient model in formal language translation theory is that of a syntax-directed translation scheme (SDTS). A fuzzy SDTS is a 5-tuple $\mathcal{T} = (V_N, V_T, W_T, s, D)$ where V_N is a set of nonterminals, V_T and W_T are alphabets, s is an initial symbol, and D is a set of double productions $A \xrightarrow{\rho} \alpha, \beta$ with $A \in V_N$, $(\alpha, \beta) \in (V_T \cup V_N)^* \times (W_T \cup V_N)$, and $\rho > 0$ valuates the translation of α into β.

Obviously, when a string is generated in V_T^*, another string is generated in W_T^* by means of a double derivation chain. A fuzzy SDTS builds the translation relation T.

DePalma and Yau [489] introduced *fractionally fuzzy grammars*. A string is derived in the same manner as in the case of a fuzzy grammar. However, the membership of a string is given by

$$\chi_G(x) = \sup_k \left(\frac{\sum_{i=1}^{n_k} g(l_k^i)}{\sum_{i=1}^{n_k} h(l_k^i)} \right) \in [0, 1]$$

with k = index of a derivation chain leading to x; n_k = length of the kth derivation chain; g and h are functions from J to $\mathbb{R}$, where J labels the productions, and $h(l) \geq g(l)$, $\forall l \in J$; and the convention $0/0 = 0$. Fractionally fuzzy grammars were used by both authors instead of fuzzy grammars in order to reduce the combinatorial aspect of parsing in a pattern recognition procedure.

Fuzzy automata can be considered as acceptors of fuzzy languages. Let $\mathcal{P} = (U, S, Y, \tilde{s}_0, \delta, \sigma)$ be a fuzzy automaton such that:

U is a finite set of inputs, $U = \{a_1, \ldots, a_m\}$.
S is a finite set of states, $S = \{q_1, \ldots, q_n\}$.
Y is a finite set of outputs, $Y = \{y_1, \ldots, y_p\}$.
$\tilde{s}_0$ is a fuzzy set on X, the fuzzy initial state.
δ is a fuzzy ternary relation on $S \times U \times S$, made up of m transition relations $\{\delta_u\}_{u \in U}$ for the states.
σ is a fuzzy relation on $S \times Y$, i.e., the output map.

When a nonfuzzy input u is processed by the automaton, the output can be symbolically written $\tilde{y} = \tilde{s}_0 \circ \delta_u \circ \sigma$ where $\circ$ is the (associative) composition of binary relations. Once a string of inputs $\theta = u_1 u_2 \cdots u_k$ has been processed by the automaton, the fuzzy output is

$$\tilde{y} = \tilde{s}_0 \circ (\delta_{u_1} \circ \delta_{u_2} \circ \cdots \circ \delta_{u_k}) \circ \sigma = \tilde{s}_0 \circ \Delta_\theta \circ \sigma.$$

Denote by Δ_θ the result of the composition $\delta_{u_1} \circ \cdots \circ \delta_{u_k}$. Δ_Λ is the identity relation.

Structural properties of fuzzy languages accepted by fuzzy automata have been investigated by Mizumoto et. al. [1729], Santos [2300]–[2315], and Kandel and Lee [1237].

3.3.2 Syntactic Pattern Recognition

The many different techniques used to solve pattern recognition problems may be grouped into two general categories—decision theoretic and syntactic. In syntactic pattern recognition, a basic set of pattern primitives is selected and forms the set of terminals of the grammar. The productions of the grammar are a list of allowable relations among the primitives. The pattern class is the set of strings generated by the pattern grammar. However, the concept of a formal grammar is often too rigid to be used for representation of real patterns, and thus fuzzy languages can handle imprecise patterns when the indeterminacy is due to inherent vagueness. The fuzziness may lie in the definition of primitives or in the physical relations among them. Thus the primitives become labels of fuzzy sets and the production rules of the grammar are weighted. The membership grade of a particular pattern in the class described by the grammar is calculated using max-min composition, that is, the grammar is fuzzy. The possibility of applying fuzzy grammars to the recognition of leukocytes and chromosomes is discussed in Lee [1482].

Chromosomes have been classified by centromeric index (ratio of arm lengths to total body length), ratios of body lengths from one chromosome to another, chromosome areas, and so on. In particular, classification has been done by applying the "rubber-mask" technique and the stochastic linguistic approach. The chromosome distortion parameters are length, width, angle, and curve, which are adjusted individually and independently in each of the four quadrants of a stereotype. Lee has studied the classification problem through the use of shape-oriented similarity measures comparing a given chromosome image to symmetrical chromosomes, median chromosomes, submedian chromosomes, and acrocentric chromosomes. His algorithm, which relates also to similarity measures introduced in Chapter 4, is based on a previous algorithm for classifying a triangle as an "approximate isosceles triangle," "approximate equilateral triangle," "approximate right triangle," "approximate isosceles right triangle," or "ordinary triangle," and an algorithm used to classify a quadrangle as an "approximate square," "approximate rectangle," "approximate rhombus," "approximate parallelogram," "approximate trapezoid," or "ordinary quadrangle" (presented by Lee in [1481]).

The following algorithm is used for classifying a chromosome image into one of three classes: "approximate median," "approximate submedian," or "approximate acrocentric." This can be done by computing χ_M, χ_{SM}, and χ_{AC} and setting a threshold δ, where δ is a parameter and $0 \leq \delta < 1$. If we compare the $\max\{\chi_M, \chi_{SM}, \chi_{AC}\}$ with δ, there are two possibilities:

1 If $\max\{\chi_M, \chi_{SM}, \chi_{AC}\} > \delta$, then there are two possibilities:
   (a) If the maximum is unique, then we choose the class corresponding to the maximum value and classify the image accordingly.
   (b) If the maximum is not unique, then we define a priority among χ_M, χ_{SM}, and χ_{AC} and classify the chromosome image accordingly.
2 Otherwise, the chromosome image is rejected as not belonging to any of the classes.

Clearly, there are three advantages of shape-oriented similarity measures. These are the following:

1 Two chromosome images may have the same shape but differ in area and dimensions and still be similar.
2 Shape-oriented similarity measures can be normalized between zero and one.
3 Shape-oriented similarity measures are invariant with respect to rotation, translation, or expansion or contraction in size.

Clearly, the perception of form embodies the automatic assignment of a top, a bottom, and sides. Thus orientation plays an important role in chromosome classification done by human beings. Due to the invariance of angle and length measurements with respect to orientation, shape-oriented chromosome classification would not be confused by the orientation of the chromosome. In this sense, shape-oriented chromosome classification is better than chromosome classification done by human beings.

We follow here an example of the application of fuzzy syntactic pattern recognition to handwritten capitals as presented by Kickert and Koppelaar [1295].

The approach is used to describe the patterns in terms of the constituent basic elements and their relationships. Using the formal grammar concept as a method of describing an algorithm for letter recognition, a prior optimization of the algorithm with regard to computing time and memory requirements can be performed.

The fact that the variability in handwritings is very high is the reason for considering a new proposal for the actual class assignment procedure based on fuzzy set concepts. In this method the rather dubious assumptions about underlying probability density functions of handwritings are avoided. More-

Syntactic and Semantic Techniques

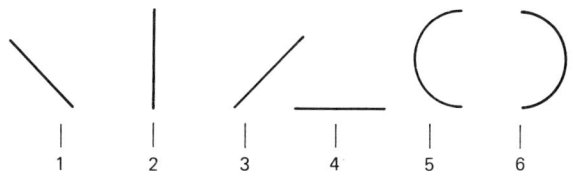

Figure 3.1 Segments ideally constituting capital letters.

over, the concept of vagueness seems to be a more appealing and convincing way of describing the variability in letters than the concept of probability.

The syntactical recognition approach consists of incorporating structural information about the pattern to be recognized and implies viewing each capital as composed of a set of basic elements. The choice of the segments constituting a capital letter is essentially heuristic. Having as one extreme the letter as a whole, the other extreme would be to view each letter as a collection of tiny small lines. An intermediate choice would be some kind of lines and arcs. The reason for not choosing a context-sensitive approach to the segmentation is that this would give a nonoptimizable program.

Kickert and Koppelarr have selected a set of segments ideally constituting capital letters; these are shown in Figure 3.1.

The letter composition process is governed by a context-free grammar. The recognition is performed sequentially: the pattern is scanned from segment to segment, and each segment is separately recognized as a member of the previously defined set of ideal letter segments. After the recognition of a segment, scanning continues to the end or the middle points of the classified segment. The recognition rules are divided according to the different places in the total pattern where a segment can occur, which resulted in eleven (arbitrarily ordered) rules, described in Table 3.1.

The set of (ideal) segments is extended with the "null" segment ε. Formally, the recognition procedure comprises a set V_T of segments ideally constituting a capital letter, a set P of (eleven) subroutines capable of detecting and recognizing the pattern segments sequentially, and a set V_N of actual pattern segments. Referring to the formalism of a regular grammar, they denote the recognition algorithm as a quadruple

$$RG = (V_N, V_T, P, S)$$

where V_N is the finite set of pattern segments, actually Boolean grids; the pattern is discretized in cells that are one or zero depending on whether they are black or not. V_T is the finite set of (ideal) segments into which the segments of V_N have to be classified $V_N = \{1_1, 1_2, \ldots, 1_6, \varepsilon\}$. P is the finite

Table 3.1

Rule Number	Application Place of Recognition Rule	Typical Application
0	Enter the pattern from middle left	
1	At the top of the segment recognized in 0	
2	In the middle of the segment of 1	A
3	At the right-hand end of the segment of 1	N, M
4	At the right-hand end of the segment of 3	M
5	At the bottom of the segment of 0	
6	At the right-hand end of the segment of 5	V, W
7	At the right-hand end of the segment of 6	W
8	In the middle of the segment of 0	
9	In the middle of the segment of 8	K
10	At the right-hand end of the segment of 8	A, H

set of eleven recognition rules (subroutines) such that $P: A \to aB$ or $A \to a$, where $A, B \in V_{N^*}$ and $a \in V_T$. S is the initial total pattern to be recognized.

From the description of the scanning and recognition procedure, it will be clear that the recognized element, parametrically viewed, consists of parameters for its position as well as parameters for its shape. A convenient set of parameters of V_N for this problem is the following: length, slope, curvature, begin, end, and connection points.

Because of the basically subjectivistic concept of a fuzzy set, the actual fuzzy membership assignment to the pattern segments is a heuristic choice. As a consequence, the set of parametrical segments V_T becomes a set of fuzzy sets. Each segment is now considered to be a fuzzy set, which itself constitutes several fuzzy sets, for example, "curved," "long," and "steep." A possible definition of this last composition is to state that the membership function of a "curved," "long," and "steep" segment is

$$\chi(x) = \min\left[\chi_{\text{"curved"}}(x); \chi_{\text{"long"}}(x); \chi_{\text{"steep"}}(x)\right]$$

(the "and" connective being defined as a min operator). Note that the recognition subroutines P, which were mappings

$$P: A \to a \text{ or } A \to aB, \quad A, B \in V_{N^*}, a \in V_T$$

Syntactic and Semantic Techniques

are now

$$\chi_a: A \to [0,1] \text{ or } A \to [0,1]B$$

with $A, B \in V_{N^*}$ (as before). However, χ is now a fuzzy set of V_T, represented by its membership function $\chi_a \in [0,1]$. This function $\chi_a(x)$ is a measure for the goodness—that is, the reliability—of the recognition of the observed pattern segment.

There are various ways of making a decision as to whether the segment is, for example, an arc. We can introduce a threshold level $\alpha \in [0,1]$ such that, if $\chi_a(x) \geq \alpha$, then x is an arc, while if $\chi_a(x) < \alpha$, x is not an arc. Another decision procedure, especially in the case of several fuzzy sets defined on the segment x, is to take that class a_j for which

$$\chi_{a_j}(x) = \max_i \chi_{a_i}(x).$$

The whole character to be recognized is considered as consisting of the intersection of all subsegments. Under the assumption that the intersection of two fuzzy sets is defined as

$$\chi_{A \cdot B}(x) = \min[\chi_A(x), \chi_B(x)],$$

the membership of the whole letter H_l will be

$$\chi_{H_l}(S) = \min[\chi_{a_0}(x_0); \chi_{a_1}(x_1); \ldots; \chi_{a_n}(x_n)]$$

Here a_j corresponds to each of the elements (fuzzy segment sets) of V_T that were recognized from the observed pattern segments x_j; S is the whole pattern constituted by $x_j (j = 1(1)n)$. The set of indices $a_0, a_1, \ldots, a_n$ forms a class description of one of the 26 letters H_l ($l = 1(1)26$). From the final 26 values $\chi_{H_l}()$ that class H_m is assigned to the pattern S for which $\chi_{H_m}(S) = \max_l \chi_{H_l}(S)$, $l = 1(1)26$.

Clearly, the essential differences between this fuzzy classification procedure and those using probabilistic concepts lies not only in its intrinsic use of the vagueness of an observed pattern but also in its different evaluation procedure, based on a fuzzy set theoretic method of inference. The simplicity of this evaluation method, only consisting of min and max operators, is certainly an advantage in computing time as well.

Comparison of this approach to the stochastic Bayesian approach, as well as criticism of the technique described above, has been discussed by Stallings [2453].

Fractionally fuzzy grammars were used be DePalma and Yau [489] also. For the recognition of handwritten characters, we follow their presentation in [489].

In syntactic pattern recognition, the patterns are strings over the terminal alphabet. These strings must be parsed in order to find the pattern classes to which they most likely belong. Many parsing algorithms require backtracking; that is, after applying some rules, it is discovered that the input string cannot be parsed successfully by this sequence of rules. Rather than starting it from the beginning again, it is desirable to reverse the action of one or more of the most recently applied rules in order to try another sequence of productions. With nonfuzzy grammars it is sufficient to keep track of the derivation tree as it is generated, with each node being labeled with a symbol from V. However, with fuzzy grammars this tree is not sufficient since the fuzzy value at the ith step is the minimum of the value at the $(i-1)$th step and the fuzzy membership of the ith rule. If this minimum was the ith rule's membership, there is no way to know the fuzzy value at the $(i-1)$th step. Thus the fuzzy value at each step must also be remembered at each node, and hence the memory requirements are greatly increased for many practical problems.

A second drawback of fuzzy grammars in pattern recognition is the fact that all string in the fuzzy language can be classified into a finite number of subsets by their membership in the language. The number of such subsets is strictly limited by the number of productions in the grammar.

To overcome these restrictions, Depalma and Yau have introduced a new method of computing the membership of a string x that can be derived by the m sequences of production rules, $r_1^k r_2^k \ldots r_{l_k}^k$, of lengths l_k, where $k = 1, 2, \ldots, m$. This leads to the following definition.

Definition 3.3.1

A *fractionally fuzzy grammar* (FFG) is a 7-tuple: FFG $= (V_N, V_T, S, P, J, g, h)$, where V_N, V_T, S, P, and J are the nonterminal alphabet, the terminal alphabet, the starting symbol, the set of productions, and a distinct set of labels on the productions as a fuzzy grammar, respectively. The functions g and h map the set of productions into the nonnegative integers such that $g(r_k) \leq h(r_i)$ for all r_i in P. A string is generated in the same manner as that by a fuzzy grammar, except that the membership of the derived string is given by

$$\chi(x) = \sup_k \frac{\sum_{j=1}^{l_k} g(r_j^k)}{\sum_{j=1}^{l_k} h(r_j^k)}$$

where $0/0$ is defined as 0.

Syntactic and Semantic Techniques

Because $0/0$ is defined as 0, it is clear that $0 \leq \chi(x) \leq 1$ for all x. It is also clear that backtracking over a rule r can now be accomplished by simply subtracting $g(r)$ and $h(r)$ from the respective running totals.

Clearly, we can divide the rules into three classes—those that strongly indicate membership in the class, those that strongly indicate membership in another class, and those that serve little purpose in separating the classes but that are traits between different classes.

The following results are due to Depalma and Yau [489].

Theorem 3.3.1 The set of all languages generated by type i fractionally fuzzy grammars properly includes the set of all languages generated by type i fuzzy grammars, where $i = 0, 1, 2,$ and 3.

Lemma 3.3.1 Let FFG be a context-sensitive fractionally fuzzy grammar and let some derivation contain the sequence

$$\cdots \to \theta_i \to \theta_{i+1} \to \cdots \to \theta_{i+k} \to \cdots,$$

where $\theta_i = \theta_{i+k}$. Then either $k \leq n^p$, where $n = |V|$ is the number of symbols in the vocabulary and $p = |\theta|$ is the length of the string θ_i, or $\theta_{i+j} = \theta_{i+m}$ for $0 \leq j < m < n^p$.

Lemma 3.3.2 Let FFG be a fractionally fuzzy grammar, and $x \in L(\text{FFG})$, which is derivable by the sequence

$$S = \theta_0 \to \theta_1 \to \theta_2 \to \cdots \to \theta_n = x.$$

If $\theta_j = \theta_k$ for $j < k$ and $0 \leq j < n$, then the membership of x in FFG is at least

$$\max \left[\frac{\sum_{m=1}^{j} g(r_{i_m}) + \sum_{m=k+1}^{n} g(r_{i_m})}{\sum_{m=1}^{j} h(r_{i_m}) + \sum_{m=k+1}^{n} h(r_{i_m})}, \frac{\sum_{m=j}^{k} g(r_{i_m})}{\sum_{m=j}^{k} h(r_{i_m})} \right].$$

Lemma 3.3.3 Let FFG be a context-sensitive fractionally fuzzy grammar with n symbols in V. Let $R_0^1 = \{S\}$. Let R_0^k be the set of all strings over V of length k that can be directly generated from a string of length less than k. Let R_j^k be the set of all strings of length k that can be directly generated from a string in R_{j-1}^k, $j = 1, 2, \ldots$. Then the set $R^k = R_0^k R_1^k \ldots R_n^k$ contains all strings over V of length k that can be generated by the FFG, and the derivation needed to generate R^k will contain all the simple derivation loops on strings of length k in L(FFG).

Depalma and Yau tested the usefulness of fractionally fuzzy grammars in pattern recognition by converting data points from handwritten characters into a string of symbols that comprised the terminal alphabet. This was accomplished by comparing each adjacent pair of points to see the relative direction traveled by the pen at that point and classifying the direction into one of eight directions, each separated by 45 degrees, with class 0 being centered at 0 degrees (the $+X$ direction) and the remaining classes being numbered 1 through 7 in a counterclockwise direction. Thus the terminal alphabet consisted of the eight octal digits, that is, $V_T = \{0, 1, 2, \ldots, 7\}$.

The individual letters were separated by an operator using an interactive graphics program. These letters then consisted of strings of octal digits whose lengths varied from 10 to about 70 characters in length. The machine was asked to separate the i's, e's, t's, and the l's without the dots on the i's and the crossings of the t's, with only regular fractionally fuzzy grammars. The grammars were generated by cut and try methods based upon the principle of class distinction and the effects of each rule on the final membership of any string generated by the rule. An important consideration, for example, was that if rule r was used, the fuzzy membership of the string would be changed in the direction toward the value $g(r)/h(r)$ by that application of rule r. Thus if $g(r)/h(r)$ was close to 1, the membership of the string would be increased and, if $g(r)/h(r)$ was close to 0, the membership of the string would be decreased. In order to make this particular technique attractive and useful, algorithms for constructing the fractionally fuzzy grammars for a given training set, to make them practical for solving pattern recognition problems, are obviously needed. This is true in spite of the fact that the results obtained by Depalma and Yau in the cursive script recognition experiment clearly demonstrate the capabilities of the idea.

3.3.3 Semantic Pattern Recognition

As mentioned in Section 3.3.2, quantitative measures of the proximity of two n-sided polygons have been investigated by Lee [1481]. The proximity indices are based on angular and dimensional comparisons. Thus, for instance, triangles can be classified into "approximate right triangle," "approximate isosceles triangle," "ordinary triangle," and so on. Siy and Chen [2403] have a similar approach in a handwritten numerical character recognition procedure. Each numeral is decomposed into primitives such as

| H | V | P | N | C | D | A | V | S | Z | OL | OR | OA | OB | OO |

Figure 3.2 Branch feature set.

horizontal lines or portions of circles. The authors use proximity measures for the (semantic) identification of the primitives. However, the structural part of Siy and Chen's procedure (graph matching) is not fuzzy.

In what follows we outline Siy and Chen's basic idea as presented in [2403].

Handwritten characters are a distorted variant of printed characters; therefore, any study of handwritten characters must start with printed characters. There are essentially three basic elements of alphanumeric characters: (1) the straight line (vertical, horizontal, and slant), (2) the circle, and (3) a portion of a circle of various orientations, all considered as fuzzy sets and shown in Figure 3.2.

Let $B = \{H, V, P, N, C, D, A, V, S, Z, OL, OR, OA, OB, OO\}$ denote the branch feature set shown in Figure 3.2.

Two patterns are said to be equivalent if their functional descriptions are the same. Thus $F(5) = H(1,2) \cdot V(1,3) \cdot D(3,4)$ defines an equivalent class in which the character 5 is a typical member. This is the same as the idea of generalization, which means that we only have to learn one typical example of character 5 in order to recognize any equivalent variants. This definition of pattern (or stimulus) equivalence is both size and position invariant, which is compatible with visual perception in man. Other variants of the character 5 of which the functional descriptions are different are $F(5') = H(1,2) \cdot P(1,3) \cdot D(3,4)$; $F(5'') = P(1,2) \cdot V(1,3) \cdot D(3,4)$. Using the fuzzy logical OR operation, we can combine the various descriptions of a character. For example, the character 5 can be represented by

$$F(5) = F(5) + F(5') + F(5'')$$

$$= H(5,2) \cdot V(1,3) \cdot D(3,4)$$

$$+ H(1,2) \cdot P(1,3) \cdot D(3,4)$$

$$+ P(1,2) \cdot V(1,3) \cdot D(3,4)$$

The decision criteria used for pattern classification are executed in two steps.

1   The branch features of the pattern to be classified and the branch features of the prototypes of each class are compared. Those prototypes that perfectly match the branch features of the pattern are retrieved.

2   The node pairs of the same branch type of the pattern to be classified and each retrieved prototype are compared. The fact that the numbering of the nodes in the pattern and prototypes may not be the same implies that an isomorphic mapping has to be found. This can be done as follows. Let $(n'_\phi(bi), n'_\theta(bi))$ and $(n_\phi(bi), n_\theta(bi))$ denote the node pair of branch type bi, $i = 1,\ldots,n$, of the pattern and prototype under consideration, respectively. The mapping

$$\Psi = \left\{ \left(n'_\phi(bi), n_\phi(bi) \right), \left(n'_\theta(bi), n_\theta(bi) \right) \mid k = 1,\ldots,n \right\}$$

is defined. When this is isomorphic, the prototype is accepted; when this is not, it is rejected.

The procedure is making use of the following labeling:

1   *Node Detection and Labeling*: A node set in the skeleton of a pattern is defined as the collection of tips (points that have one neighbor), corners (points that have two neighbors and where an abrupt change of line direction occurs), and junctions (points that have three or more neighbors).
2   *Branch Detection and Labeling*: A branch is a line segment connecting a pair of adjacent nodes. A branch of which the length is less than threshold S_B is considered extraneous and consequently removed by declaring its nodes equivalent.

In classifying branches, two sources of fuzziness are attributed to the measure of (1) straightness and (2) orientation.

1   *Measure of Straightness*: The measure of straightness of a branch is determined by fitting a straight line with the minimum least squares error, and the branch is represented by the line with the least squares error. A measure of straightness for a noncircular branch is defined by

$$\chi_{SL} = 1 - S/S_T, \quad \text{if } S < S_T$$
$$= 0, \quad \text{if } S \geq S_T$$

where S_T is the threshold least squares error. A given branch is classified as a portion of a circle, if $0 \leq \chi_{SL} < 0.5$; a straight line, if $0.5 < \chi_{SL} \leq 1$; either, if $\chi_{SL} = 0.5$.

2 *Measure of Orientation*: If the classification of a branch to the class of a straight line, portion of a circle, or a circle with node is known, then the measure of orientation can be used to further characterize this branch:
   (a) *Straight Line Class*: This class contains the sets H, V, P, N.
   (b) *Portion of Circle Class*: This class contains the sets C, D, A, V, S, Z. These sets can be further grouped as vertical (C, D), horizontal (A, V), or either (S, Z).
   (c) *Circle with Node Class*: This class contains the sets OL, OR, OA, OB.

The experimental data used in this case shows 98.4 percent correct classification and has illustrated that fuzzy logic can be applied to the feature extraction of handwritten numerical characters. Dominant prototypes that are functional representations of branch features and node features can be learned and stored for each character. For pattern classification the functional representation of the incoming pattern is compared with those of the prototypes. The recognition system proposed was successfully simulated on a digital computer using the IEEE pattern recognition data base.

Kotoh and Hiramatsu [1418] have proposed an approach to semantic pattern recognition where a feature is viewed as a fuzzy partition of the pattern space. Namely, if the possible values of feature y are given by the fuzzy members or the linguistic values $x_1, x_2, \ldots, x_k$, then these values realize a fuzzy partition provided that the orthogonality condition

$$\sum_{i=1}^{k} \chi_{x_i}[y(q)] = 1, \quad \forall q \in X,$$

where $y(q)$ is the specific feature of q, is satisfied. A fuzzy pattern class is expressed as a logical formula of feature values that correspond to different features. To deal with this model, operations such as refinement and unification of fuzzy-valued features related to intersection and union of fuzzy sets, respectively, have been introduced in [1418]. Two pseudocomplementations of feature values are defined; these differ from the usual fuzzy set complementation; for example, "medium or large" and "medium and large" are the two pseudocomplements of "small." In this approach, each object is evaluated with respect to a fuzzy pattern class by means of a fuzzy logical expression that is specific to this pattern class. However, an opaque algorithm cannot always be reduced to the computation of a fuzzy logical formula. As a specific example, let F be a fuzzy pattern class defined by the fuzzy feature values $F_1, F_2, \ldots, F_r$ where F_i is a fuzzy value of feature i. An object q is thus characterized with respect to the class F by k membership values $\chi_{F_i}(y_i(q))$ denoted $\chi_i(q)$ for convenience. $\chi_F(q)$ is then constructed by aggregating the $\chi_i(q)$ in some subjective manner.

Among the several fuzzy pattern classes $\{F^j\}_{j=1}^t$ that q can be assigned to, we assign q to class s such that

$$\chi_{F^s}(q) = \max_j \{\chi_{F^j}(q)\}$$

provided that $\chi_{F^s}(q)$ is sufficiently large. A related idea can be found in Kaufmann's work [1255], in which he has considered fuzzy perceptions as a recognition tool. The linguistic approach of Zadeh [2977] is using linguistic values for features, such that the relationships between $\chi_F(q)$ and the feature fuzzy values $y_j(q)$ are expressed as an $(r + 1)$ary fuzzy relation R_F on $X_1 \times \cdots \times X_r \times [0, 1]$ where X_j is the universe of $y_j(q)$. R_F is specific to the fuzzy pattern class F and is derived from a relational tableau having n lines and $(r + 1)$ columns such that p_i^j is a linguistic feature value for $j \leq r$ and p_i^{r+1} is a linguistic truth value. A first way of calculating $\chi_F(q)$ is to explicitly construct

$$R_F = \bigcup_{i=1,n} \bigcap_{j=1,r+1} p_i^j.$$

By knowing the linguistic feature values $y_i(q)$, $i = 1,\ldots,r$, of an object q, $\chi_F(q)$ is obtained by max-min composition

$$\chi_F(q) = [y_1(q) \times \cdots \times q_r(q)] \circ R_F.$$

Another way of determining $\chi_F(q)$ is to build a branching questionnaire by viewing each column j of the relational tableau as a set of possible answers to a question concerning feature j. Analogously, Chang and Pavlidis [347] discuss certain theoretical aspects of fuzzy decision trees to be presented in the next section.

3.4 RECOGNITION VIA FUZZY DECISION TREE

The fuzzy decision tree discussed by Chang and Pavlidis [347] is a tree such that each nonleaf node i has a k-tuple decision function f_i from X to $[0, 1]^k$ and k ordered sons. Each nonleaf son j of a node i corresponds to a question determined by the answer to the preceding question i. $f_i(q; j)$ valuates the branch from i to j. More specifically, a fuzzy decision function f_x at node x is a real-valued unary k-tuple function $k \geq 2$,

$$f_x: X \rightarrow [0, 1]^k,$$

where X is the input (e.g., a digitized picture I, or a voice spectrograph S)

and the k-tuple is the labels (decision values) $v(x_i)$ of the outgoing branches (x, x_i), $i = 1, \ldots, k$ where x_i is the ith son of node x. A 0-1 decision function f_x is a fuzzy decision function that can assume only integer values 0, 1:

$$f_x: X \to \{0, 1\}^k,$$

with exactly one element of the k-tuple equal to 1.

Definition 3.4.1 [347]

A decision tree T_r is a tree with root r such that each nonleaf node i has a corresponding k-tuple decision function f_i and k ordered sons $i_1, \ldots, i_k$. A fuzzy decision tree is one with fuzzy decision functions, and a 0-1 decision tree is one with 0-1 decision functions. A decision tree is binary if and only if all the decision functions are double functions. A complete (or completely balanced) k-ary decision tree is a decision tree in which the decision function at each nonleaf node is k-ary and each path from the root to a leaf has the same length.

Definition 3.4.2 [347]

The decision path (x, y) is the path of the decision tree from node x to node y. The decision path x is the decision path (root, x). The decision path $(w, x, \ldots, y, z)$ is the path from node w via nodes $x, \ldots, y$ to node z.

Definition 3.4.3 [347]

The (decision) value $V(x)$ of a decision path x is the product of the decision values (labels) of the branches composing it, that is,

$$V(x) = \bigwedge_{y \in \text{path } x} v(y).$$

where $\wedge$ is the real-number product in the prob model or is the minimum function as defined in the max-min model.

It has been shown [347] that, given the same decision tree T with the same decision values, the 0-1, fuzzy max-min, and fuzzy prob criteria may all lead to different decisions. By definition, two decision trees are equivalent if and only if they give the same decisions.

It is interesting to note, however, that the max-min decision tree and the prob decision tree are mutually reducible to one another in $0(n)$ time, where n is, as usual, the number of decision classes.

An example of a decision tree is given in Figure 3.3. A guaranteed bottom algorithm for general fuzzy decision trees has time complexity $0(n)$

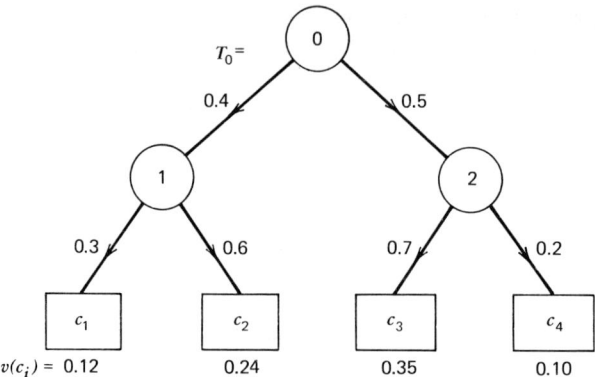

Figure 3.3 Fuzzy decision tree.

as a least upper bound. The branch-bound-back track (BBB) algorithm presented in [347] usually has the complexity $O(\log n)$ and still gives the optimal solution. It is interesting to note that this approach differs from Zadeh's [2981], in which a decision tree (branching questionnaire) is characteristic of a pattern class and the leaves are the p_i^{r+1}. Moreover, in Chang and Pavlidis' model, the truth values $f_i(q; j)$ are numerical and not linguistic. Some of the interesting and useful results of Chang and Pavlidis are the following.

Theorem 3.4.1 [347] Given any binary 0-1 decision tree T that divides a bounded m-dimensional space into k subspaces (i.e., classification into k classes with m parameters), we can always construct a corresponding fuzzy decision tree T' arbitrarily closely equivalent to T (by choice of the scale r). In other words, the classification hyperplanes induced by T' may be made as close to those induced by T as desired.

For the linear classifier-decision tree theorem, we have the following.

Theorem 3.4.2 [347] For any general two-parameter two-class linear classifier, if $Ax_1 + Bx_2 > C$, then Class 1, and if $Ax_1 + Bx_2 < C$, then Class 2, where $C \geqslant 0$. For a bounded (open or closed) two-dimensional sample space S, there always exists an equivalent fuzzy decision tree T' using trivial comparisons alone (equivalent in the sense that T' always gives exactly the same classification as the linear classifier).

The efficient BBB algorithm for fuzzy decision trees has been used in [347], demonstrating the advantages of the former over 0-1 decision trees when handwritten numeral recognition is needed. For more details on the BBB algorithm and its use in pattern recognition, the interested reader is referred to [347] and [348].

3.5 FUZZY PARTITION VIA RUSPINI

In one of the earliest works on fuzzy techniques in pattern recognition, Ruspini [2244] has introduced, in 1969, the notion of a fuzzy partition to represent the clusters in a data set. A fuzzy partition is a family of fuzzy sets $F_1, \ldots, F_m$ on X such that

$$\forall x \in X, \quad \sum_{i=1}^{m} \chi_{F_i}(x) = 1.$$

Ruspini states that, "The advantage of fuzzy set representation in cluster analysis is that stray points or points isolated between clusters as well as other types of uncertainties may be classified as such" (Ruspini [2244]).

According to Ruspini [2245], the problem of fuzzy clustering may be stated as follows. Assume a finite data set X and a positive real-valued function δ (the distance or dissimilarity function), whose domain is X^2, such that:

1 $\forall x \in X, \delta(x, x) = 0$.
2 $\forall x, y \in X, \delta(x, y) = \delta(y, x)$.

Find a fuzzy partition $F_1, \ldots, F_m$, where m is *a priori* known, such that close elements in X (in the sense of δ) will have similar classification and dissimilar elements will have different classification. The classification of an element x is the vector $C(x) = [\chi_{F_1}(x) \ldots \chi_{F_m}(x)]$. One of the possible ways of satisfying the above requirement is to select the function $C(x)$ so as to minimize some suitably defined functional. Let us outline Ruspini's idea for constructing such a functional.

Let v be a functional from $[0, 1]^m \times [0, 1]^m$ to $\mathbb{R}^+$ such that $v(a, a) = 0$ and $v(a, b) = v(b, a)$, and let f be a positive nondecreasing not identically zero real function of one real variable satisfying $f(0) = 0$; then the function C should be selected such that

$$\forall x, y \in X, \quad v(C(x), C(y)) = f(\delta(x, y)).$$

Generally, this equation has no solution. It is then relaxed into a minimization problem: Find C minimizing

$$\sum_{x, y \in X} w(x)w(y)[v(C(x), C(y)) - f(\delta(x, y))]^2$$

where w is an appropriate weighting function. Usually, v is taken as a Euclidean distance. Various forms of f have been tried and discussed by Ruspini [2244]–[2250], where many experimental results are provided. A slightly different approach using association measures is described in [2251]. The association measure between a point x and a fuzzy set F on X is taken as the inverse of a weighted average distance between x and F, where the average distance between x and F is defined as

$$d(x, F) = \frac{1}{|F|} \sum_{i=1}^{|X|} \chi_F(x^i) \, \delta(x, x^i).$$

The basic idea is that the membership value of x in a fuzzy cluster F_j varies in proportion with the inverse of the average distance between x and F.

In a recent paper [2251], Ruspini presents some of his ideas regarding recent developments in fuzzy cluster analysis. We follow [2251].

Definition 3.5.1

Let g be a fuzzy set in X. Then g is called a fuzzy r-cluster (or a cluster, for short) iff:

1 $g(x) = 1$ implies that $g(y) = r(x, y)$.
2 $|g(x) - g(y)| \leq 1 - r(x, y)$.

Condition (1) of Definition 3.5.1 extends both conditions of the definition of conventional cluster.

Condition (2) is essentially a continuity condition having no counterpart in the conventional case. This condition is the simplest formalism available to indicate that the fuzzy classifications of similar points should be similar. Note that no value other than 1 should be used as a factor of the difference $1 - r(x, y)$ on the right-hand side as it is desirable that no condition of relationship between the classifications for x and y be imposed whenever $r(x, y) = 1$.

The following proposition is easily derived from the above definition.

Proposition 3.5.1 If g and h are fuzzy clusters, then they are either identical or their cores are disjoint.

Definition 3.5.2

Let C be a family of fuzzy subsets of X such that:

1. Every point of X is in the core of some member of C.
2. Every member of C has a nonempty core.

Then C is said to be a fuzzy coverage of X.

Definition 3.5.3

Let r be a reflexive, symmetric fuzzy relation in X. If the inequality

$$|r(x, y) - r(x, z)| \leq 1 - r(y, z)$$

is satisfied for all x, y, and z in X, then r is called a likeness relation in X.

Theorem 3.5.1 Let C be a fuzzy coverage of X. In order for every element of that coverage to be a fuzzy cluster with nonempty core, it is necessary and sufficient that:

1. r be a likeness relation.
2. c be the fuzzy quotient of X by r.

It is clear that this theorem really means that: "A necessary and sufficient condition for the existence of a fuzzy r-cluster coverage is that r be a likeness relation. In that case the coverage is unique and its elements are the fuzzy sets with membership function $r(x, \cdot)$, for all x in X."

The significance of this existence requirement is made clearer by the following result.

Proposition 3.5.2 If r is a likeness relation, then its complement $d = 1 - r$ is a bounded pseudometric in X.

Essentially, this result indicates that, for r to be a likeness relation, its complement must be a (pseudo)metric in X. Thus the requirement of likeness is not only weaker than that of equivalence (as all equivalence relations are likeness relations), but it is also weaker than that of fuzzy similarity, which has complements that must satisfy the stronger ultrametric inequality.

$$d(x, y) \leq \max(d(x, z), d(z, y)), \quad x, y, z \text{ in } X.$$

In a similar way, let P denote an element of the complete complemented lattice L. If the value $r(x, y)$ of the similarity between any two points of X is equated with the degree of truth (in a [0, 1] scale) of the statement: "It is desirable to classify x and y in the same cluster," then the similarity function r defines a truth function

$$T: B_X \to [0, 1]$$

such that if

$$P(x, y) \equiv \text{"It is desirable to classify } x \text{ and } y \text{ in the same cluster"},$$

then

$$T(P(x, y)) = r(x, y).$$

If the domain of the function T is extended to encompass all of L by means of the formulas of the modified Lukasiewicz's Aleph-1 logic:

$$T(\bar{P}) = 1 - T(P)$$

$$T(P \cap Q) = [T(P) + T(Q) - 1]^+$$

$$T(P \cup Q) = \min(T(P) + T(Q), 1)$$

$$T(P \to Q) = \min(1, 1 - T(P) + T(Q)) = T(\bar{P} \cup Q),$$

then requiring that the relation induced by common cluster membership be transitive is equivalent to asserting that the compound statement

If ["It is desirable to classify x and y in the same cluster"

and "It is desirable to classify y and z in the same cluster"],

then "It is desirable to classify x and z in the same cluster"

is always true (i.e., has a truth value of 1) for any x, y, and z in X; in a formal way we get

$$\min[1, 1 - T(P(x, y) \cap P(y, z)) + T(P(x, z))] = 1$$

or

$$T(P(x, z)) \geq [T(P(x, y)) + T(P(y, z)) - 1]^+ ,$$

which implies

$$r(x,z) \geq [r(x,y) + r(y,z) - 1]^+.$$

To quote Ruspini [2251]:

> In summary, the requirement that similarity relations be likeness relations (i.e., complements of pseudometrics) is equivalent to the transitivity of the clustering relation. Thus, under the interpretation provided by the modified Aleph-1 multivalued logic, the clustering relation is an equivalence relation if and only if the underlying similarity structure is given by a likeness relation. This is the natural extension of the corresponding result for conventional sets which stated that the clustering relation was an equivalence relation if and only if the underlying similarity structure was given by a conventional equivalence relation. The substantial difference in this case is that, under the multivalued logic interpretation, meaningful solutions exist for a much larger class of problems than for the conventional case.

Clearly, if fuzzy logic forms are being used instead of the modified Aleph-1 logic, that is,

$$T(\bar{P}) = 1 - T(P)$$

$$T(P \cap Q) = \min(T(P), T(Q))$$

$$T(P \cup Q) = \max(T(P), T(Q))$$

$$T(P \to Q) = \max(1 - T(P), T(Q)) = T(\bar{P} \cup Q),$$

then the transitivity of the clustering relation is equivalent to the equation

$$r(x,z) = \sup_{y} [\min(r(x,y), r(y,z))],$$

which defines fuzzy equivalence or similarity relations.

The prototype interpolation introduced in [2251], based on the extension to $F(X)$ of a similarity function r, can be thought of as being equivalent to the definition of the similarity between imaginary centroids of each fuzzy agglutinate.

By definition of an agglutination, it is meant that an extension r^*, defined in $F(X)^2$, is sought for the similarity relation r, defined in X^2. In other words, it is desirable to consider fuzzy subsets of X as possible amalgamates and appropriate extensions of r are sought so as to provide a meaningful similarity relation between fuzzy sets. Any such similarity function will then define the similarity between the fuzzy amalgamates as a weighted average of the similarities between points of x, the weight values being dependent on the degrees of membership of each point to the respective agglutinates.

Definition 3.5.4

By centroid of a fuzzy subset, it is meant the point x_f defined by

$$x_f = \frac{1}{\Omega(f)}\left(\sum yf(y)\right)$$

where the sums are over the support of f.

Definition 3.5.5

Let X be a subset of a Euclidean space R^n and let the similarity function r be the complement of the Euclidean distance d in X. Then the extension r^* of r must be equal to the complement of the Euclidean distance between centroids of fuzzy subsets of R (i.e., equivalent, through the fuzzy set to centroid mapping, to the Euclidean distance in the convex hull of X).

Addition of Euclidean consistency as a goal assures that, in the Euclidean case, the extension will be consistent with our own intuitive notion of distance between centroids in Euclidean spaces. Moreover, the importance of the inclusion of Euclidean consistency as a goal is due to the following results, which link it with likeness relations and therefore with the existence of fuzzy clusters.

Theorem 3.5.2 Let r be a similarity relation in $X \subset R^n$. If r is the complement of the Euclidean distance r in X, then its unique extension r^* satisfying the Euclidean consistency property is given by

$$(d^*(f, g))^2 = -\tfrac{1}{2}\sum\sum[(\phi(x) - \gamma(x))(d^2(x, y))(\phi(y) - \gamma(y))]$$

where

$$\phi(x) = \frac{f(x)}{\Omega(f)}$$

$$\gamma(x) = \frac{g(x)}{\Omega(g)},$$

where

$$d^* = 1 - r^*, \qquad d = 1 - r,$$

and where all sums are over the set union of the supports of f and g.

Theorem 3.5.3 Let f and g be fuzzy subsets of a finite reference set X. Let r be a similarity relation in X and let $Q^*(f, g)$ be defined by

$$Q^*(f, g) = -\tfrac{1}{2} \sum \sum \left[(\phi(x) - \gamma(x))(d^2(x, y))(\phi(y) - \gamma(y)) \right],$$

where ϕ and γ are defined above and

$$d = 1 - r,$$

such that all the sums are over the set union of the supports of f and g. In order for $Q^*(f, g)$ to be nonnegative for all pairs of (nonempty core) fuzzy subsets f and g in $F(X)$, it is necessary and sufficient that r be a likeness relation.

If r is a likeness relation, then Q^* defines a similarity relation r^* in $F(X)$ by

$$r^*(f, g) = 1 - Q^*(f, g)^{1/2}.$$

Further, r^* is itself a likeness relation in $F(X)$.

Clearly, likeness relations not only assure the existence of clusters in X but, further, they can be extended as likeness relations of the set $F(X)$ of all nonempty core fuzzy subsets of X. The extension r^* can then be used (by forming the fuzzy set theoretic quotient of $F(X)$ by r^*) to define fuzzy clusters having fuzzy subsets of X as typical elements (i.e., generalized "centroids").

We feel that this important axiomatic development allowed the identification of relations between the concepts of *clusters* and cluster prototype and is very important to future development of fuzzy techniques in pattern classification. Other interesting axiomatic developments regarding the above and the concepts of clusterability and hierarchical taxonomy have been discussed by Ruspini [2247], and provide some theoretical guidelines that enhance the understanding of taxonomical structures as well as the use of fuzzy set concepts in unsupervised classification of data samples.

3.6 FUZZY ISODATA

In many problems related to pattern analysis, the researcher is interested not only in the clustering of data but also in the cluster centers as representative elements of the cluster or its typical value. This has been achieved by the ISODATA algorithm.[†]

[†] Ball, G. H., and Hall, D. J., (1967). A clustering technique for summarizing multivariate data," *Behav. Sci.*, **12**, 153–155.

The steps of the algorithm are:

1. Read initial cluster means, splitting parameters, lumping parameters, and the minimum number of points a cluster may have.
2. Group each sample of the feature space to its nearest (Euclidean) cluster center.
3. After all samples have been grouped, compute the coordinate values of each cluster center.
4. If any "split threshold" is exceeded, split that cluster into two clusters. The "split threshold" is one of the *a priori* established thresholds.
5. If a split occurred in the previous step, regroup the data and compute new coordinates for the cluster centers.
6. If any cluster has too few members (where "too few" is established *a priori*), eliminate that cluster.
7. If a cluster was eliminated in the previous step, regroup (by nearest neighbor or any other relevant method) and compute new coordinates for the cluster centers.
8. If the distance between the two cluster centers is smaller than some predetermined distance, combine the two clusters.
9. If any two clusters were combined in the previous step, compute the new coordinates for the cluster error.
10. Repeat steps 2 through 9 until steady state occurs or until a maximum number of iteratives is exceeded.

We can use the least square error criterion for the classification, namely

$$J = \sum_{j=1}^{m} \sum_{x \in X_j} d^2(x, \bar{x}_j)$$

where $\bar{x}_j = \sum_{x \in X_j} x = $ centroid of X_j.

Even though ISODATA is widely used, we would like to raise several objections to it:

1. Spurious fixed points.
2. Possible split clusters.
3. Complications with "goodness" of technique (do we really get clusters?).
4. Need of a tie breaking rule.

Moreover, ISODATA assumes that the data to be clustered is linearly operable and requires that the following information be specified:

1. The number of initial clusters.
2. The initial cluster means.
3. The split threshold.
4. The lump threshold.
5. The minimum number of elements to constitute a cluster.

As far as computer requirements, ISODATA requires that the entire sample space be kept in memory while most algorithms require only that the data be stored on a mass storage unit (disc, drum, tape). This imposes a significant sample space restriction on the ISODATA algorithm.

A slight improvement can be obtained if we extend the criterion function to its fuzzy form by redefining J as

$$J = \sum_{j=1}^{m} \sum_{x \in X_j} d^2(x, \bar{x}_j) \chi_j^k(x)$$

where $k \geq 1$. This is really an extension of least squares error, which reduces to the nonfuzzy case where

$$\chi_j(x) = \begin{cases} 0 \\ 1 \end{cases}$$

and not a number in the closed interval $[0, 1]$.

Northouse and Fromm[†] have suggested the following technique for the threshold calculations regarding ISODATA:

1. Compute $\bar{x}_i$, the grand mean value of the ith dimension for the entire sample population consisting of N samples.

2. Compute s_i, the standard deviation of the ith dimension for the entire sample population.

3. Define the starting vector as Y_k. There will be 2^n such vectors, each having n elements.

$$Y_k = (y_{k1}, y_{k2}, \ldots, y_{kn}), \quad k = 1, 2, \ldots, 2^n$$

[†]Northouse, R. A., and Fromm, F. R., (1972). CLASS: Non-parametric clustering of large data problems, TA-AI-72-2, Univ. of Wisconsin, Milwaukee.

where

$$y_{ki} = \bar{x}_i + \text{sgn}\left\{\text{sine}\left(\frac{2}{2^i}k - \frac{2}{2^{i+1}}\right)\right\}s_i$$

$$\bar{x}_i = \frac{1}{N}\sum_{j=1}^{N} x_{ij} \quad \text{and} \quad s_i = \left[\sum_{j=1}^{N} \frac{(x_{ij} - \bar{x}_i)^2}{(N-1)}\right]^{1/2}, \quad i = 1, 2, \ldots, N.$$

Include $\bar{X}$, the grand mean value of the entire sample population, as the $(2^n + 1)$st point.

$$Y_{2^n+1} = \bar{X}$$

Perhaps one of the most important parameters in the clustering process is the splitting parameter. This parameter determines when a cluster should be broken into two or more clusters. This parameter is usually static; that is, it does not change during the clustering process. It may be calculated dynamically by the following procedure.

For each successive iteration k, we can determine the new splitting threshold as

$$S_k = S_{k-1} + \frac{(1 - S_0)}{\gamma}$$

where $S_0 \cong 0.6$ (by experimentation) and $\gamma =$ maximum number of iterations.

To determine if a cluster is to be split, we can use a nearest neighbor criterion.

We need to first calculate the cluster means η_i^k for each cluster k in every dimension i where

$$\eta_i^k = \frac{1}{N^k}\sum_{p=1}^{N^k} x_i^{kp}, \quad i = 1, 2, \ldots, n$$

where x_i^{kp} is the ith element of the pth sample in cluster k and N^k is the number of samples in cluster k. From this we can calculate the average distances: D_i^{k1} is the average distance of points $> \eta_i^k$ in the ith dimension from the mean η_i^k; and D_i^{k2} is the average distance of points $< \eta_i^k$ in the ith dimension from the mean η_i^k where

$$D_i^{k1} = \frac{1}{N^{k1}}\sum_{p=1}^{N^{k1}} \left(x_i^{kp} - \eta_i^k\right)$$

and

$$D^{k2} = \frac{1}{N^{k2}} \sum_{p=2}^{N^{k2}} \left(x_i^{kp} - \eta_i^k\right)$$

for N^{k1} = number of points in cluster k with $x_i^{kp} > \eta_i^k$ and N^{k2} = number of points in cluster k with $x_i^{kp} < \eta_i^k$. D^{k1} is the average distance of the N^{k1} points right of η_1^k from cluster k, while D^{k2} is the average distance of the N^{k2} points left of η_1^k from cluster k. Now define

$$a^{k1} = \max \frac{D_i^{k1}}{\min x_i^{k1}}, \qquad \max x_i^{k1} \neq 0$$

and

$$a^{k2} = \max \frac{D_i^{k2}}{\min x_i^{k2}}, \qquad \min x_i^{k2} \neq 0$$

for $k = 1, 2, \ldots$ (number of clusters) and $i = 1, 2, \ldots$ (number of dimensions).

For any cluster that has

$$a^{k1} > S \quad \text{or} \quad a^{k2} > S,$$

that cluster is split in the dimension of the corresponding maximum D_i^{kp}.

Another possible cluster reconfiguration is the lumping of two or more clusters when they become too close to one another. The related problem is defining the minimum distance allowed before clusters are combined. This distance, τ, is known as the lumping parameter.

Algorithms such as ISODATA need to specify this parameter *a priori*. It was found empirically, however, that this threshold could be dynamically calculated from several parameters, such as the dimension of the data, the present number of clusters, and the average minimum distance between clusters. It was found experimentally that τ could then be represented by

$$\tau = \frac{\sqrt{n}}{3m} \sum_{k=1}^{m} D_k$$

where n is the dimension of the sample space, m is the present number of clusters, and D_k is the average distance between cluster k and the other

$(m-1)$ clusters, that is

$$D_k = \min \|H_{kj}\|, \quad k, j = 1, 2, \ldots, m, k \neq j$$

where $H_{kj} = \eta^k - \eta^j$ (distance between kth and jth clusters). η^i is the mean vector $(\eta^i_1, \eta^i_2, \ldots, \eta^i_n)$ for cluster i and

$$\|H_{kj}\| = \sqrt{\sum_{i=1}^{\eta} (\eta^k_i - \eta^j_i)^2}.$$

Hence $m^{-1}\sum_{k=1}^{m} D_k$ is the average minimum distance between any cluster and its nearest neighbor cluster.

Now for any $D_k < \tau$, the kth cluster is eliminated, and its members are reassigned to neighboring clusters by the nearest neighbor policy.

Let $\mathcal{F} = (F_1, \ldots, F_m)$ be a hard (i.e., nonfuzzy) partition of X. $\text{conv}(F_i)$ denotes the convex hull of F_i in $V = \mathbb{R}^n$.

The subsets F_i of a nonfuzzy partition of X are said to be compact well-separated (CWS) clusters iff for all i, j, and k with $j \neq k$, any pair (x, y) with x in F_i and y in $\text{conv}(F_i)$ are closer together as measured by d than any pair (u, v) with u in F_j and v in $\text{conv}(F_k)$. This property can be quantified by the index

$$\beta(m, \mathcal{F}) = \frac{\min\limits_{1 \leq i \leq m} \min\limits_{\substack{1 \leq j \leq m \\ j \neq i}} d(F_i, \text{conv}(F_j))}{\max\limits_{1 \leq i \leq m} \text{diam}(F_i)}$$

The following result is due to Dunn [598]: X can be partitioned into m CWS clusters relative to d iff

$$\bar{\beta}(m) = \max_{\mathcal{F}} \beta(m, \mathcal{F}) > 1$$

($\mathcal{F}$ belongs to the set of m-partitions of X). The problem of finding an $\mathcal{F}$ such that $\beta(m, \mathcal{F}) = \bar{\beta}(m)$ is difficult. The above index is usually replaced by the simpler criterion

$$J(\mathcal{F}, v) = \sum_{i=1}^{m} \sum_{x \in F_i} d(x, v_i)^2$$

where v is an m-tuple of elements of $\text{conv}(X)$ called the cluster centers and

Fuzzy Isodata

d is now supposed to be induced by an inner product:

$$d(x, y) = \left[(x - y)^t M(x - y)\right]^{1/2}.$$

M is called a sample covariance matrix. Usually, M is taken as the identity. $J(\mathcal{F}, v)$ can be interpreted as the average least squares error of assimilating the elements of F_i to v_i, for all $i = 1, m$. The problem becomes: find $\mathcal{F}^*$ and v^*, for a given m, such that

$$J(\mathcal{F}^*, v^*) = \min_{\mathcal{F}} \inf_{v \in \text{conv}(X)} J(\mathcal{F}, v).$$

Dunn [595] and Bezdek [172]–[181] have relaxed J to allow fuzzy partitions as global minima because in the nonfuzzy ISODATA inferences drawn from the partition can be dangerous if it is not well known in advance that CWS clusters are actually present. The following discussion is based on the work by Bezdek [172]–[182] and Bezdek and Castelaz [183].

Let $X = \{x_1, x_2, \ldots, x_n\} \subset \mathbb{R}^s$ be a finite data set in feature space $\mathbb{R}^s$, let c be an integer $2 \leq c < n$, and let V_{cn} denote the vector space of all real $(c \times n)$ matrices over $\mathbb{R}$, equipped with the usual scalar multiplication and vector addition. A nonfuzzy c-partition of X is conveniently represented by a matrix $U = [u_{ik}] \in V_{cn}$, the entries of which satisfy

$$u_{ik} \in \{0, 1\}, \quad 1 \leq i \leq c, 1 \leq k \leq n$$

$$\sum_{i=1}^{c} u_{ik} = 1, \quad 1 \leq k \leq n$$

$$\sum_{k=1}^{n} u_{ik} > 0, \quad 1 \leq i \leq c.$$

The set of all such matrices is denoted by M_c:

$$M_c = \{U \in V_{cn} \mid u_{ik} \text{ satisfies the above } \forall i, k\},$$

and by M_{c0}, the superset of M_c obtained by allowing zero rows in U. To interpret $U \in M_c$ as a hard c-partition of X, regard the ith row of U, say $U_{(i)} = (u_{i1}, u_{i2}, \ldots, u_{in})$, as exhibiting (the values of) the characteristic function $u_i: X \to \{0, 1\}$ defined by

$$u_i(x_k) \triangleq u_{ik} = \begin{cases} 1; & x_k \in i\text{th subset } Y_i \text{ partitioning } X \\ 0; & \text{otherwise.} \end{cases}$$

The subsets $\{Y_i\}$ are the set-theoretic realization of the characteristic functions $\{u_i\}$, and because they are isomorphic descriptions, we may call either Y_i or u_i the ith hard cluster, or subset of X, in the c-partition U.

Next, let $v = (v_1, v_2, \ldots, v_c)$, where $v_i \in \mathbb{R}^s$ for $1 \leq i \leq c$; let $\|\cdot\|_E$ be the Euclidean norm on $\mathbb{R}^s$; and define the functional $J_1: (M_c \times \mathbb{R}^{cs}) \to \mathbb{R}$ by

$$J_1(U, v) = \sum_{k=1}^{n} \sum_{i=1}^{c} (u_{ik}) \|x_k - v_i\|_E^2.$$

If the range of each u_{ik} is extended to the closed interval $[0, 1]$, the resultant function $u_{ik}: X \to [0, 1]$ becomes a membership function and u_{ik} is called a fuzzy subset or fuzzy cluster in X. Here $u_{ij} = u_i(x_j)$ is called the grade of membership of x_j in fuzzy set u_i. Generalizing the above by this device produces the set

$$M_{fc} = \left\{ U \in V_{cn} \,\middle|\, u_{ij} \in [0, 1] \,\forall i, j; \, \sum_{i=1}^{c} u_{ij} = 1 \,\forall j; \, \sum_{j=1}^{n} u_{ij} > 0 \,\forall i \right\}.$$

Each $U \in M_{fc}$ is called a fuzzy c-partition of X; M_{fc} is a fuzzy c-partition space associated with X.

Physically, partitions of X that lie in M_{fc} enjoy the advantage of allowing each individual in the data the choice of belonging unequivocally to one and only one partitioning subset (e.g., a progenitor), or of sharing membership in several partitioning subsets (e.g., a hybrid). In either instance total membership of each subject in the data is restricted by the column sum condition to unity.

To generate c-partitions of X that are fuzzy in the above sense, let L be the linear hull or span of the data, $L = \text{span}(X)$, and let L_c denote the c-fold Cartesian product of L with itself. For each real number $m \in [1, \infty)$, define the real-valued functional $J_m: M_{fc} \times L_c \to R$ by

$$J_m(U, v) = \sum_{k=1}^{n} \sum_{i=1}^{c} (u_{ik})^m \|x_k - v_i\|^2, \quad 1 \leq m < \infty.$$

where $\|\cdot\|$ is any differentiable norm on R^s, and the vector $v = (v_1, v_2, \ldots, v_c)$ has components the vectors $v_i \in L$. These v_i's are interpreted as cluster centers or prototypes of the c fuzzy clusters defined by their companion U matrix, and play a fundamental role in our development. J_m is a weighted, least squares objective function: for $U \in M_c$, $J_m = J_1$ for all m, and J_1 is the classical within-groups sum of squares objective function. As

Fuzzy Isodata

mentioned, the fuzzy ISODATA clustering algorithms arise from necessary conditions for local minima of the constrained nonlinear programming problem associated with

$$\text{minimize } \{J_m(U, v)\} \text{ over } M_{fc} \times L_c.$$

Fuzzy ISODATA Algorithm (q = 2)

1. Choose a fuzzy partition $u(\cdot)$ of k nonempty membership functions, $u_i(\cdot) \neq 0$, $1 \leq i \leq k$ (k is a fixed integer between 2 and the number n of elements in X).
2. Compute the k weighted means,

$$v_i = \sum_{x \in X} (u_i(x))^2 \cdot \frac{x}{\sum_{x \in X} (u_i(x))^2}, \quad 1 \leq i \leq k.$$

3. Construct a new partition, $\hat{u}(\cdot)$, according to the following scheme. Let $I(x) = \{1 \leq i \leq k \mid v_i = x\}$; if $I(x)$ is not empty, let $\hat{i}$ be the least integer in $I(x)$ and put

$$\hat{u}_i(x) = \begin{cases} 1, & i = \hat{i} \\ 0, & i \neq \hat{i} \end{cases}$$

for $1 \leq i \leq k$; otherwise, if $I(x)$ is empty (the usual case), put

$$\hat{u}_i(x) = \sum_{i=1}^{k} \frac{1/\|x - v_i\|^2}{1/\|x - v_i\|^2}.$$

4. Compute some convenient measure, δ, of the defect between $u(\cdot)$ and $\hat{u}(\cdot)$. If δ is less than a specified threshold, T, then stop; otherwise, put $u(\cdot) = \hat{u}(\cdot)$ and go to step 2.

If X is binary, the optimal cluster centers $\{v_i\}$ are related to X, since if X is binary, then the components of each v_i satisfy:

$$0 \leq v_{ij} \leq 1, \quad \forall i, j$$

$$v_{ij} = 0 \Leftrightarrow x_{pj} = 0, \quad p = 1, 2, \ldots, n \Leftrightarrow v_{kj} = 0, \forall k$$

$$v_{ij} = 1 \Leftrightarrow x_{pj} = 1, \quad p = 1, 2, \ldots, n \Leftrightarrow v_{kj} = 1, \forall k.$$

The v_{ij} are called the feature centers of cluster center v_i. The implications for feature selection are these: $v_{ij} = 0$, iff none of the data possesses feature j so it is uncharacteristic of the mixed populations, and is minimal as a subclass descriptor. On the other hand, $v_{ij} = 1$, iff all members of the data have feature j; in this instance j is a maximal descriptor of all subclasses. In either situation feature j will be totally irrelevant to subclass discrimination. If $v_{ij} \to 0$, feature j becomes uncharacteristic of class i; conversely, as $v_{ij} \to 1$, we presume that feature j is strongly indicative of class i individuals.

To use the v_{ij} for identification of optimal features for subclass discrimination, define the following measure of cluster center separation:

$$f_{ij} = (|v_{i1} - v_{j1}|, |v_{i2} - v_{j2}|, \ldots, |v_{is} - v_{js}|),$$

$1 \leq i \leq c$; $1 \leq j \leq c$. Since $f_{ij} = f_{ji}$ and $f_{ii} = \emptyset$ is the zero vector, our attention can be confined to the $c(c-1)/2$ f's with $1 \leq i \leq c-1$ and $i+1 \leq j \leq c$. Thus for the components $f_{ij,k}$ of f_{ij},

$$0 \leq f_{ij,k} < 1, \quad \forall i, j, \text{ and } k$$

$$f_{ij,k} = 0 \Leftrightarrow \begin{cases} 1 & \text{either all vectors in } X \text{ have feature } k, \text{ or} \\ 2 & \text{none of the vectors in } X \text{ has feature } k \end{cases}$$

When $f_{ij,k} = 0$, feature k is apparently useless as a measure of variability between subclass i and j. $f_{ij,k}$ cannot attain the value 1 because v_{ik} and v_{jk} are simultaneously either 0 or 1. However, when $0 < v_{ij} < 1$, $\forall i$ and j, $f_{ij,k}$ can become arbitrarily close to 1 as more and more members of one class possess feature k while very few opposite subclass members have it. Accordingly, it seems plausible to infer from $f_{ij,k} \to 1$ that k is an optimal discriminator for subclasses i and j, and hence the number $\{f_{ij,k} | k = 1, 2, \ldots, s\}$ rank—by their magnitudes—the relative utility of the s features as separators of classes i and j.

For $c = 2$ the components of f_{12} rank the features, and any number of features from 1 to s can be selected directly. For $c > 2$, the procedure is more complex, for then the best features for pairwise discrimination are generally a function of i and j. In this case, moreover, we cannot select less than M features, $c \leq 2^M$, for separation of binary-valued data. A natural way to extend the method for c subclasses is to average the values of $f_{ij,k}$ over the $c(c-1)/2$ pairs (i, j) to obtain an overall average efficiency of feature k:

$$f_k = \left\{\frac{2}{(c)(c-1)}\right\} \sum_{j=i+1}^{c} \sum_{i=1}^{c-1} f_{ij,k},$$

which indicates the relative ability of feature k to separate all distinct pairs of the c subclasses. Since $0 \leq f_k < 1$, $\forall k$, our procedure is to calculate the numbers $(f_1, f_2, \ldots, f_s)$ and order them in decreasing magnitude; the optimal features are those having the highest values.

Following Bezdek [181], we assume that $m > 1$, that the cn distances $d_{ik} = \|x_k - v_i\|^2$ are always positive, and that the norm is any inter product induced norm. Clearly, J_m descends weakly on the iterates generated by fuzzy ISODATA, that is,

$$J_m(U^{(k+1)}, v^{(k+1)}) \leq J_m(U^{(k)}, v^{(k)}), \quad k = 0, 1, 2, \ldots .$$

Thus it can be shown that J_m is a descent function under other constraints that yield the result that under some very simple hypotheses, fuzzy ISODATA algorithms will converge, even though the convergence is in a somewhat weaker sense than the type of convergence usually established by fixed-point theorems.

As asserted by Bezdek [181], it is entirely conceivable that the fuzzy ISODATA algorithms are themselves a subfamily of a more general set of procedures designed to minimize functionals such as

$$J_m(U, p) \doteq \sum_{k=1}^{n} \sum_{i=1}^{c} (u_{ik})^m d_{ik}, \quad 1 \leq m < \infty$$

where $U \in M_{fc}$, and p is a parameterization of the (cn) "distances" between the n x_k's and the c fuzzy subsets $\{u_i\}$ in U involving variables other than v_i. Independent of the change, $J_m(U, p)$ will descend strictly on iterates of the operator involved. Although this method fails to establish convergence of J_m at $m = 1$, it may very well apply to generalizations of fuzzy ISODATA in the other direction.

Furthermore, it was shown [181] that feature selection with fuzzy ISODATA yielded optimal subsets of features, which resulted in no appreciable loss of classifier performance and upheld results of other studies. These facts seem to recommend further study of the fuzzy techniques and related methods.

CHAPTER FOUR

Inexact Hierarchical Classification

4.1 INTRODUCTION

Classification procedures that use labeled samples as training devices are said to be supervised. On the other hand, those procedures that use unlabeled samples are denoted as unsupervised learning and clustering procedures, that is, we investigate what can be achieved when all we have is a collection of samples without any knowledge of their classification.

There are several reasons for the interest in such procedures. First, the labeling of a large data set of patterns is quite costly compared to an unsupervised classifier that has a very limited labeled set of samples as its initial training data. Second, the designer may alter the classification approach by gaining some insight into the nature of the data, during the early stages of the classification. Finally, improved performance can be achieved if dynamic changes in the characteristics of the patterns are tracked. A classifier running in an unsupervised mode is capable of tracking such changes, as a function of time, in the nature of the patterns. Among the classical techniques in unsupervised clustering, we have the maximum likelihood approach; ISODATA procedure; unsupervised Bayesian procedures; use of criterion functions for clustering, such as the sum-of-squared error criterion, related minimum variance criteria, or scattering criteria, graph-theoretic methods; and obviously hierarchical clustering, a class of techniques that range from agglomerative (bottom-up, clumping) clustering as well as nearest neighbor and further neighbor algorithms to divisive (top-down, splitting) procedures. Excellent discussions of these classical techniques can be found in the influential books by Anderberg (1973),

Sneath and Sokal (1974), Sokal and Sneath (1963), Duda and Hart (1973), and Van Ryzin (1977).†

Almost all of the research on pattern recognition has examined the problem of recognizing an isolated pattern that is embedded in a homogeneous background. Typically, the pattern is further simplified in that it is resolved and projected onto a two-dimensional matrix, which enormously reduces the amount of potential information in the input pattern. This is usually the case with cluster anlysis, which has the objective of classifying experimental data in a certain number of categories where the elements of each category should be as similar as possible and dissimilar from those of other categories. This implies the existence of a measure of distance or similarity between the elements to be classified. The number of such categories may be fixed beforehand or may be a consequence of some constraints imposed on them.

In what follows, our attention is focused primarily on exploring the mathematical properties of inexact matrices and using them in pattern recognition. In most classification problems, subjective information plays an important role, and thus our analysis is data-free and analytic in nature. Namely, no data set of particular patterns is assumed and we make no attempt in the present content to discuss the possible applications of inexact matrices to related problem areas such as decision processes, system modeling, or approximation theory.

4.2 FUZZY RELATIONS

The concepts of fuzzy graphs and fuzzy relations are natural generalizations of crisp groups and relations, using fuzzy sets, and in many cases the extension principle.

4.2.1 Fuzzy Graphs

The following exposition of fuzzy graphs is due to Rosenfeld [2230].

A fuzzy graph $\tilde{G}$ is a pair $(\tilde{V}, \tilde{E})$ where $\tilde{V}$ is a fuzzy set on V and $\tilde{E}$ is a fuzzy relation on $V \times V$ such that

$$x_{\tilde{E}}(v, v') \leq \min(\chi_{\tilde{V}}(v), \chi_{\tilde{V}}(v')).$$

---
†See the bibliography at the end of Chapter 1.

The above inequality expresses the idea that the strength of the link between two vertices cannot exceed the degree of "importance" or of "existence" of the vertices. In other words, $\tilde{E}$ is a fuzzy relation on $\tilde{V} \times \tilde{V}$ in the sense that dom($\tilde{E}$) and ran($\tilde{E}$) are contained in $\tilde{V}$. However, in some situations it may be desirable to relax this inequality.

Classical concepts and definitions pertaining to graphs have been extended to fuzzy graphs, as we see here.

A path whose length is n in a fuzzy graph is a sequence of distinct vertices $v_0, v_1, \ldots, v_n$ such that $\chi_{\tilde{E}}(v_{i-1}, v_i) > 0$, $\forall i = 1, n$. The strength of the path is $\min_i \chi_{\tilde{E}}(v_{i-1}, v_i)$, $1 \leq i \leq n$, and $\chi_{\tilde{V}}(v_0)$ if $n = 0$. A strongest path joining two vertices v_0 and v_n has a strength $\chi_{\tilde{E}}^{\hat{}}(v_0, v_n)$ where $\hat{\tilde{E}}$ is the transitive closure of $\tilde{E}$.

In an ordinary graph, the distance between two vertices is the length of the shortest path linking them. A set U of vertices is called a cluster of order k iff:

1 $\forall v, v' \in U, d(v, v') \leq k$.
2 $\forall v \notin U, \exists v' \in U, d(v, v') > k$.

Here $d(v, v')$ denotes the distance between v and v'. When $k = 1$, a k-cluster is called a clique, that is a maximum complete subgraph. In a fuzzy graph, a nonfuzzy subset U of V is called a fuzzy cluster of order k if

$$\min_{\substack{v \in U \\ v' \in U}} \chi_{\tilde{E}^k}(v, v') > \max_{v \notin U}\left[\min_{v' \in U} \chi_{\tilde{E}^k}(v, v')\right]$$

where $\tilde{E}^k$ is the kth power of $\tilde{E}$.

The following definitions assume that the set of vertices is not fuzzy and $\tilde{E}$ is symmetrical $[\chi_{\tilde{E}}(v, v') = \chi_{\tilde{E}}(v', v)]$. The degree of a vertex v is $dg(v) = \Sigma_{v' \neq v}\chi_{\tilde{E}}(v, v')$. The minimum degree of G is $\delta(G) = \min_v dg(v)$.

G is said to be λ-degree connected (Yeh and Bang [2934]), iff:

1 $\forall v, v' \in V, \chi_{\tilde{E}}(v, v') \neq 0$ (if $v \neq v'$).
2 $\delta(G) \geq \lambda$.

A λ-degree component of G is a maximal λ-degree connected subgraph of G. For any $\lambda > 0$, the λ-degree components of a fuzzy graph are disjoint.

As in the theory of nonfuzzy sets, the notion of fuzzy graphs may be explained in terms of fuzzy relations.

4.2.2 *n*-ary Fuzzy Relations

The following is based on Zadeh [2954].

A fuzzy mapping from $X = \{x\}$ to $Y = \{y\}$ is a fuzzy set on $X \times Y$ with membership function $\chi_\xi(x, y)$. A fuzzy function $\xi(y)$ is a fuzzy set on X with membership function

$$\chi_{\xi(y)}(x) = \chi_\xi(x, y).$$

Its inverse $\xi^{-1}(x)$ is a fuzzy set on Y with

$$\mu_{\xi^{-1}(x)}(y) = \chi_\xi(x, y).$$

Based on the above, we can define fuzzy (binary) relation η as a fuzzy collection of ordered pairs. Thus a fuzzy relation from $X = \{x\}$ to $Y = \{y\}$ is a fuzzy subset of $X \times Y$ characterized by a membership function χ_R that associates with each pair (x, y) its grade-membership $\chi_R(x, y)$ in R.

Let $X_1, \ldots, X_n$ be n universes. An n-ary fuzzy relation R in $X_1 \times \cdots \times X_n$ is a fuzzy set on $X_1 \times \cdots \times X_n$.

Let $v = (v_1, \ldots, v_n)$ be a variable on $X = X_1 \times \cdots \times X_n$. A fuzzy restriction $R(v)$ is a fuzzy relation R that acts as an elastic constraint on the values, elements of X, that may be assigned to a variable.

The projection of a fuzzy relation R on $X_{i_1} \times \cdots \times X_{i_k}$, where $(i_1, \ldots, i_k)$ is a subsequence of $(1, 2, \ldots, n)$, is a relation on $X_{i_1} \times \cdots \times X_{i_k}$ defined as

$$\text{proj}[R; X_{i_1}, \ldots, X_{i_k}] = \int_{X_{i_1} \times \cdots \times X_{i_k}} \frac{\sup_{x_{j_1}, \ldots, x_{j_i}} \chi_R(x_1, \ldots, x_n)}{(x_{i_1}, \ldots, x_{i_k})}$$

where $j_1, \ldots, j_i$ is the subsequence complementary to $i_1, \ldots, i_k$ in $1, \ldots, n$.

If R is a fuzzy set in $X_{i_1} \times \cdots \times X_{i_k}$, then its cylindrical extension in $X_1 \times \cdots \times X_n$ is a fuzzy set $c(R)$ on $X_1 \times \cdots \times X_n$ defined by

$$c(R) = \int_{X_1 \times \cdots \times X_n} \frac{\chi_R(x_{i_1}, \ldots, x_{i_k})}{(x_1, \ldots, x_n)}.$$

Let R and S be two fuzzy relations on $X_1 \times \cdots \times X_r$ and $X_s \times \cdots \times X_n$, respectively, with $s \leq r + 1$: the join of R and S is $c(R) \cap c(S)$, where $c(R)$ and $c(S)$ are cylindrical extensions on $X_1 \times \cdots \times X_n$.

In terms of their cylindrical extensions, the composition of two fuzzy relations R and S, respectively, on $X_1 \times \cdots \times X_r$ and on $X_s \times \cdots \times X_n$ with

$s \leqslant r$ is expressed by

$$R \circ S = \text{proj}[c(R) \cap c(S); X_1 \times \cdots \times X_{s-1} \times X_{r+1} \times \cdots \times X_n].$$

$R \circ S$ is a fuzzy relation in the symmetrical difference of the universes of R and S.

An n-ary fuzzy restriction $R(v_1,\ldots,v_n)$ is said to be separable iff $R(v_1,\ldots,v_n) = R(v_1) \times \cdots \times R(v_n)$ where $\times$ denotes the Cartesian product and $R(v_i)$ is the projection of R on X_i, that is,

$$\chi_R(x_1,\ldots,x_n) = \min_i \chi_{\text{proj}[R;\, x_j]}(x_j) \,|\, j = 1,\ldots,n.$$

Clearly, given a function from $X_1 \times \cdots \times X_n$ to Y and a fuzzy restriction R on the arguments of f, the extension principle becomes

$$f(A_1,\ldots,A_n) = \int_{X_1 \times \cdots \times X_n} \frac{\min(\chi_{A_1}(x_1),\ldots,\chi_{A_n}(x_n), \chi_R(x_1,\ldots,x_n))}{f(x_1,\ldots,x_n)}.$$

where A_i is a fuzzy set on X_i.

The extension principle can be written

$$\chi_B(y) = \sup_{\substack{x_1,\ldots,x_n \\ y = f(x_1,\ldots,x_n)}} \min(\chi_{A_1}(x_1),\ldots,\chi_{A_n}(x_n))$$

where $B = f(A_1,\ldots,A_n)$. By denoting $R = c(A_1) \cap \cdots \cap c(A_n) = A_1 \times \cdots \times A_n$ and letting S be the ordinary relation defined by $\chi_S(x_1,\ldots,x_n, y) = 1$ iff $y = f(x_1,\ldots,x_n)$, we have $B = R \circ S$, and the extension principle appears as a particular case of composition of fuzzy relations. When a restriction T on $(x_1,\ldots,x_n)$ is added, B becomes $B = (R \cap T) \circ S$.

Let R be a fuzzy relation on $X \times Y$. The domain of R, denoted $\text{dom}(R)$, and the range of R, denoted $\text{ran}(R)$, are, respectively, defined by

$$\chi_{\text{dom}(R)}(x) = \sup_y \chi_R(x, y), \qquad \forall x \in X$$

and

$$\chi_{\text{ran}(R)}(y) = \sup_x \chi_R(x, y), \qquad \forall y \in Y.$$

Yeh [2933] has presented the following definitions.

R is: ε-determinate iff $\forall x \in X$, $\exists$ at most one $y \in Y$, such that $\chi_R(x, y) \geqslant \varepsilon$; ε-productive iff $\forall x \in X$, $\exists y$, $\chi_R(x, y) \geqslant \varepsilon$; an ε-function iff R is both

Fuzzy Relations

ε-determinate and ε-productive (a 1-function is an ordinary function when restricted to its 1-cut); ε-onto iff $\forall y \in Y, \exists x \in X, \chi_R(x, y) \geq \varepsilon$; ε-injective iff R is an ε-function and R^{-1} is ε-determinate; ε-bijective iff R and R^{-1} are both ε-functions.

If $X = Y$, then R is: reflexive iff $\forall x \in X, \chi_R(x, x) = 1$; ε-reflexive iff $\forall x \in X, \chi_R(x, x) \geq \varepsilon$; weakly reflexive iff $\forall x \in X, \forall y \in X, \chi_R(x, x) \geq \chi_R(x, y)$.

R is symmetric iff $\forall x \in X, \forall y \in X, \chi_R(x, y) = \chi_R(y, x)$.

Definition 4.2.1

Let R and S be a binary fuzzy relation such that $R \subseteq X \times Y$ and $S \subseteq Y \times Z$. We define the *composition* of R and S as

$$\chi_T = \chi_{R \circ S}(x, y) = \sup_{y \in Y} \min(\chi_R(x, y), \chi(y, z)), \quad \forall x \in X, \forall z \in Z.$$

There are some properties that are common to binary relations. They can be proven without difficulty. Let U be an extra relation on $Z \times W$ and T on $Y \times Z$. Then we can find these properties:

1 *Associativity*: $R \circ (S \circ U) = (R \circ S) \circ U$.
2 *Distributivity over Union*: $R \circ (S \cup T) = (R \circ S) \cup (R \circ T)$.
3 *Weak Distributivity over Intersection*: $R \circ (S \cap T) \subseteq (R \circ S) \cap (R \circ T)$.
4 *Monotonicity*: $S \subseteq T$ implies $R \circ S \subseteq R \circ T$.
5 *Symmetrization*: $R \circ R^{-1}$ is a weakly reflexive and symmetric relation on $X \times X$.

A nonzero fuzzy relation Q on X is weakly reflexive and symmetric iff there is a universe Y and a fuzzy relation R on $X \times Y$ such that $Q = R \circ R^{-1}$.

Clearly, $\chi_{R \circ S}(x, z)$ is the strength of a set of chains linking x to z, each chain of form x-y-z, such that the strength of such a chain is that of the weakest link. The strength of the relation between x and z is that of the strongest chain between x and z.

When the related universes X and Y are finite, a fuzzy relation R on $X \times Y$ can be represented as a matrix $[R]$ whose generic term $[R]_{ij}$ is $\chi_R(x_i, y_j) = r_{ij}$, $i = 1, n$, $j = 1, m$, where $|X| = n$ and $|Y| = m$.

The composition of finite fuzzy relations can thus be viewed as a matrix product. With $[S]_{jk} = s_{jk}$, $k = 1, p$, $p = |Z|$,

$$[R \circ S]_{ik} = \sum_j r_{ij} s_{jk}$$

where Σ is in fact the operation max and product the operation min.

Since $R \circ S$ can be written $\text{proj}[c(R) \cap c(S); X \times Z]$ where R and S are, respectively, on $X \times Y$ and $Y \times Z$, other compositions can be defined. For example, changing min to $*$, we define $R \boxdot S$ as

$$\chi_{R \boxdot S}(x, z) = \sup_{y} (\chi_R(x, y) * \chi_S(y, z)).$$

If $*$ is associative, and nondecreasing with respect to each of its arguments, the sup-$*$ composition satisfies associativity, distributivity over union, and monotonicity.

Let R be a fuzzy relation on $X \times X$; R is max-min transitive iff $R \circ R \subseteq R$, or more explicitly

$$\forall (x, y, z) \in X^3, \quad \chi_R(x, z) \geq \min(\chi_R(x, y), \chi_R(y, z)).$$

Other transitivities, associated with other kinds of composition of fuzzy relations, can be defined. Generally, R is said to be max-$*$ transitive iff $R \boxdot R \subseteq R$.

A specific case of max-$*$ transitive is where $a * b$ is given by:

1. $a \stackrel{.}{*} b = \max(0, a + b - 1)$ (bold intersection).
2. $a \square b = \frac{1}{2}(a + b)$ (arithmetic mean).
3. $a \vee b = \max(a, b)$ (union).
4. $a \stackrel{+}{*} b = a + b - ab$ (probabilistic sum).

Bandler and Kohout [125] have defined the triangle product of two crisp relations, R and S, as

$$(R \triangleleft S)_{ik} = \frac{1}{N_j} \sum_j (R_{ij} \to S_{jk})$$

when N_j is the number of elements and $\to$ is a fuzzy implication operator. The triangle product is nonsymmetric and in essence represents the (mean) degree to which the fuzzy afterset $a_i R$, defined by $A_i R = \{b \in B \mid a_i R b, a \in A, R \in B(A \to B)\}$, is contained in the fuzzy foreset Sc_j, defined by $Sc_j = \{a \in A \mid a S c_j, c_j \in C, S \in B(A \to C)\}$. Namely, it is the mean degree to which being related by R to a_i implies relating by S to c_k. Clearly, the afterset aR of $a \in X_1$ is the fuzzy subset of X_2 consisting of those $y \in X_2$ to which a is related, each with its degree, thus producing

$$\chi_{aR}(y) = \chi_R(a, y), \quad R \subseteq X_1 \times X_2.$$

Similarly, the foreset Sc of $c \in X_3$ is a fuzzy subset of X_2 consisting of those

Fuzzy Relations 137

$y \in X_2$ that are related to c, each with its degree of intensity given by

$$\chi_{S_c}(y) = \chi_S(y, c), \quad S \subseteq X_2 \times X_3.$$

An example of the application of this concept as a tool for analysis and synthesis of the behavior of complex natural and artificial systems has been illustrated by Bandler and Kohout in [125].

A special kind of fuzzy relation, based on a new composition, is called the α-composite fuzzy relation, and has been introduced by Sanchez [2276]–[2291]. Formally, let $Q \subseteq X \times Y$ and $R \subseteq Y \times Z$ be two fuzzy relations. We define $T = Q \, \alpha \, R$, $T \subseteq X \times Z$, the α-composite fuzzy relation of Q and R, by

$$(Q \, \alpha \, R)(x, z) = \min_y [Q(x, y) \, \alpha \, R(y, z)]$$

where $y \in Y$, for all $(x, z) \in X \times Z$.

The following results are quite trivial:

1. If $b, d \in L$ where L is a fixed complete Brouwerian lattice[†] and if $b \leq d$, then $a \, \alpha \, b \leq a \, \alpha \, d$, $\forall a \in L$.
2. If $R_1, R_2 \subseteq Y \times Z$ and if $R_1 \subseteq R_2$, then

$$Q \, \textcircled{a} \, R_1 \subseteq Q \, \textcircled{a} \, R_2, \quad \forall Q \subseteq X \times Y,$$

where

$$(Q \, \textcircled{a} \, R_i)(x, z) = \min_y [Q(x, y) \, \alpha \, R_i(y, z))$$

and

$$y \in Y, \quad \forall (x, z) \in X \times Z.$$

As an example of this concept, consider a set of medical patients P, having a set of symptoms S, with a set of diagnoses D. Given Q, $Q \subseteq P \times S$, a fuzzy relation between patients and symptoms, and T, $T \subseteq P \times D$, a fuzzy relation between patients and diagnoses, under the assumption that the range of these fuzzy relations is a Brouwerian lattice, we can investigate the concept of medical knowledge observed from Q and T. The expression

---

[†]A Brouwerian lattice L is a lattice in which, for any given elements x and y, the set of all Z in L such that $x \wedge z \leq y$, where $x \wedge z$ denotes the greatest lower bound of x and z, contains a greatest element denoted $x \, \alpha \, y$.

of this medical knowledge can be computed by determining the greatest fuzzy relation $R \subseteq S \times D$ such that

$$R(s, d) \to [Q(p, s) \to T(p, d)]$$

for all $s \in S$, $d \in D$, and $p \in P$.

Since a Brouwerian logic is a propositional calculus, that is a lattice with 0 and I, in which

$$P \to Q = I, \quad \text{iff } P \leq Q,$$

$$P \to (Q \to R) = (P \wedge Q) \to R, \quad \forall P, Q, R.$$

We can conclude that $R(s, d)$ can be determined as above if and only if

$$[R(s, d) \wedge Q(p, s)] \to T(p, d)$$

or

$$R(s, d) \wedge Q(p, s) \leq T(p, d).$$

Hence $\forall (p, d) \in P \times D$,

$$\bigvee_s [R(s, d) \wedge Q(p, s)] \leq T(p, d)$$

or simply, $R \circ Q \subseteq T$.

However, it is clear that the greatest R, such that $R \circ Q \subseteq T$, is equal to

$$R_{\max} = Q^{-1} \, @ \, T,$$

which gives us the medical knowledge associated with Q and T.

Given, now, a patient p_i with a known fuzzy set of symptoms $Q_{p_i}(s)$, $\forall s \in S$, the composition $R_{\max} \circ Q_{p_i} = T_{p_i}$ expresses the "closest" or the "most reliable" fuzzy set of diagnoses presented by p_i.

The same idea can be applied to the mechanism discussed in Section 3.2. In general, the correspondence from $f(p)$, p being a natural object, to $\chi_C(p)$ is relational. In other words, C is assumed to be characterized by a relational tableau R with linguistic entries that allow us to set translation rules of fuzzy propositions related to inexact, ill-defined, and vague dependencies. The translation of R into a fuzzy relation R_r is quite simple. Thus given a fuzzy set A related to the attributes, the composition $R_r \circ A$

corresponds to the linguistic interpolation yielding new grades of membership for C, which is one of the basic problems raised in Chapter 3.

4.3 SIMILARITY RELATIONS

Especially important in science are the relations that are reflexive, symmetric, and transitive; these relations are known in classical set theory as equivalence relations. The concept of a *similarity* relation is essentially a generalization of the concept of an equivalence relation. More concretely, we can formulate the following definition.

Definition 4.3.1

A *similarity relation* S, is a fuzzy relation in X that is reflexive, symmetric, and transitive.

Thus let x_i and x_j be elements of X, and let $x_s(x_i, x_j)$ denote the grade-membership of the ordered pair (x_i, x_j) in S. Then S is a similarity relation in X iff $\forall x, y, z \in X$:

1. $\chi_s(x, x) = 1$ (reflexive).
2. $\chi_s(x, y) = \chi_s(y, x)$ (symmetric).
3. $\chi_s(x, z) \geq \max_y [\min(\chi_s(x, y), \chi_s(y, z))]$ (transitive).

Clearly, this definition of transitivity is the explicit expression of general transitivity under the max-min composition denoted by $\circ$; namely,

$$S \supset S \circ S.$$

It should be noted that, if we use a different definition of composition, for example, a max-product, denoted by $*$, then

$$S \supset S * S.$$

or, more explicitly,

$$\chi_s(x, z) \geq \max_y [\chi_s(x, y) * \chi_s(y, z)].$$

Clearly, the *n-step fuzzy relation* defined as

$$\chi_s^n(x, y) \triangleq \sup_{x \in \Omega} \{\min[\chi_s(x, x_1), \ldots, \chi_s(x_{n-1}, y)]\}$$

where $\mathbf{x} = (x_1, x_2, \ldots, x_{n-1}) \in \hat{X}$ [$\hat{X}$ is the $(n-1)$-fold Cartesian product of X with itself where $x, y \in X$] implies that for all $x, y \in X$ and all $n \geq 1$,

$$0 \leq \chi_s^n(x, y) \leq \chi_s^{n+1}(x, y) \leq 1.$$

Consequently, $\lim_{n \to \infty} \chi_s^n(x, y) \triangleq \bar{\chi}_s(x, y)$ exists by the monotone convergence principle; namely, for every $\varepsilon > 0$, there is an integer n such that $|\chi_s^n(x, y) - \bar{\chi}_s(x, y)| < \varepsilon$ for $n > N$. Since the sequence is nondecreasing and is bounded from above and from below, we can conclude that existence of the limit is due to the following theorem.

Theorem 4.3.1 A bounded nondecreasing sequence $\{a_i\}_i$ has a limit; the limit is the smallest number that is not less than any a_j.

Definition 4.3.2

Let x and y be two elements of X, and let $\chi_s^n(x, y)$ be the n-step fuzzy relation as defined above. Then we define the *propinquity* $\bar{\chi}(x, y)$ in $[0, 1]$ such that

$$\bar{\chi}(x, y) = \lim_{n \to \infty} \chi_s^n(x, y).$$

Definition 4.3.3

Let $x, y \in X$. Then x and y are said to have a threshold relation $(xR_T y)$ iff $\bar{\chi}_s(x, y) \geq T$.

Theorem 4.3.2 $\forall x, y, z \in X, [\bar{\chi}_s(x, z) \geq \min[\bar{\chi}_s(x, y), \bar{\chi}_s(y, z)]]$.

Proof

1. $xR_T x, \forall T \in [0, 1]$, since $1 = \chi_s(x, x) \leq \bar{\chi}_s(x, x) \leq 1, x \in X$.
2. $xR_T y$ iff $yR_T x$, since

$$\lim_{n \to \infty} \chi_s^n(x, y) = \bar{\chi}_s(x, y) = \bar{\chi}_s(y, x) = \lim_{n \to \infty} \chi_s^n(y, x).$$

3. $xR_T y \wedge yR_T z \to xR_T z$, since

$$\bar{\chi}_s(x, z) \geq \min[\bar{\chi}_s(x, y), \bar{\chi}_s(y, z)]. \qquad \text{Q.E.D.}$$

Clearly, we can associate with every relation an appropriate matrix to represent the relation, and hence we can classify the patterns using the partition induced by the threshold relation. This relation is similar to the idea of α-level-sets of a fuzzy relation R, denoted by R_α (and is a nonfuzzy set in $X \times Y$), which gives rise to the notion of relation matrices.

Similarity Relations

Before proceeding, we investigate some of the important properties of fuzzy or *inexact matrices*. The results are a generalization of Boolean matrix theory and some known results on lattice matrices.

When the related universes X and Y are finite, a fuzzy relation R on $X \times Y$ can be represented as a matrix $[R]$ whose generic term $[R]_{ij}$ is $\chi_R(x_i, y_j) = r_{ij}$, $i = 1, n, j = 1, m$, where $|X| = n$ and $|Y| = m$.

The composition of finite fuzzy relations can thus be viewed as a matrix product, with $[S]_{jk} = s_{jk}$, $k = 1, p, p = |Z|$,

$$[R \circ S]_{ik} = \sum_j r_{ij} s_{jk}$$

where Σ is in fact the operation max and product the operation min.

Let R be a fuzzy relation on $X \times X$ where $|X| = n$. The mth power of a fuzzy relation is defined as $R^m = R \circ R^{m-1}$, $m > 1$, and $R^1 = R$. The following propositions are discussed:

1. The power of R either converges to idempotent R^c for a finite c or oscillates with finite period (if R^m does not converge, then it must oscillate with a finite period since $|X|$ is finite and the composition is deterministic and cannot introduce numbers not in R originally).
2. If $\forall i, j, \exists k$ such that $r_{ij} \leq \min(r_{ik}, r_{kj})$, then R converges to R^c where $c \leq n - 1$.

Other results in more particular cases can be found in the next section.

Since $R \circ S$ can be written $\text{proj}[c(R) \cap c(S); X \times Z]$ where R and S are, respectively, on $X \times Y$ and $Y \times Z$, other compositions may be introduced by modifying the operator used for the intersection.

Changing min to $*$, we define $R \circledast S$ through

$$\chi_{R \circledast S}(x, z) = \sup_y (\chi_R(x, y) * \chi_S(y, z)).$$

Zadeh [2954] proved that, when $*$ is associative and nondecreasing with respect to each of its arguments, the sup-$*$ composition satisfies associativity, distributivity over union, and monotonicity.

Examples of such operators are product and bold intersection.

We may encounter another kind of alternative compositions, inf-max compositions. The following property holds: $\overline{R \circ S} = \overline{R} \mathbin{\bar{\circ}} \overline{S}$ where $\bar{\circ}$ denote inf-max composition.

In the following section we present a more detailed discussion of fuzzy relations presented by inexact matrices.

4.4 INEXACT MATRICES

In order to use the notion of a proximity relation for cluster analysis, we must develop a reevaluation procedure that will produce the transitive-closure of the proximity matrix (the matrix describing all proximity relations between its elements).

Motivated by this idea, we investigate the properties of inexact (fuzzy) matrices under the operations of fuzzy logic. It should be noted that some of the results can be obtained in many other mathematical configurations since the following discussion is basically a generalization of Boolean matrix theory.

It is also interesting to note that there are several aspects of inexact matrix theory that are interesting in their own right. However, our discussion is centered on those properties that can be applied toward the solution of the classification problem of static patterns under the subjective measure of similarity.

Let Z be our inexact (fuzzy) algebra, which is a distributive lattice with unique identities under $+$ and $*$, e_+ and e_*, respectively. Consider Z_{pq}, the complete set of $p \times q$ matrices with elements in Z. Two elements $S = [s_{ij}]$ and $T = [t_{ij}]$ in Z_{pq} are regarded equal iff $s_{ij} = t_{ij}$ for all i and j. We now define the following compositions in Z_{pq}.

Definition 4.4.1

1. Sum: $S + T = W$ iff $w_{ij} = s_{ij} + t_{ij}$ for all i and j.
2. Lattice Product: $S * T = W$ iff $w_{ij} = s_{ij} t_{ij}$ for all i and j.
3. $S \geqslant T$ iff $s_{ij} \geqslant t_{ij}$ for all i and j.
4. $S = \bar{T}$ iff $s_{ij} = \bar{t}_{ij}$ for all i and j.

Evidently, the system $[Z_{pq}, +, *]$ is a lattice with the universal element E_* (all entries e_*) and the zero element E_+ (all entries e_+).

Scalar Multiplication: $W = rS = Sr$ iff $w_{ij} = rs_{ij}$, $s \in Z_{pq}$, $r \in Z$.

Matrix Product: $W = ST$ iff $w_{ij} = \Sigma_{k=1}^{q} s_{ik} t_{kj}$, $s \in Z_{pq}$, $T \in Z_{qm}$, and $W \in Z_{pm}$, where the symbol Σ is used to denote the sum with respect to the operation $+$(max) of Z.

Transpose: $W = S^t$ iff $w_{ij} = s_{ji}$, $s \in Z_{pq}$, $W \in Z_{qp}$.

Permanent: $\operatorname{per}(S) = |S| = \sum\limits_{i}[\prod\limits_{j}(s_{ji_j})]$, where $S \in Z_{pp}$, $|S| \in Z$, and the summation is taken over all permutations $(i_1, i_2, \ldots, i_p)$.

Inexact Matrices

Adjoint: $W = \operatorname{adj} S$ iff $w_{ij} = S_{ij}$ where $S \in Z_{pp}$, $W \in Z_{pp}$, and S_{ij} is the cofactor of s_{ji} in $|S|$.

It is interesting to note that Z_{pp} forms a lattice-ordered semigroup with the matrix product as third binary composition, and the matrix I (all diagonal entries e_* and all others e_+) serves as unity.

A vector is called *inexact* iff all its entries are elements of Z. A matrix is called *inexact* iff all its rows are inexact vectors. Since our interest is mainly in *inexact symmetric matrices* we refer to them just as inexact matrices. Thus for the set of all $p \times p$ inexact symmetric matrices Z_{pp}, we have the following definitions and propositions.

Definition 4.4.2

Let $A = [a_{ij}]$ and $B = [b_{ij}]$ be two $p \times p$ inexact matrices.

1. $A + B = [\max(a_{ij}, b_{ij})]$.
2. $AB = [\max_k \{\min(a_{ik}, b_{kj})\}]$.
3. $A \leq B$ exists iff $a_{ij} \leq b_{ij}$, $\forall i, j$.

It is easy to prove that $A \leq B$ implies $SA \leq SB$ and $AT \leq BT$, $\forall S, T \in Z_{pp}$. $A \leq B$ evidently implies $A + B = B$ and conversely. We immediately verify that the set of $p \times p$ inexact matrices is a monoid under matrix multiplication. Namely, the set of $p \times p$ inexact matrices is closed under inexact matrix multiplication and the unit $p \times p$ matrix (all diagonal entries 1 and all other 0) is inexact.

Definition 4.4.3

An inexact matrix A is called *constant* if all its rows are equal.

The following propositions are immediate consequences of the above definitions.

Proposition 4.4.1 If A is a constant inexact matrix and B an inexact matrix of the same order, then AB and BA are constant inexact matrices.

Proof

1. By Definition 4.4.3, a_{ik} is the same for all i. By definition, $AB = [\max_k \{\min(a_{ik}, b_{kj})\}]$ and thus the elements in AB are independent of i and therefore AB is constant. This is true regardless of whether A and B are symmetric or not.
2. $BA = [\max_k \{\min(b_{ik}, a_{kj})\}]$. $B \geq I$ implies $b_{ii} = 1$, $\forall i, i = 1, 2, \ldots, n$.

Thus since a_{kj} is independent of k, denote a_{kj} by a_j and

$$\max_k \{\min(b_{ik}, a_{kj})\} = \min\{a_j, \max_k(b_{ik})\}.$$

However, $\max_k(b_{ik}) = 1$ and thus a_j is the minimum term regardless of k.
Q.E.D.

It should be noted that, if $B \neq I$, the second part is not valid.

EXAMPLE 4.4.1 Let

$$A = \begin{bmatrix} 0.5 & 0.3 \\ 0.5 & 0.3 \end{bmatrix}$$

and

$$B = \begin{bmatrix} 0.1 & 0.4 \\ 0.7 & 0.1 \end{bmatrix};$$

then

$$PA = \begin{bmatrix} 0.4 & 0.3 \\ 0.5 & 0.3 \end{bmatrix},$$

which is clearly not constant.

Proposition 4.4.2 Let $A = [a_{ij}]$ be a constant inexact matrix and $B = [b_{ij}]$ an inexact matrix. If for all i, $\max_j[b_{ij}] \geq [a_{ij}]$, then $BA = A$.

Proof For all i we have the following:

$$\max_j [b_{ij}] \geq [a_{ij}],$$

then

$$\max_k \{\min(b_{ik}, a_{kj})\} = \min\{a_j, \max_k(b_{ik})\} = a_j. \qquad \text{Q.E.D.}$$

Corollary 5.5.1

If A is a constant inexact matrix, then $A^2 = A$ (i.e., inexact constant matrices are idempotent).

Theorem 4.4.1 Let A be a $p \times p$ inexact matrix. Then the sequence $A, A^2, A^3, \ldots$ is ultimately periodic.

Inexact Matrices 145

Proof Let $p = \{x_1, x_2, \ldots, x_m\}$ be the set of all the inexact elements that occur in matrix A. Then the number of different matrices that can be obtained by multiplying A is at most m^{p^2}, which is clearly finite.   Q.E.D.

Proposition 4.4.3 Let A be a $p \times p$ matrix and let $B = A + I$, where $+$ denotes the operation max. Then $B \leq B^2 \leq \cdots \leq B^{p-1} = B^p = B^{p+1} = \cdots$. Let $\tilde{A} = \sup_k A^k$ where A is a $p \times p$ inexact matrix. From Proposition 4.4.3 it is clear that, if $B = A + I$, then $\tilde{B} = B^{p-1}$.

Proof We prove a stronger result claiming that there exists an integer $q \leq p - 1$ such that

$$B \leq B^2 \leq \cdots \leq B^q = B^{q+1} = \cdots .$$

For simplicity of the proof, we denote the operation min by concatenation. Let $B = (a_{ij})$ and $B^h = (b_{ij})$. Then

$$B^{h+1} = \left(\sum_{k=1}^{p} b_{ik} a_{kj} \right) = \left(b_{ij} + \sum_{k=1}^{p} b_{ik} a_{kj} \right)$$

since $a_{jj} = 1$. Thus

$$b_{ij} \leq \sum_{k=1}^{p} b_{ik} a_{kj} \quad \text{and} \quad B^h \leq B^{h+1}.$$

This shows $B \leq B^2 \leq \cdots$ We must show that this chain does not strictly increase indefinitely. It will be sufficient to show that $B^{p-1} = B^p$. Since we already know $B^{p-1} \leq B^p$, we need only show $B^p \leq B^{p-1}$. We are using the fact that $\leq$ is a partial ordering and hence antisymmetric.

Consider an off-diagonal element of B^p. It is of the form

$$\sum_{k_{n-1}=1}^{p} \sum_{k_{n-2}=1}^{p} \cdots \sum_{k_1=1}^{p} a_{ik_1} a_{k_1 k_2} \cdots a_{k_{p-2} k_{p-1}} a_{k_{p-1} j}.$$

Clearly, there are $p - 1 + 2 = p + 1$ subscripts, so not all of them can be distinct (Dirichlet's principle). Consider a term of the above form and suppose there exists an integer s such that $j = k$. The term is of the form

$$a_{ik_1} a_{k_1 k_2} \cdots a_{k_{s-1} j} a_{j k_{s+1}} \cdots a_{k_{p-1} j}$$

but, since $ab \leq a$, this is contained in the term

$$a_{ik_1}\cdots a_{k_{s-1}j},$$

which is the $i-j$ entry of B^s and is contained in the $i-j$ entry of B^{p-1}. The previous argument is symmetric if $i = k_s$, for some integer s.

The remaining case occurs if there exist integers r and s such that $k_s = k_r$. Assuming $s < r$, we get

$$a_{ik_1}\cdots a_{k_{s-1}k_r}a_{k_rk_{s+1}}\cdots a_{k_{r-1}k_r}a_{k_rk_{r+1}}\cdots a_{k_{p-1}j},$$

which is contained in

$$a_{ik_1}\cdots a_{k_{s-1}k_r}a_{k_rk_{r+1}}a_{k_rk_{r+1}}\cdots a_{k_{p-1}j}.$$

This term is contained in the ijth entry of B^{p-1}. Therefore $B^p \leq B^{p-1}$ and $B^p = B^{p-1}$. Q.E.D.

Proposition 4.4.4 Let A be a $p \times p$ inexact matrix and let $B = A + I$ where I is of the same order. Then $\tilde{B} = \operatorname{adj} B$.

Before proving Proposition 4.4.4, the following discussion is quite helpful.

Let B be a symmetric inexact matrix with diagonal of 1's. Then we can consider an inexact (fuzzy) undirected finite graph G having a *primitive connection matrix* B corresponding to the grade memberships of the edges of G, where the columns and rows of B represent the vertices of G.

For any primitive connection matrix, we define the *characteristic inexact matrix* or *inexact transmission matrix* $\psi(B) = [x_{ij}]$ such that x_{ij} is the inexact transmission function of the two-terminal system connecting vertex i to j. It is clear that $\psi(B)$ is a symmetric matrix, since the graph is an undirected one, and thus $x_{ij} = x_{ji}$, $\forall i, j$, and $x_{ii} = 1$, $\forall i$.

During the process of computing the characteristic inexact matrix, simplifications of the inexact structures are possible.

Lemma 4.4.1 Let B be a square inexact transmission matrix of order p. Then there exists an integer $q \leq p - 1$ such that

$$B^q = B^{q+1} = \cdots = \psi(B).$$

Proof of Lemma 4.4.1 Let $B = [b_{ij}]$. The ij entry of B^2 is

$$\sum_{k=1}^{p} b_{ik}b_{kj},$$

Inexact Matrices

and this term has the grade of membership of

$$\max_k \left[\min(b_{ik}, b_{kj}) \right]$$

iff there is a direct path between vertices i and j or there is a path from i to j through one intermediate vertex. Extending this argument to B^i, it is clear that no path requires more than $p-2$ intermediate vertices, since there are only p vertices and internal loops are excluded. Hence the entry of B^{p-1} has the grade of membership of $\max_{\text{subterms}}\{ij \text{ terms of } B^{p-1}\}$ iff i and j are connected, namely $B^{p-1} = \psi(B)$. Q.E.D.

Proof of Proposition 4.4.4 Let G be the inexact undirected finite graph having a primitive connection matrix B. By using network analysis techniques, we can prove that the transmission matrix $T_G = [t_{ij}]$ of graph G is given by $T_g = \text{adj } B$, that is, $B_{ii} = 1$, and the transmission t_{ij} from v_i to v_j, $i \neq j$, equals B_{ij}. The fact that $B_{ii} = 1$ follows immediately from $b_{ii} = 1$. Now we consider any pair (i,j), $i, j = 1, 2, \ldots, p$ with $i \neq j$.

Let S be the set of permutations

$$s = \begin{pmatrix} 1 \ldots p \\ s_1 \ldots s_p \end{pmatrix}$$

with $s_j = i$. Let $s = u \cdot v \cdots$ be the decomposition of s into cyclic permutations, such that u is of the form

$$u = (j, i, \ldots, j)$$

and let U be the set of all such permutations u. Also let

$$B_s = b_{1s_1}, \ldots b_{ps_p}.$$

We have $B_s = B_u \cdot B_v \cdots$ and thus $B_u \geq B_s$. Since $U \subseteq S$, we get

$$\sum_{s \in S} B_s = \sum_{u \in U} B_u.$$

Evidently, $\sum_{u \in U} B_u = b_{ji} t_{ij}$, whereas $\sum_{s \in S} B_s$ consists of all terms of $|B|$ containing b_{ji}.

Hence t_{ij} is the cofactor of b_{ji} in $|B|$, that is, $t_{ij} = B_{ij}$, which implies $T_G = \text{adj } B$.

Furthermore, this proof shows that if B and $\psi(B)$ are the *inexact primitive connection* and *transmission* matrices of some graph G, respectively, then $\psi(B) = \text{adj } B$.

Proposition 4.4.5 Let B be a $p \times p$ inexact matrix such that $B = A + I$. Then:

1. $\psi(\psi(B)) = \psi(B) \, [\tilde{\tilde{B}} = \tilde{B}]$.
2. $B \leq \psi(B) \, [\tilde{B} \geq B]$.
3. $\psi(B^2) = \psi(B) \, [\tilde{B}^2 = \tilde{B}]$.

Proof

1. $\psi(B) = B^{p-1} = B^p = \cdots = B^{p^2}$; $\psi(\psi(B)) = \psi(B^p) = (B^p)^p = B^{p^2} = B^p = \psi(B)$.
2. $B \leq B^{p-1} = \psi(B)$.
3. $\psi(B^2) = (B^2)^{p-1} = B^{2p} = \cdots = B^p = B^{p-p} = \psi(B)$. Q.E.D.

Theorem 4.4.2 Let A and B be $p \times p$ inexact matrices and let I be a $p \times p$ unit matrix. If $C = A + I$ and $D = B + I$, then

$$\widetilde{C+D} = \widetilde{\tilde{C}\tilde{D}} = \widetilde{\tilde{D}\tilde{C}}.$$

Proof $\tilde{C} \geq C$ and $\tilde{D} \geq I$ imply $\tilde{C}\tilde{D} \geq C$. Similarly, $\tilde{C}\tilde{D} \geq D$ and therefore $\tilde{C}\tilde{D} \geq C + D$, which implies $\widetilde{\tilde{C}\tilde{D}} \geq \widetilde{C+D}$. However, $\tilde{C} \leq \widetilde{C+D}$ and $\tilde{D} \leq \widetilde{C+D}$ imply $\tilde{C}\tilde{D} \leq \widetilde{C+D}$. The result follows immediately since $\widetilde{\tilde{C}\tilde{D}} \leq \widetilde{\widetilde{C+D}} = \widetilde{C+D}$. Q.E.D.

Theorem 4.4.3 Let A and B be $p \times p$ inexact matrices and let $C = A + I$ and $D = B + I$, when I is the $p \times p$ unit matrix. Then

$$C \leq \tilde{D} \quad \text{iff} \quad \tilde{C} \leq \tilde{D}.$$

Proof $C \leq \tilde{C}$ implies that, if $\tilde{C} \leq \tilde{D}$, then $C \leq \tilde{D}$. $C \leq \tilde{D}$ implies $\tilde{C} \leq \tilde{\tilde{D}} = \tilde{D}$. Q.E.D.

4.5 APPLICATIONS TO CLUSTERING

Once we describe the clustering problem as one of finding natural groupings in a data set, we have to investigate measures of similarity between samples, as well as the evaluation of a partitioning of a set of samples into clusters. It is this partitioning that we are interested in when we discuss the subject of hierarchical clustering, which is the topic of this section.

Applications to Clustering

Let us consider a sample space of k samples that we want to partition into q classes. The first stage is to partition the data set into k clusters, each containing exactly one sample. Then we partition the sample space into $k-1, k-2, \ldots, j$ clusters where at level j of the sequence we have

$$q = n - j + 1.$$

Thus level one corresponds to k clusters and level k to one. Given any two samples x_1 and x_2, at some level they will be grouped together in the same cluster. If the sequence has the property that, whenever two samples are in the same cluster at level k, they remain together at all higher levels, then the sequence is said to be a *hierarchical clustering*. Examples of hierarchical clustering appear in biological taxonomy, where individuals are grouped into species, species into genera, genera into families, and so on.

For every hierarchical clustering there is a corresponding tree, called a *dendrogram*, that shows how the samples are grouped.

Hierarchical clustering procedures are divided into two distinct classes, agglomerative and divisive. Agglomerative (bottom up, clumping) procedures start with q singleton clusters and form the sequence by successively merging clusters. Divisive (top down, splitting) procedures start with all of the samples in one cluster and form the sequence by successively splitting clusters. The computation needed to go from one level to another is usually simpler for the agglomerative procedures. However, when there are many samples and we are interested in only a small number of clusters, this computation will have to be repeated many times. At any level the distance between nearest clusters can provide the dissimilarity value for that level. It should be noted that we have not said how to measure the distance between two clusters. The considerations here are much like those involved in selecting a criterion function. Basic distance measures include:

$$d_{\min}(\rho_i, \rho_j) = \min_{\alpha \in \rho_i, \beta \in \rho_j} \|\alpha - \beta\|$$

$$d_{\max}(\rho_i, \rho_j) = \max_{\alpha \in \rho_i, \beta \in \rho_j} \|\alpha - \beta\|$$

$$d_{\text{avg}}(\rho_i, \rho_j) = \frac{1}{k_i k_j} \sum_{\alpha \in \rho_i, \beta \in \rho_j} \sum \|\alpha - \beta\|$$

$$d_{\text{mean}}(\rho_i, \rho_j) = \|M_i - M_j\|.$$

All of these measures have a minimum-variance flavor, and they usually yield the same results if the clusters are compact and well-separated.

However, if the clusters are close to one another, or if their shapes are not basically hyperspherical, quite different results can be obtained.

Interestingly enough, if we define the dissimilarity between two clusters by

$$\delta_{\min}(\rho_i, \rho_j) = \min_{\alpha \in \rho_i, \beta \in \rho_j} \delta(\alpha, \beta)$$

or

$$\delta_{\max}(\rho_i, \rho_j) = \max_{\alpha \in \rho_i, \beta \in \rho_j} \delta(\alpha, \beta),$$

then the hierarchical clustering procedure will induce a distance function or the given set of n samples. Furthermore, the ranking of the distances between samples will be invariant to any monotonic transformation of the dissimilarity values.

The graph-theoretic principle involves the selection of a threshold distance. Once the threshold distance d_0 is selected, two elements are said to be in the same cluster if the distance between them is less than d_0. This procedure can easily be generalized to apply to arbitrary similarity measures. Suppose that we pick a threshold value d_0 and say that α is similar to β if $s(\alpha, \beta) > d_0$.

$$S_{ij} = \begin{cases} 1, & \text{if } s(\alpha_i, \alpha_j) > d_0 \\ 0, & \text{otherwise} \end{cases} \quad i, j = 1, \ldots, n.$$

This matrix defines a *similarity graph* in which nodes correspond to points and an edge joins node i and node j if and only if $S_{ij} = 1$.

The clusterings produced by the single-linkage algorithm and by a modified version of the complete-linkage algorithm are readily described in terms of this graph. With the single-linkage algorithm, two samples α and β are in the same cluster if and only if there exists a chain $\alpha_1, \alpha_2, \ldots, \alpha_k$ such that α is similar to α_1, α_1 is similar to α_2, and so on for the whole chain. Thus this clustering corresponds to the connected components of the similarity graph. With the complete-linkage algorithm, all samples in a given cluster must be similar to one another, and no sample can be in more than one cluster. If we drop this second requirement, then this clustering corresponds to the maximal complete subgraphs of the similarity graph, the "largest" subgraphs with edges joining all pairs of nodes. In general, the clusters of the complete-linkage algorithm will be found among the maximal complete subgraphs, but they cannot be determined without knowing the unquantized similarity values.

It is clear that the nearest-neighbor algorithm could be viewed as an algorithm for finding a minimal spanning tree. Conversely, given a minimal spanning tree, we can find the clusterings produced by the nearest-neighbor algorithm. Removal of the longest edge produces the two-cluster grouping, removal of the next longest edge produces the three-cluster grouping, and so on. This amounts to an inverted way of obtaining a divisive hierarchical procedure, and suggests other ways of dividing the graph into subgraphs. For example, in selecting an edge to remove, we can compare its length to the length of other edges incident upon its nodes. Let us say that an edge is inconsistent if its length ξ is significantly larger than $\bar{\xi}$, the average length of all other edges incident on its vertices.

When the data points are strung out into long chains, a minimal spanning tree forms a natural skeleton for the chain. If we define the diameter path as the longest path through the tree, then a chain will be characterized by the shallow depth of branching off the diameter path. In contrast, for a large, uniform cloud of data points, the tree will usually not have an obvious diameter path, but rather several distinct, near-diameter paths. For any of these, an appreciable number of nodes will be off the path. While slight changes in the locations of the data points can cause major rerouting of a minimal spanning tree, they typically have little effect on the recognition process.

To estimate subjectively the resemblance between pairs of data points, we adopt the convention of arranging data for numerical classification in the form of a matrix. Each entry $[ij]$ in such a matrix is the score of the proximity relation (subjective similarity) between data points i and j. It should be noted that the numerical values in the proximity matrix are in general only quantitatively descriptive numbers whose significance cannot be evaluated, in general, by conventional statistical techniques, and thus are determined subjectively.

Since the proximity relation is not necessarily transitive, we must utilize the theory of inexact matrices as described in Section 4.4 in order to formulate a transitive closure structure that enables us to separate the data set into mutually exclusive clusters that are, in essence, equivalent classes under a certain threshold.

Definition 4.5.1

Let R be an inexact relation on data set D and let r be a subset of D. R is said to refine r iff xRy implies that $x \in r$ iff $y \in r$. Symbolically,

$$xRy \to (x \in r \leftrightarrow y \in r).$$

Theorem 4.5.1 Let $T_1 \geq T_2$. Then R_{T_1} refines R_{T_2}.

Proof $xR_{t_1}y \to \bar{\chi}_s(x, y) \geq T_1 \geq T_2 \to xR_{T_2}y.$ Q.E.D.

It is clear that, if R_T is a threshold relation induced by $\chi_s(x, y)$, R'_T is a threshold relation induced by $\chi'_s(x, y)$, and $\chi_s(x, y) \leq \chi'_s(x, y)$ for all $x, y \in \Omega$, then R_T refines R'_T.

It is our assumption that if $x \neq y$ then $\bar{\chi}_s(x, y) \in [0, 1)$ and thus the function $\bar{\eta}_s(x, y) = 1 - \bar{\chi}_s(x, y)$ acts as a distance function. This is very obvious, since:

1 $\bar{\eta}_s(x, y) > 0$ for $x \neq y$ and $\bar{\eta}_s(x, x) = 0$.
2 $\bar{\eta}_s(x, y) = \bar{\eta}_s(y, x)$.
3 $\bar{\eta}_s(x, z) \leq \bar{\eta}_s(x, y) + \bar{\eta}_s(y, z)$
   as
   $\bar{\mu}_s(x, z) \geq \min[\bar{\chi}_s(x, y), \bar{\chi}_s(y, z)] \geq \bar{\chi}_s(x, y) + \bar{\chi}_s(y, z) - 1$.

We assume for the simplicity of the analysis that we deal with only a finite number of patterns, and hence we consider our threshold relation on finite sets only.

Theorem 4.5.2 Let $x_1, x_2, \ldots, x_n \in \Omega$ where n is a finite number. Then

$$R_T = \psi(R_T^*) = (R_T^*)^{n-1}$$

where

$$xR_T^* y \quad \text{iff } \chi_s(x, y) \geq T$$

and

$$\psi(R_T^*) = \sup_i (R_T^*)^i = \text{adj}(R_T^*).$$

Proof

1 Clearly, $\psi(R_T^*)$ refines R_T since $\psi(R_T^*) = \cup_j (R_T^*)^j$ and if $x[\psi(R_T^*)]y$ then $\exists x_1, x_2, \ldots, x_{n-1} \in \Omega$ such that $\chi_s(x, x_1) \geq T, \ldots, \chi_s(x_{n-1}, y) \geq T$ and hence $\chi_s^n(x, y) \geq \min[\chi_s(x, x_1), \ldots, \chi_s(x_{n-1}, y)] \geq T$, which implies that $\bar{\chi}_s(x, y) \geq \chi_s^n(x, y) \geq T$ and thus $xR_T y$.

2 Assume $xR_T y$. Then

$$\chi_s(x, y) = \chi_s^{n-1}(x, y)$$

$$= \max_{x_1, x_2, \ldots, x_{n-2} \in \Omega} \{\min[\chi_s(x, x_1), \ldots, \chi_s(x_{n-1}, y)]\} \geq T.$$

Applications to Clustering

Therefore

$$\exists x_1, x_2, \ldots, x_{n-2} \in \Omega$$

such that

$$\chi_s(x, x_1) \geq T$$
$$\chi_s(x_1, x_2) \geq T$$
$$\vdots$$
$$\chi_s(x_{n-2}, y) \geq T$$

and thus we have

$$x R_T^* x_1, x_1 R_T^* x_2, \ldots, x_{n-2} R_T^* y$$

$$\Rightarrow x(R_T^*)^{n-1} y \text{ exists}$$

$$\Rightarrow x[\psi(R_T^*)] y \text{ exists}.$$

3 Let $R_T^* = [\chi_{ij}]$. The ij entry of $(R_T^*)^2$ is $\sum_{k=1}^n \chi_{ik} \chi_{kj}$, and this term has the grade membership of

$$\max_k \left[\min(\chi_{ik}, \chi_{kj})\right]$$

iff there is a direct path between vertices i and j or there is a path from i to j through one intermediate vertex. Extending this argument to $(R_T^*)^i$, it is clear that no path requires more than $n - 2$ intermediate vertices, since there are only n vertices and internal loops are excluded. Hence the ij entry of $(R_T^*)^{n-1}$ has the grade membership of $\max_{\text{subterms}}\{ij \text{ terms of } (R_T^*)^{n-1}\}$ iff i and j are connected, namely, $(R_T^*)^{n-1} = \psi(R_T^*)$. Q.E.D.

Corollary 4.5.1 $R_T^* = \psi(R_T^*) \leftrightarrow (R_T^*)^2 = R_T^*$.

Based on the above result, algorithm A is presented to compute $\psi(R_T^*)$.

Proof of Corollary 4.5.1 Assume $(R_T^*)^2 = R_T^*$. By induction, $R_T^* = (R_T^*)^{n-1}$ and $(R_T^*)^{n-1} = \psi(R_T^*)$. Conversely, if

$$R_T^* = \psi(R_T^*)$$

then

$$R_T^* = (R_T^*)^{n-1}$$

$$(R_T^*)^2 = (R_T^*)^n = \psi(R_T^*) = R_T^*. \qquad \text{Q.E.D.}$$

Algorithm A

Given the matrix R_T^* constructed from the inexact patterns $x_1, x_2, \ldots, x_n$, generate the matrix $(R_T^*)^l = (R_T^*)^{l+1}$ for some l, under the operation of fuzzy matrix multiplication.

The repeated matrix multiplication makes algorithm A unattractive from an efficiency viewpoint. Algorithm B achieves the same result, and requires only a single scan over the matrix. In fact, algorithm B works correctly on a wider range of input, since it is not required that the diagonal elements of the input matrix $= 1$, as with algorithm A.

Algorithm B

1 Label vertices of R_T^* by the integers $1, \ldots, n$.
2 Generate the matrix R_T^*.
3 DO $K = 1$ TO n
4   DO $I = 1$ TO n
5   IF $R_T^*(I, K) \neq 0$ THEN
6   DO $J = 1$ TO n
7          $R_T^*(I, J) = \max(R_T^*(I, J), \min(R_T^*(I, K), R_T^*(K, J)))$
8   END
9   END
10 END

The basic idea is to scan down *column* K, and for each nonzero element encountered (say in row I), each element in *row* I (say element $R_T^*(I, J)$] is possibly improved by comparing $R_T^*(I, J)$ to $\min(R_T^*(I, K), R_T^*(K, J))$. A rigorous proof of correctness is achieved by attaching the following inductive assertion W between statements 7 and 8:

$$(W) \quad R_T^*(I, J) = G(I, J, K)$$

where $G(I, J, K) \stackrel{\text{def}}{=} \max\{\min(\text{all chains from } I \text{ to } J \text{ such that each intermediate element has a label} \leq K)\}$.

Before proving that assertion W is true whenever control leaves step 7, it is noted that the relation

$$R_T^*(I, J) = G(I, J, n)$$

is the desired relation at the termination of the algorithm since $G(I, J, n) = \max\{\min(\text{all chains from } I \text{ to } J)\}$.

Assertion W is proved by induction on K.

1 $K = 1$: The first time W is reached, K has the value 1, and analysis shows that $R_T^*(I, J) = \max(R_{T_0}^*(I, J), \min(R_{T_0}^*(I, 1), R_{T_0}(1, K)))$ where $R_{T_0}^*$ represents the original matrix, and the right-hand side of the equation is equal to $G(I, J, 1)$.

2 Assume $R_T^*(I, J) = G(I, J, K)$, $1 \leq K < n$. Then we have to show

$$R_T^*(I, J) = G(I, J, K + 1).$$

There are two subcases to consider. If $G(I, J, K + 1)$ does not involve element $K + 1$, then no change is made to the matrix and the desired result is true. If $G(I, J, K + 1)$ does involve element $K + 1$, then we can guarantee that element $K + 1$ appears only once, since loops do not increase the max of any chain. Thus we can break the optimal chain into two subchains, $R_T^*(I, K + 1)$ and $R_T^*(K + 1, J)$. Since both subchains involve intermediate elements numbered $\leq K$, the inductive hypothesis applies to each subchain, and the desired result follows.

It is interesting to note that, during the process of computing the characteristic fuzzy matrix, minimization of the fuzzy structures is possible. In general, we cannot apply binary minimization techniques and thus more specific methods, directed toward the minimization of fuzzy functions, should be developed.

Generally, an indexed collection

$$\lambda = \{D_\alpha; \alpha \in J\}$$

of subsets of a set S satisfying

$$S = \bigcup_{\alpha \in J} D_\alpha$$

$$D_\beta \cap D_\gamma = \emptyset \quad \text{for all } \beta \neq \gamma \in J$$

is said to be a *partition* of S.

Clearly, we can apply the notions discussed above to the partition of our pattern set into disjoint classes, which depends on the *a priori* assigned threshold.

EXAMPLE 4.5.1 Let $X = \{x_1, x_2, x_3, x_4\}$ and let $\chi_s(x_i, x_j)$, $i, j = 1, 2, 3, 4$ be as follows:

$$R_T^* = \begin{array}{c} \\ x_1 \\ x_2 \\ x_3 \\ x_4 \end{array} \begin{array}{cccc} x_1 & x_2 & x_3 & x_4 \\ \left[\begin{array}{cccc} 1 & 0 & T_2 & T_1 \\ 0 & 1 & T_4 & T_3 \\ T_2 & T_4 & 1 & \overline{T}_1 \\ T_1 & T_3 & \overline{T}_1 & 1 \end{array}\right] \end{array}.$$

The minimized form of $\psi(R_T^*)$ is represented by

$\psi(R_T^*) =$

$$\begin{array}{c} \\ x_1 \\ x_2 \\ x_3 \\ x_4 \end{array} \begin{array}{cccc} x_1 & x_2 & x_3 & x_4 \\ \left[\begin{array}{cccc} 1 & T_2T_4 + T_1T_3 + \overline{T}_1T_1T_4 + T_1T_2T_3 & T_2 + T_1\overline{T}_1 + T_1T_3T_4 & T_1 + \overline{T}_1T_2 + T_2T_3T_4 \\ T_2T_4 + T_1T_3 + T_1\overline{T}_1T_4 + \overline{T}_1T_2T_3 & 1 & T_4 + \overline{T}_1T_3 + T_1T_2T_3 & T_3 + \overline{T}_1T_4 + T_1T_2T_4 \\ T_2 + T_1\overline{T}_1 + T_1T_3T_4 & T_4 + \overline{T}_1T_3 + T_1T_2T_3 & 1 & \overline{T}_1 + T_1T_2 + T_3T_4 \\ T_1 + \overline{T}_1T_2 + T_2T_3T_4 & T_3 + \overline{T}_1T_4 + T_1T_2T_4 & \overline{T}_1 + T_1T_2 + T_3T_4 & 1 \end{array}\right] \end{array}$$

Let $T_1 = 0.3$, $T_2 = 0.5$, $T_3 = 0.6$, and $T_4 = 0.9$. Then

$$\psi(R_T^*) = \begin{array}{c} \\ x_1 \\ x_2 \\ x_3 \\ x_4 \end{array} \begin{array}{cccc} x_1 & x_2 & x_3 & x_4 \\ \left[\begin{array}{cccc} 1 & 0.5 & 0.5 & 0.5 \\ 0.5 & 1 & 0.9 & 0.7 \\ 0.5 & 0.9 & 1 & 0.7 \\ 0.5 & 0.7 & 0.7 & 1 \end{array}\right] \end{array}$$

and we have the partitions

$$R_{T=1} = \{[x_1], [x_2], [x_3], [x_4]\}$$

$$R_{0.7 < T \leq 0.9} = \{[x_1], [x_2, x_3], [x_4]\}$$

$$R_{0.5 < T \leq 0.7} = \{[x_1], [x_2, x_3, x_4]\}$$

$$R_{0 \leq T \leq 0.5} = \{[x_1, x_2, x_3, x_4]\}.$$

Applications to Clustering

EXAMPLE 4.5.2 Let

$$\chi_s(x, y) = \frac{1}{1 + |x - y|}, \qquad x, y \in \mathbb{N}$$

$R_T^* = $

	0	1	2	3	4	5	6	7	8	...
0	1	$\frac{1}{2}$	$\frac{1}{3}$	$\frac{1}{4}$	$\frac{1}{5}$	$\frac{1}{6}$	$\frac{1}{7}$	$\frac{1}{8}$	$\frac{1}{9}$	...
1	$\frac{1}{2}$	1	$\frac{1}{2}$	$\frac{1}{3}$	$\frac{1}{4}$	$\frac{1}{5}$	$\frac{1}{6}$	$\frac{1}{7}$	$\frac{1}{8}$	...
2	$\frac{1}{3}$	$\frac{1}{2}$	1	$\frac{1}{2}$	$\frac{1}{3}$	$\frac{1}{4}$	$\frac{1}{5}$	$\frac{1}{6}$	$\frac{1}{7}$	...
3	$\frac{1}{4}$	$\frac{1}{3}$	$\frac{1}{2}$	1	$\frac{1}{2}$	$\frac{1}{3}$	$\frac{1}{4}$	$\frac{1}{5}$	$\frac{1}{6}$	...
4	$\frac{1}{5}$	$\frac{1}{4}$	$\frac{1}{3}$	$\frac{1}{2}$	1	$\frac{1}{2}$	$\frac{1}{3}$	$\frac{1}{4}$	$\frac{1}{5}$	...
5	$\frac{1}{6}$	$\frac{1}{5}$	$\frac{1}{4}$	$\frac{1}{3}$	$\frac{1}{2}$	1	$\frac{1}{2}$	$\frac{1}{3}$	$\frac{1}{4}$	...
6	$\frac{1}{7}$	$\frac{1}{6}$	$\frac{1}{5}$	$\frac{1}{4}$	$\frac{1}{3}$	$\frac{1}{2}$	1	$\frac{1}{2}$	$\frac{1}{3}$	...
7	$\frac{1}{8}$	$\frac{1}{7}$	$\frac{1}{6}$	$\frac{1}{5}$	$\frac{1}{4}$	$\frac{1}{3}$	$\frac{1}{2}$	1	$\frac{1}{2}$	...
8	$\frac{1}{9}$	$\frac{1}{8}$	$\frac{1}{7}$	$\frac{1}{6}$	$\frac{1}{5}$	$\frac{1}{4}$	$\frac{1}{3}$	$\frac{1}{2}$	1	...
⋮	⋮	⋮	⋮	⋮	⋮	⋮	⋮	⋮	⋮	⋱

$\psi(R_T^*) = $

	0	1	2	3	4	5	6	7	8	...
0	1	$\frac{1}{2}$	$\frac{1}{2}$	$\frac{1}{2}$	$\frac{1}{2}$	$\frac{1}{2}$	$\frac{1}{2}$	$\frac{1}{2}$	$\frac{1}{2}$	...
1	$\frac{1}{2}$	1	$\frac{1}{2}$	$\frac{1}{2}$	$\frac{1}{2}$	$\frac{1}{2}$	$\frac{1}{2}$	$\frac{1}{2}$	$\frac{1}{2}$	...
2	$\frac{1}{2}$	$\frac{1}{2}$	1	$\frac{1}{2}$	$\frac{1}{2}$	$\frac{1}{2}$	$\frac{1}{2}$	$\frac{1}{2}$	$\frac{1}{2}$	...
3	$\frac{1}{2}$	$\frac{1}{2}$	$\frac{1}{2}$	1	$\frac{1}{2}$	$\frac{1}{2}$	$\frac{1}{2}$	$\frac{1}{2}$	$\frac{1}{2}$	...
4	$\frac{1}{2}$	$\frac{1}{2}$	$\frac{1}{2}$	$\frac{1}{2}$	1	$\frac{1}{2}$	$\frac{1}{2}$	$\frac{1}{2}$	$\frac{1}{2}$	...
5	$\frac{1}{2}$	$\frac{1}{2}$	$\frac{1}{2}$	$\frac{1}{2}$	$\frac{1}{2}$	1	$\frac{1}{2}$	$\frac{1}{2}$	$\frac{1}{2}$	...
6	$\frac{1}{2}$	$\frac{1}{2}$	$\frac{1}{2}$	$\frac{1}{2}$	$\frac{1}{2}$	$\frac{1}{2}$	1	$\frac{1}{2}$	$\frac{1}{2}$	...
7	$\frac{1}{2}$	$\frac{1}{2}$	$\frac{1}{2}$	$\frac{1}{2}$	$\frac{1}{2}$	$\frac{1}{2}$	$\frac{1}{2}$	1	$\frac{1}{2}$	...
8	$\frac{1}{2}$	$\frac{1}{2}$	$\frac{1}{2}$	$\frac{1}{2}$	$\frac{1}{2}$	$\frac{1}{2}$	$\frac{1}{2}$	$\frac{1}{2}$	1	...
⋮	⋮	⋮	⋮	⋮	⋮	⋮	⋮	⋮	⋮	⋱

Clearly, for $1 \geq T_1 \geq T_2 \geq 0$, R_{T_1} refines R_{T_2}; therefore, for every monotone nonincreasing finite sequence of thresholds

$$0 \leq T_j \leq T_{j-1} \leq \cdots \leq T_2 \leq T_1 \leq 1,$$

we can obtain a corresponding j-level hierarchy of clusters $J_i = \{$equivalence classes of R_{T_i} in $X \mid 1 \leq i \leq j\}$. For each i, J_i is a partition of X and every class in J_{i+1} is the union of some nonempty class of subsets in J_i. It is interesting to note that if we define recursively

$$\Gamma_1 = R_T^*$$

$$\Gamma_n = \Gamma_{n-1}^2,$$

then

$$\Gamma_j = (R_T^*)^{2^j}$$

and since, by Theorem 4.5.2,

$$\psi(R_T^*) = (R_T^*)^{n-1},$$

we can obtain the following relationship: $2^j \geq n - 1$ will determine $\Gamma_j[\bar{\chi}_s(x, y)]$, or $j \geq \lceil \log_2(n - 1) \rceil$ will determine $\Gamma_j[\bar{\chi}_s(x, y)]$. It should be noted that this is not a necessary condition since we can find certain cases for smaller j's to imply

$$\Gamma_j[\bar{\chi}_s(x, y)].$$

It is also interesting to note that we can present the same structure through a simple use of graph theory.

Definition 4.5.2

An *inexact graph* IG is a pair $[V, S]$ where V is a set of data vertices, and S is a proximity relation on V. A vertex v_i is said to be T-accessible from another vertex v_j for some $0 < T \leq 1$, if and only if

$$\chi_s(v_i, v_j) \geq T.$$

IG is called *strongly T-connected* if and only if every pair of vertices are mutually accessible.

A *T-cluster* in V is a maximal subset U of V such that each pair of elements in U is mutually T-accessible. Thus the construction of T-clusters of V is really nothing but constructing all maximal strongly T-connected subgraphs of IG.

This method can be related to existing nonfuzzy techniques that make use of link-node plots or spatial-dependence rearrangements of matrices.

Applications to Clustering

The link-node plot relates directly to the graph-theoretic approach using minimal spanning trees. The rearrangement of matrices represents a transformation on a matrix giving numbers of variables differing significantly between pairs of categories. The transformation usually brings similar rows and columns together and so makes the organization of the matrix more obvious. In a large data set, an algorithmic approach that systematically rearranges rows and corresponding columns in order to optimize $|j - k| A^2(j, k)$, where j and k are the row and the column, respectively, and $A(j, k)$ is the number of different variables between category j and category k, will accomplish such a rearrangement. In the following subsections, we investigate several closely related techniques in which the rules of the game have been altered in order to allow the reader to envision far more options in matrix clustering than have been available before.

4.5.1 Optimistic Transitions

Following Santos and Wee [2315], the matrix composition under the operation "max-min" is known as a *pessimistic* grade of transition. Here investigate several different structures and their applicability to the pattern classification problem.

First, we call the matrix composition with the grade of transition under the operation "min-max" an *optimistic* model. We denote it as

$$C = A \# B \leftrightarrow c_{ij} = \left[\min_k \left\{ \max(a_{ik}, b_{kj}) \right\} \right].$$

We can construct the properties of this operation in a way similar to the construction in the "max-min" case. The most interesting results are:

1. $A\#(B\#C) = (A\#B)\#C$.
2. $A\#A^\circ = A^\circ \#A = A$ where $A^\circ = [a^\circ_{ij}]$ and

$$a^\circ_{ij} = \begin{cases} 0 & \text{if } i = j \\ 1 & \text{if } i \neq j. \end{cases}$$

3. $A^\alpha \# A^\beta = A^{\alpha + \beta}$.
4. $(A^\alpha)^\beta = A^{\alpha\beta}$.
5. If $A < C$ and $B < D$, then $A\#B < C\#D$.

It is clear that Theorem 4.5.2 does not hold under the operation #, as can be seen from the following examples.

EXAMPLE 4.5.3 Let

$$R_T^* = \begin{bmatrix} 1 & 0 & 0.5 & 0.3 \\ 0 & 1 & 0.9 & 0.6 \\ 0.5 & 0.9 & 1 & 0.7 \\ 0.3 & 0.6 & 0.7 & 1 \end{bmatrix}.$$

Then

$$(R_T^*)_\#^2 = R_T^* \# R_T^* = \begin{bmatrix} 0 & 0.6 & 0.7 & 0.6 \\ 0.6 & 0 & 0.5 & 0.3 \\ 0.7 & 0.5 & 0.5 & 0.5 \\ 0.6 & 0.3 & 0.5 & 0.3 \end{bmatrix}$$

$$(R_T^*)_\#^3 = (R_T^*)_\#^2 \# R_T^* = \begin{bmatrix} 0.6 & 0 & 0.5 & 0.3 \\ 0 & 0.6 & 0.6 & 0.6 \\ 0.5 & 0.6 & 0.7 & 0.6 \\ 0.3 & 0.6 & 0.6 & 0.6 \end{bmatrix}$$

$$(R_T^*)_\#^4 = (R_T^*)_\#^3 \# R_T^* = \begin{bmatrix} 0 & 0.6 & 0.6 & 0.6 \\ 0.6 & 0 & 0.5 & 0.3 \\ 0.6 & 0.5 & 0.5 & 0.5 \\ 0.6 & 0.3 & 0.5 & 0.3 \end{bmatrix}$$

$$(R_T^*)_\#^5 = (R_T^*)_\#^4 \# R_T^* = \begin{bmatrix} 0.6 & 0 & 0.5 & 0.3 \\ 0 & 0.6 & 0.6 & 0.6 \\ 0.5 & 0.6 & 0.6 & 0.6 \\ 0.3 & 0.6 & 0.6 & 0.6 \end{bmatrix}$$

$$(R_T^*)_\#^{2n} = (R_T^*)_\#^4$$

$$(R_T^*)_\#^{2n+1} = (R_T^*)_\#^5$$

for $n > 2$.

However, we can prove the following theorem.

Theorem 4.5.3 Let $x_1, x_2, \ldots, x \in \Omega$ where n is a finite number. Then

$$\hat{R}_T = \hat{\psi}(\hat{R}_T^*) = (\hat{R}_T^*)_\#^{n-1}$$

where $x_i R_T^* x_j$ iff $\eta_s(x_i, x_j) \leq T$, $i, j \in \{1, 2, \ldots, n\}$ and $\hat{\psi}(\hat{R}_T^*) = \mathrm{adj}(\hat{R}_T^*)$ under $\#$ operation.

Proof By Theorem 4.5.2 and De Morgan's laws. Q.E.D.

Applications to Clustering

Clearly, $\hat{\psi}(\hat{R}_T^*) = E - \psi(R_T^*)$ where E is an $n \times n$ matrix of 1's.

EXAMPLE 4.5.4 Let

$$\hat{R}_T^* = \begin{bmatrix} 0 & 1 & 0.5 & 0.7 \\ 1 & 0 & 0.1 & 0.4 \\ 0.5 & 0.1 & 0 & 0.3 \\ 0.7 & 0.4 & 0.3 & 0 \end{bmatrix}$$

$$(R_T^*)^2 = \hat{\psi}(\hat{R}_T^*) = \begin{bmatrix} 0 & 0.5 & 0.5 & 0.5 \\ 0.5 & 0 & 0.1 & 0.3 \\ 0.5 & 0.1 & 0 & 0.3 \\ 0.5 & 0.3 & 0.3 & 0 \end{bmatrix}.$$

Hence the partitions described in this section can be performed directly on the matrix representation of the distance relation.

4.5.2 Max-Product Structures

Another possibility is to derive a max-product relation instead of the max-min relation, namely,

$$\chi_s(x, z) = \max_y [\chi_s(x, y) \cdot \chi_s(y, z)]$$

and thus

$$C = A \,\S\, B \leftrightarrow c_{ij} = \left[\max_k (a_{ik} \cdot b_{kj})\right]$$

where

$$A^\circ = [a_{ij}^\circ] \quad \text{and} \quad a_{ij}^\circ = \begin{cases} 1 & \text{if } i = j \\ 0 & \text{if } i \neq j. \end{cases}$$

The transitive closure in this case is defined similarly as

$$\tilde{\psi}(\tilde{R}_T^*) = \tilde{R}_T^* \cup (\tilde{R}_T^*)^2 \cup (\tilde{R}_T^*)^3 \cup \cdots$$

where

$$(\tilde{R}_T^*)^k = \underbrace{\tilde{R}_T^* \,\S\, \tilde{R}_T^* \,\S\, \cdots \,\S\, \tilde{R}_T^*}_{k \text{ times}}, \quad k = 1, 2, 3, \ldots.$$

Since $\forall \alpha, \beta \in [0, 1]$, $\alpha \cdot \beta \leq \min(\alpha, \beta)$, it is clear that

$$R_T^* \circ R_T^* \subset R_T^* \Rightarrow \tilde{R}_T^* \S \tilde{R}_T^* \subset \tilde{R}_T^*$$

where

$$(R_T^*)^2 = R_T^* \circ R_T^*$$

represents

$$\chi_{R_T^* \cdot R_T^*}(x, z) = \max_y \left\{ \min \left[\chi_{R_T^*}(x, y), \chi_{R_T^*}(y, z) \right] \right\}$$

and

$$(\tilde{R}_T^*)^2 = \tilde{R}_T^* \S \tilde{R}_T^*$$

represents

$$\chi_{\tilde{R}_T^* \cdot \tilde{R}_T^*}(x, z) = \max_y \left[\chi_{\tilde{R}_T^*}(x, y) \cdot \chi_{\tilde{R}_T^*}(y, z) \right].$$

It should be noted that, just for consistency, we have used $\tilde{R}_T^*$ instead of R_T^* since they are representing precisely identical proximity relations. Theorem 4.5.2 is satisfied also under the max-product operation.

EXAMPLE 4.5.5 Let

$$\tilde{R}_T^* = \begin{bmatrix} 1 & 0 & 0.5 & 0.3 \\ 0 & 1 & 0.9 & 0.6 \\ 0.5 & 0.9 & 1 & 0.7 \\ 0.3 & 0.6 & 0.7 & 1 \end{bmatrix}$$

$$(\tilde{R}_T^*)^2 = \begin{bmatrix} 1 & 0.45 & 0.5 & 0.35 \\ 0.45 & 1 & 0.9 & 0.63 \\ 0.5 & 0.9 & 1 & 0.7 \\ 0.35 & 0.63 & 0.7 & 1 \end{bmatrix}$$

$$(\tilde{R}_T^*)^3 = \begin{bmatrix} 1 & 0.45 & 0.5 & 0.35 \\ 0.45 & 1 & 0.9 & 0.63 \\ 0.5 & 0.9 & 1 & 0.7 \\ 0.35 & 0.63 & 0.7 & 1 \end{bmatrix} = \hat{\psi}(\tilde{R}_T^*).$$

It is trivial to verify that the rates of convergence of both the max-min and max-product operations are identical, but the results of the transitive-closure matrix are different.

Applications to Clustering

Similarly, we could repeat the process with a *min-sum* operation, which would produce the following result:

$$\chi_{\text{min-sum}}(x, y) = 1 - \chi_{\text{max-product}}(x, y), \quad \text{for all } (x, y) \text{ in } \tilde{\psi}(\tilde{R}_T^*).$$

It is clear that $\tilde{\psi}(\tilde{R}_T^*) \subset \psi(R_T^*)$ since

$$\max_y \{\min[\chi_s(x, y), \chi_s(y, z)]\} \geq \max_y [\chi_s(x, y) \cdot \chi_s(y, z)]$$

and

$$\hat{\psi}(\hat{R}_T^*) \subset \psi_{ms}(R_T^*)ms$$

where $\psi_{ms}(R_T^*)ms$ is the transitive closure under the min-sum relation.

Regarding the graph-theoretic approach, we can use the inexact graph to construct the *T*-clusters in *V*. This can be illustrated by the following example.

EXAMPLE 4.5.6 Let R_T^* be as in Example 4.5.1, and let the following graph G represent $\psi(R_T^*)$:

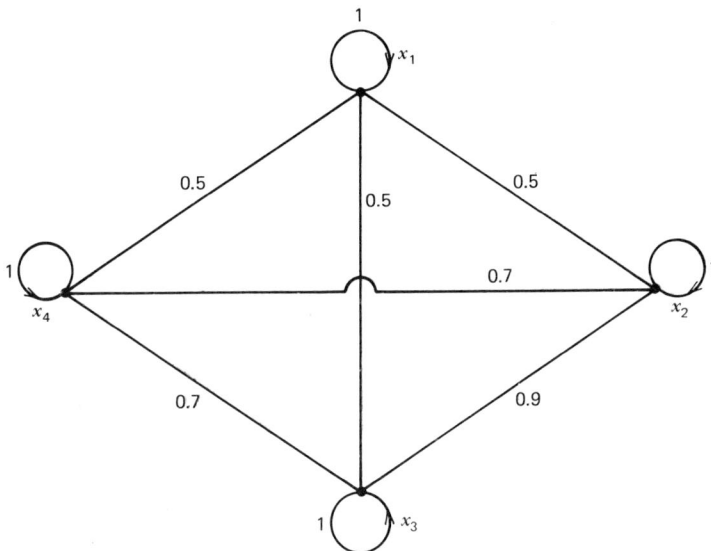

On every level we can construct the algebraic sum of the proximity values of the edges to a specific vertex. Thus

$$W(x_3) = W(x_2) = 3.1$$

$$W(x_4) = 2.9$$

$$W(x_1) = 2.5,$$

and we have the partitions

$$R_{T \leq 2.5} = \{[x_1, x_2, x_3, x_4]\}$$

$$R_{2.5 < T \leq 2.9} = \{[x_1], [x_2, x_3, x_4]\}$$

$$R_{2.9 < T \leq 3.1} = \{[x_1], [x_2, x_3], [x_4]\}$$

$$R_{T > 3.1} = \{[x_1], [x_2], [x_3], [x_4]\},$$

which are identical with those of Example 4.5.1.

We can extend these ideas to present a solution to the classification problem of time-varying patterns, since the literature on pattern recognition and classification to date has not considered the notion of patterns as central to the problem of classification. In fact, traditionally, the patterns have been assumed to be a set of time-invariant features.

Let pattern x be presented at intervals of Δt. By assumption, the pattern represents a dynamic system and we can introduce a time function $x(t)$ defined as

$$x(t) = x(k)$$

for

$$(k-1)\Delta t \leq t < k\Delta t, \quad k = 1, 2, \ldots.$$

Let $X(t)$ be the space of fuzzy sets describing all possible properties η at time t, where property η is a combinational structure of dynamic and history (denoted as D/H). The subset $B(t)$ is the set of reasonable alternatives at time t. Then a characteristic function for $B(t)$ is given by the mapping

$$\chi_{B(t), t} : X(t) \to [0, 1].$$

Moreover, each membership function $\chi_{B(t), t^*}$ represents the ranking given

Applications to Clustering

to all fuzzy sets at time t^* with respect to the set of criteria and constraints imposed at time t, which have produced the subset $B(t)$ of $X(t)$. Thus for $t^* > t$ classification is considered at a later time with respect to previous criteria and constraints.

Definition 4.5.3

A *dynamic fuzzy relation* (DFR), $R(t, t^*)$, is a fuzzy subset of Cartesian product

$$X(t) \times X(t^*) = \{\langle x(t), x(t^*)\rangle | x(t) \in X(t) \wedge x(t^*) \in X(t^*)\}$$

with the characteristic function

$$\chi_{R(t,t^*)}: X(t) \times X(t^*) \to [0, 1].$$

Clearly, $\chi_{R(t,t^*)}$ can be used as some measure of strength of relationship over a period of time. The operations described for static relations can be applied here in the time context.

EXAMPLE 4.5.7 Let $R(t, t^*)$ and $R(t^*, t^{**})$ be two time-dependent DFR's. Namely, $R(t, t^*) \subseteq X(t) \times X(t^*)$ and $R(t^*, t^{**}) \subseteq X(t^*) \times X(t^{**})$. The composition $R(t, t^*) \circ R(t^*, t^{**}) \subseteq X(t) \times X(t^{**})$ can be defined by the characteristic function

$$\chi_{R(t,t^*) \circ R(t^*,t^{**})}(\alpha, \beta) = \max_{\gamma} \left\{ \min\left[\chi_{R(t,t^*)}(\alpha, \gamma), \chi_{R(t^*,t^{**})}(\gamma, \beta)\right]\right\}$$

for $\alpha \in X(t)$ and $\beta \in X(t^{**})$.

Definition 4.5.4

The fuzzy elements $\alpha(t)$ and $\gamma(t)$, such that $\alpha(t) \in X(t)$ and $\gamma(t) \in X(t^*)$, are said to be λ-*compatible* with respect to $R(t, t^*)$ iff $\chi_{R(t,t^*)}[\alpha(t), \gamma(t)] \geq \lambda$.

Obviously, we can now collect all λ-*compatible* pairs $[\alpha(t), \lambda(t)]$ into a well-defined pattern.

Definition 4.5.5

Let $\alpha(t) \in X(t_1)$ and $\gamma(t) \in X(t_2)$ at distinct times t_1, t_2 such that $t_1 < t_2$. Then $\alpha(t)$ and $\gamma(t)$ are chained iff for any $t \in [t_1, t_2]$ there exists $\delta(t) \in X(t)$ such that $\min\{\chi_{R(t_1,t)}[\alpha(t), \delta(t)], \chi_{R(t,t_2)}[\delta(t), \gamma(t)]\} > 0$.

Theorem 4.5.4 Let $\alpha(t) \in X(t_1)$ and $\gamma(t) \in X(t_2)$ be λ-compatible with respect to $R(t_1, t) \circ R(t, t_2), \forall t \in [t_1, t_2]$, then $\alpha(t)$ and $\gamma(t)$ are chained.

Proof Clearly, by the definitions of λ-compatible and fuzzy composition,

$$\max_{\delta(t)} \left\{ \min\left[\chi_{R(t_1, t)}[\alpha(t), \delta(t)], \chi_{R(t, t_2)}[\delta(t), \gamma(t)] \right] \right\} \geq \lambda, \quad \forall t \in [t_1, t_2].$$

Therefore, there exists $\delta(t) \in X(t)$ such that

$$\min\left\{ \chi_{R(t_1, t)}[\alpha(t), \delta(t)], \chi_{R(t, t_2)}[\delta(t), \gamma(t)] \right\} \geq \lambda > 0$$

since $0 < \lambda \leq 1$. (The case $\lambda = 0$ is not interesting, since it shows only a zero measure of connectiveness between elements.) Q.E.D.

Theorem 4.5.5 Let $B(t) = [b_{ij}(t)]$ be a $p \times p$ fuzzy matrix at a given time t, and let $b_{ii}(t) \geq \max[b_{ij}(t), b_{ji}(t)]$, $\forall i, j \mid i, j \in \{1, 2, \ldots, p\}$. Then we have

$$\forall i, j \left[b_{ij}(t) = \max_k \left\{ \min\left[b_{ik}(t), b_{kj}(t) \right] \right\} \right]$$

iff

$$\forall i, j, k \left[b_{ij}(t) \geq \min\{ b_{ik}(t), b_{kj}(t) \} \right].$$

Proof

1 Clearly, because the maximum is taken over all k, the left-hand side implies the right-hand side.

2 The statement

$$\forall i, j, k \left[b_{ij}(t) \geq \min\{ b_{ik}(t), b_{kj}(t) \} \right]$$

clearly implies that

$$\forall i, j \left[b_{ij}(t) \geq \max_k \left\{ \min\left[b_{ik}(t), b_{kj}(t) \right] \right\} \right].$$

Also

$$\max_k \left\{ \min\left[b_{ik}(t), b_{kj}(t) \right] \right\} \geq \min\left[b_{ii}(t), b_{ij}(t) \right].$$

However,

$$b_{ii}(t) \geq \max\left[b_{ij}(t), b_{ji}(t) \right] \quad \forall i, j$$

Applications to Clustering

implies

$$b_{ij}(t) = \min[b_{ii}(t), b_{ij}(t)].$$

This can be easily seen from the following:

$$\min[b_{ii}(t), b_{ij}(t)] = b_{ii}(t) \cdot b_{ij}(t) = b_{ij}(t) + b_{ji}(t) \cdot b_{ij}(t) = b_{ij}(t).$$

Hence

$$\max_{k} \{\min[b_{ik}(t), b_{kj}(t)]\} \geq b_{ij}(t)$$

and thus the equality is established. Q.E.D.

Experiments using these techniques for the recognition of tropical cyclones have been discussed in Kandel [1209].

In conclusion, the basis for the similarity matrix is the assumption that pairwise, similarity coefficients between elements, or vectors in a vector space, can be subjectively found; based on these values we can partition the data set into fuzzy partitions ranging between the *conjoint partition* (one category) and disjoint partition (singletons). The pairwise similarity measures can be obtained with the help of some feature extraction process that we assume to be available. It is important to note that the selection of this function (or the features) is a subjective process that is quite fuzzy by itself. The reason for that is that we can always argue that the *similarity measure* is "good" or "bad" and that is is never objective and never the best possible. Therefore, the partition suggested by this criterion should compensate for such possible perturbations in the similarity values; that is, the partition proposed should detect the *internal structure* of the given data set, and put less emphasis on the absolute values of the similarity measures.

Once the grouping has been achieved, we can evaluate the partition, especially in cases where the number of categories is unknown *a priori*. However, it should be noted again that data clustering is not a rigorously or uniformly defined concept. The division of a data set into classes whose elements are similar in some sense is confronted with decompositions into subsets exhibiting a cohesiveness that is not entirely related to point-to-point similarity.

The value judgement of the user is the ultimate criterion for evaluating the meaning of the classification. If using these techniques produces an answer of value, no more need be asked of them.

4.6 CONVEX DECOMPOSITIONS

The "hard" partition described in previous sections, defined by the nondegenerate fuzzy c-partition space

$$P_{fc} = \left\{ U \in V_{cn} \mid \chi_{ik} \in [0,1], \forall i, k; \sum_{i=1}^{c} \chi_{ik} = 1, \forall k; \sum_{k=1}^{n} \chi_{ik} > 0, \forall i \right\},$$

has given rise to the work by Ruspini [2244]–[2251], Bezdek [172]–[182], Woodbury and Clive [2854], and related papers.

Bezdek and Harris [189] have defined another kind of a fuzzy similarity relation R, given by

$$\text{max-}\Delta \text{ transitive if and only if } r_{ij} \geq \bigvee_{l=1}^{n} \left((r_{il} + r_{lj} - 1) \vee 0 \right), \qquad \forall i, j,$$

where the max-Δ transitivity is implied by max-prod transitivity, since $(a + b - 1) \vee 0 \leq ab$ for $a, b \in [0, 1]$, so

$$r_{ij} \geq \bigvee_{l=1}^{n} (r_{il} \cdot r_{lj}) \geq \bigvee_{l=1}^{n} (r_{il} + r_{lj} - 1) \vee 0.$$

Clearly, we have the result that $R \in R_\Delta$ if and only if the function $d: X \times X \to [0, 1]$ defined by $d(x_i, x_j) = 1 - r_{ij}$ is a pseudometric.

If we define $\text{conv}(R_k)$ to be the convex hull of hard equivalence relations, it is easy to show [189] that $\text{conv}(R_3) = R_\Delta$ and for $k > 3$ we have

$$\text{conv}(R_k) \subset R_\Delta,$$

where $\subset$ indicates proper subset.

At present, the only clustering procedure based on fuzzy relations appears to be the one described in [594], [1243], [2521], [2954]. A concise summary of this method follows: beginning with a reflexive, symmetric fuzzy relation matrix R, its transitive closure $\bar{R}$ is obtained by any of three algorithms: $(\vee \cdot \wedge)$ composition iteration, $R \leq R^2 \leq R^3 \leq \cdots \leq R^q = \bar{R} = R^k \in R_\wedge$, $\forall k \geq q \leq n - 1$, Zadeh [2954], or Tamura et al. [2521]; a column-row scanning algorithm, Kandel and Yelowitz [1243]; or Prim's minimal spanning tree algorithm, Dunn [594].

Once $\bar{R}$ is obtained, its entries are used to define a nested sequence of hard relations $R_{ij} \in R_n$ by thresholding at levels in-between successive values of $\bar{r}_{ij}$. Thus we might have from the entry $\bar{r}_{ij}$ the hard relation $x_p R_{ij} x_q \Leftrightarrow \bar{r}_{pq} \geq \bar{r}_{ij}, \forall p, q$. In this fashion we may construct a nested sequence of hard equivalence relations (therefore hard c-partitions of X) in $X \times X$, which ultimately yield a partition tree or dendogram. While this method appeared at first to be quite novel, it was shown by Dunn in [594] that, because max-min transitivity is equivalent to the ultra-metric inequality, the resultant hierarchies of hard clusters were in fact a subset of single-linkage hierarchies, from a well-known graph-theoretic method for hard clustering.

As an alternative, if R conv(R_n), Bezdek and Harris [189] use the convex decomposition or clustering as follows: suppose $R = \sum_{k=1}^{p} c_k R_k$. Each $R_k \in R_n$ is isomorphic to a hard c-partition of X, say $U_k \in P_c$. Note that c, the number of clusters in X, is in general function of k, so there will be no hope that $\sum_{k=1}^{p} c_k U_k$ is well-defined, although when it is, the resultant U lies in P_{fc}. Thus from $R = \sum_{k=1}^{p} c_k R_k$ there follows the sequence

$$\left\{ (c_k, U_k) \mid U_k \in P_c \forall 1 \leq k \leq p; \sum_{k=1}^{p} c_k = 1 \right\}.$$

Since the decomposition $\sum c_k R_k$ exhibits the "percentage" of each R_k needed to build up fuzzy relation R, they interpret c_k as an indicator of the relative merit of the associated U_k as a c-partition of X. This also provides a method for choosing c, the number of clusters most likely to exhibit substructure in X; finally, we may observe that the partitions $\{U_k\}$ generated this way are not nested hierarchically.

In short, the max-Δ transitivity is equivalent to the triangle inequality, and is necessary for fuzzy similarity relations admitting a convex decomposition by equivalence relations. Furthermore, each similarity relation of this type induces a pseudometric in $X \times X$. For relations constructed from fuzzy c-partitions of X via $U^T(\Sigma \cdot \wedge)U$, this pseudometric is half of the l_1 distance between membership vectors (for points in the data) for the fuzzy partition used. Examples of applications to cluster analysis and cluster validity seem to support the contention that the space of similarity relations characterized by max-Δ transitivity is an important one for applications in pattern recognition. A complete characterization of the convex hull of hard equivalence relation space, together with an implementable decomposition algorithm, will provide a new clustering method based on fuzzy similarity relations, which seems to hold great promise according to Bezdek and Harris [189].

Pseudosimilarity relations have been used also for decision-making in fuzzy environments by Asai et al. [68]. Some preliminary results on ranking fuzzy sets using linguistic preference relations have been reported by Efstathiou and Tong [613]. We follow their presentation.

Definition 4.6.1

A fuzzy relation R on the set of objects X is a *preference order* denoted P, if it is antireflexive, antisymmetric, and transitive.

This definition of a preference order implies:

1. There is no preference for x over x.
2. If x is preferred to y, then y is not preferred to x.
3. If x is preferred to y and y is preferred to z, then x is preferred to z.

The transitivity used here is that of the classical form: namely xRy and yRz imply xRz.

A linguistic preference relation L on a set of objects X is one in which the preference for one element of X over another is expressed linguistically rather than as a precise number in the interval [0, 1]. Thus we are concerned with the properties of relations in which preference has the form

"A is slightly preferred to B"

or

"C is strongly preferred to D."

A more compact representation of this information is

$$L(A, B) = \text{"slightly,"} \qquad L(C, D) = \text{"strongly."}$$

A first requirement of L is that it should be antireflexive. In linguistic terms, this means simply that $L(A, A) = $ "no preference." We also feel that L should be antisymmetric in that, if $L(A, B)$ is any preference whatsoever, then $L(B, A) = $ "no preference." However, the question of transitivity is somewhat problematic. If we consider the example

$$L(A, B) = \text{"slightly"}$$

$$\underline{L(B, C) = \text{"slightly"}}$$

$$L(A, C) = \text{"?"}$$

then "?" should be something stronger than "slightly" and at the same time be more vague. A long chain of such reasoning would not be expected to produce answers as precise as those at the start of the chain.

Convex Decompositions

An important issue in linguistic classification is how to derive the entries in the linguistic relation. Clearly, when clustering fuzzy sets, many subjective criteria are involved such as the decision-maker's attitude. Clustering may be performed with reference to criteria that are obvious to the decision-maker, but difficult to measure mathematically. Furthermore, their very subjectivity and vagueness makes a linguistic approach seem all the more reasonable. These criteria are derived from the form of the fuzzy sets and the relationships between them. In doing so, the sets must be interpreted as possibility distributions, which give enough information for the classification alternatives.

Efstathiou and Tong [613] have developed a set of interactive algorithms for producing a linguistic preference between a pair of sets. The complete L is obtained by repeated pairwise comparisons within the set of alternatives. The criteria we use are:

1 Overlap of support sets.
2 Separation of peak values.
3 Trade-off between better and worse outcomes.
4 Proximity to threshold values.
5 Relative heights of sets.

The algorithms are designed so that the first questions are the most discriminatory and lead immediately to firm statements of preference. Thus nonoverlapping support sets indicate a definite rank order and no more questions need be asked. But, as the analysis proceeds in less clear-cut cases, the resulting preference is less strong. In this way overlapping support sets, peak values that are not well-separated, and one set's higher possibility of good outcomes indicate a marginal preference.

Apart from the questions of normativeness and meaning of fuzzy sets, related to every subjective approach to clustering or preference, most of these techniques also require a careful understanding of uncertainty. For example, the ratings are uncertain, and so there must be uncertainty in asserting that A is preferred to B. However, before any action can be taken, there is some point at which the decision must be made precise and one of the following three cases must be stated as true: either A is preferred to B, A and B are preferentially equivalent, or B is preferred to A. A criticism sometimes levelled at fuzzy set theory in general is that among all the talk of imprecision and vagueness, numbers must be produced. It may be appropriate here to introduce the fuzzy set precision at the same point as the classification precision, and dispense with the qualifying adverbs and their connotations of degrees of belief.

In conclusion of this entire topic, we would like to mention the work by Cao Hongxing and Chen Guofan [296], who have used fuzzy partition based on imprecise relations to partition weather processes. The procedure by which synoptic situations are classified into some typical patterns has been in use in weather prediction for a long time. The authors have applied the principles of fuzzy similarity to the partition of daily 500 mb circulation processes during the winter-spring period in 1972. The objective of their study has been to group all circulations into different patterns that possess clear features so as to provide a basis for weather forecasting based on the principles of fuzzy relations and cluster analysis.

CHAPTER FIVE

Applications and Performance Measures

5.1 PHONETIC AND PHONEMIC LABELING OF SPEECH

Considerable effort has been made in the last decade to develop systems capable of understanding continuous speech. De Mori and Laface present one such model for assigning phonetic and phonemic labels to speech segments, using fuzzy algorithms [481].

The system executes fuzzy algorithms that assign degrees of worthiness to structured interpretations of syllabic segments extracted from the signal of a spoken sentence. The knowledge source is a series of syntactic rules whose syntactic categories are phonetic and phonemic features detected by a precategorical and a categorical classification of speech sounds, as illustrated in Figure 5.1.

The approach proposed in [481] is based on the detection of phonetic categories related to some distinctive features, represented by a tree of features. A node represents a feature that can be hypothesized only if the feature corresponding to the father has been previously hypothesized. Thus, for example, the feature "affricate" can be hypothesized only for nonsonorant consonants; such features will be called "precategorical classification features" (PCF's).

The phonetic features are syntactic categories related to the lower level description of the acoustic features by a "branching questionnaire" of the type proposed by Zadeh [2978]. Thus PCF is assigned to a speech segment after answering a composite classificational question $Q \triangleq B$, where the bodies of component questions Q_j, $j = 1, 2, \ldots, n$, are fuzzy sets $B_1, B_2, \ldots, B_n$ involved in an analytic representation of B. The fuzzy sets $B_1, B_2, \ldots, B_n$ are linguistic variables defined over the range of acoustic

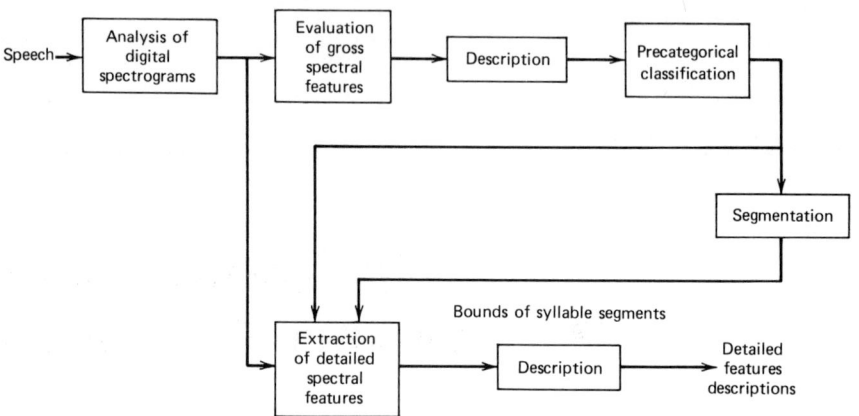

Figure 5.1 Block diagram: extraction and description of spectral features.

measurements that phoneticians have found useful for characterizing distinctive features.

Let Q_i, $i = 1, 2, \ldots, K$, be the composite question related to the ith PCF to be hypothesized. The answers to component questions Q_{ij} of the composite question Q_i are fuzzy linguistic variables (e.g., high, medium, low) whose membership functions are computed from diagrams stored in the phonetic source of knowledge and that are defined over the universe of an acoustic measurement. The answers to the composite question Q_i are related by syntactic rules to the answers to the component questions Q_{ij}; thus each answer to Q_i can be associated with a membership function whose value is computed under the control of semantic rules from the values of membership functions associated with each answer to the component questions.

This use of fuzzy algorithms models to some extent the fact that most of the acoustic-phonetic properties of speech sounds are only known with a degree of vagueness, for example, the signal energy is high for vowels, nonsonorant consonants have high-frequency components, and in unvoiced stops there is an interval of silence followed by some noise.

A fuzzy linguistic variable, representing a judgment that can be expressed after the inspection of some acoustic parameters, is defined by a fuzzy restriction

$$R(X, u)$$

where u is a generic value of an acoustic parameter and X is the subjective judgment. $\chi_{R(X)}(u)$ is a grade of membership function defining $R(X)$ as a fuzzy subset of u.

Phonetic and Phonemic Labeling of Speech

The fuzzy variables form a basis for a composite question Q. An answer to Q may be interpreted as a specification of the grade of evidence of a phonetic feature in a speech segment. This grade is a function of the grades of membership of the speech segment with every fuzzy variable.

For example, the classification of a consonant as sonorant or nonsonorant is performed after the detection of vocalic intervals and is obtained as an answer to a composite question:

$Q_1 \triangleq$ is the interval corresponding to the acoustic pattern p nonsonorant?

The question Q_1 has six component questions:

$Q_{11} \triangleq$ how high is R_v? (R_v is the ratio between the energy in the 200–900 Hz band and the energy in the 5–10 kHz band.)

$Q_{12} \triangleq$ how low is the minimum dip in R_v with respect to the values of R_v in the preceding and following vowels?

$Q_{13} \triangleq$ what is the minimum value of S with respect to the value assumed by S on the silences? (S is the total energy of a spectrum.)

$Q_{14} \triangleq$ what is the duration of the consonant?

$Q_{15} \triangleq$ what is the minimum dip in the signal?

$Q_{16} \triangleq$ what is the maximum dip in R_v?

Each question may admit some possible fuzzy answers. With the answers to these atomic questions, we can use an inferred set of fuzzy rules for defining the syntactic categories sonorant and nonsonorant that correspond to the answers to Q_1. The inference methodology is described in [484]. In that system learning consists of varying the "grammaticalities" of the rules and selecting a set of rule memberships that ensure the absence of type-1 errors and minimizes the number of type-2 errors, where we define:

$$\text{poss}[\langle \text{sonorant} \rangle \text{ is in } p] = \chi_{\langle \text{sonorant} \rangle}$$

and

$$\text{poss}[\langle \text{nonsonorant} \rangle \text{ is in } p] = \chi_{\langle \text{nonsonorant} \rangle},$$

and the following types of errors are considered:

1. *Type-1 error*: $\chi_{\langle \text{sonorant} \rangle} = 1$ for a nonsonorant sound or $\chi_{\langle \text{nonsonorant} \rangle} = 1$ for a sonorant sound.
2. *Type-2 error*: $\chi_{\langle \text{sonorant} \rangle} > \chi_{\langle \text{nonsonorant} \rangle}$ for a nonsonorant sound or $\chi_{\langle \text{nonsonorant} \rangle} > \chi_{\langle \text{sonorant} \rangle}$ for a sonorant sound.

This means that a wrong hypothesis never appears as certainly true and that the wrong hypothesis appears to be more plausible than the true hypothesis in a minimum number of situations.

Learning is performed by an interactive program that allows the operator to vary the grammaticalities of the rules in given intervals and with given steps. For each assignment of the grammaticalities, the program gives the number of type-1 errors and type-2 errors as well as the input samples for which an error has been found. The final assignment of the grammaticalities is done subjectively after experiments on all coarticulations, on the basis of the objective measurements of the number of type-1 and type-2 errors obtained with different assignments of grammaticalities.

Phoneme labeling of pseudosyllables is performed by a fuzzy algorithm consisting of a branching questionnaire. The generation of a hypothesis about the presence of a phoneme is the answer to a composite question associated with a membership function.

Phoneme classification is a context-dependent operation except for non-reduced vowels that can be recognized with no knowledge of the context. Vowels belong to sonorant segments, for which formants are tracked. Emission of hypotheses about vowels is primarily based on the analysis of the plot of the second formant frequency versus the first formant frequency.

Classification of sonorant sounds into phonemes requires a preliminary recognition of the features liquid or nasal. This operation as well as the phonemic transcription is controlled by rules that may be context-dependent. Such rules are functions of answers to the questions of a branching questionnaire. Similar to the phonetic case, it is based on a hierarchy of concurrent processes or steps of fuzzy algorithms, whereby the acoustic rendering of phonemic features are combined in a hierarchy of rules. To get the right hypothesis with the highest degree of confidence, the effectiveness of these rules is measured by their grammaticalities.

The rules and their grammaticalities represent the knowledge of a phonetician who interprets a spectrogram on the basis of his or her experience, which is subjective, but is based on objective facts and relations established by experiment. The experimental results obtained in [481] suggest that the rules used for precategorical classification depend neither on the talker nor the context.

Some other rules (for example, those used for the classification of nasals) are context-dependent but give results that are talker-independent.

By the use of fuzzy algorithms, we can account for the imprecision of the knowledge used in interpreting acoustic patterns and for the vagueness that may arise when a pattern is described in terms of acoustic features. Since the algorithms are flexible, advantage can be taken of the redundancy of features to get the best evidence assigned to the right hypothesis in spite of

Speech Identification

the vagueness and imprecision of the data and rules used to generate it. The success of the system clearly depends on the number of acoustic features used, their significance, detectability, and noninteraction.

5.2 SPEECH IDENTIFICATION

In a series of papers, Pal and Majumder [1962], [1964], [1966] have discussed the application of fuzzy sets to vowel and speaker recognition. This section is based on their approach to speech identification.

Consider an unknown pattern

$$X = \begin{bmatrix} x_1 \\ x_2 \\ \vdots \\ x_i \\ \vdots \\ x_n \end{bmatrix}$$

where x_i denotes the measured ith feature of the event, represented by a point in the multidimensional vector space Ω_x, consisting of m ill-defined pattern classes $C_1, C_2, \ldots, C_j, \ldots, C_m$. Let $R_1, R_2, \ldots, R_j, \ldots, R_m$ be the reference vectors where R_j associated with C_j contains h_j number of prototypes such that

$$R_j^{(l)} \in R_j, \quad l = 1, 2, \ldots, h_j.$$

The pattern X can then be assigned to be a member of that class to which it shows maximum similarity as measured by the following algorithms:

Algorithm 1

Define a membership function $\chi_j(X)$ associated with pattern X for the jth class as

$$\chi_j(X) = \left[1 + \left(\frac{d(X, R_j)}{E} \right)^F \right]^{-1.0}$$

where E is an arbitrary positive constant, F is any integer, $d(X, R_j)$ is the distance between X and R_j, and

$$d(X, R_j) = \wedge_l \left[\sum_i \left(W_{ji}^{(l)} (x_i - R_{ji}^{(l)}) \right)^2 \right]^{0.5}$$

where $\wedge$ denotes minimum. The above expression represents the minimum value of the weighted distances of an unknown pattern X from all its expected values in class C_j and $W_{ji}^{(l)} (|W_{ji}^{(l)}| \leq 1)$ corresponds to the lth prototype in C_j and denotes the magnitude of the weighting coefficient along the ith coordinate.

Constants E and F have the effect of altering the fuzziness of a set. The χ-value defining the grade of membership of X in C_j as shown by this equation is unity for $d(X, R_j) = 0$, zero for $d(X, R_j) = \infty$, and increases with decreasing the value of $d(X, R_j)$. Thus an unknown pattern X is recognized to be a member of the kth class if

$$\chi_k(X) = \vee_j \{\chi_j(X)\}$$

where $\vee$ denotes maximum and $j, k = 1, 2, \ldots, m$.

Algorithm 2

Let $p_1, p_2, \ldots, p_i, \ldots, p_n$ be the n properties each of which represents some aspects of the unknown pattern X and has value only in the interval $[0, 1]$ such that

$$X = \{p_1, p_2, \ldots, p_i, \ldots, p_n\}$$

where

$$p_i = \left(1 + \left| \frac{\bar{x}_i - x_i}{E} \right|^F \right)^\alpha$$

where $\alpha = -1$ and $\bar{x}_i$ is the ith reference constant determined from representative events of all the classes. Constants E and F have the same effect as in the previous method in affecting the fuzziness of a set.

If there are h number of prototypes in a class C_j, each reference point may then be represented as

$$R_j^{(l)} = \{p_{1j}^{(l)}, p_{2j}^{(l)}, \ldots, p_{ij}^{(l)}, \ldots, p_{nj}^{(l)}\}$$

where $p_{ij}^{(l)}$ denotes the degree to which property p_i is possessed by the lth

Speech Identification

prototype in C_j. Then the *similarity vector* $S_j(X)$ for the pattern X with respect to the jth class has the form

$$S_j(X) = \{s_{1j}, s_{2j}, \ldots, s_{ij}, \ldots, s_{nj}\}$$

$$s_{ij} = \frac{1}{j} \sum_i s_{ij}^{(l)}$$

$$s_{ij}^{(l)} = \left(1 + W \left|1 - \frac{p_i}{p_{ij}^{(l)}}\right|\right)^{-2Z}$$

where the numerical value of s_{ij} denotes the grade of similarity of the ith property with that of C_j. W is any positive constant dependent on each of the properties, and Z is an arbitrary integer. With the knowledge of all the similarity vectors, we can decide $X \in C_k$ if

$$|S_j(X)| < |S_k(X)|, \quad k, j = 1, 2, \ldots, m; k \neq j.$$

The recognition of the vowel as well as the specific speaker is based on the sequential recognition scheme [1962] described in Figure 5.2.

Prototype points chosen for vowel identification are the average of the coordinate values corresponding to the entire set of samples in a particular class. The features with increasing variance have been weighted with decreasing values of the weighting coefficients, and the inverse of the standard deviation of the formants as weighting coefficients were studied. Although shorter and longer types of vowels /i/, /u/, /e/, and /o/ are treated as the same group, they were given individual reference vectors and weighting coefficients computed over their respective set of events. Properties corresponding to each of the features were computed with $E = 100$, $F = 2$, and using

$$\bar{x}_i = \bigvee_q \frac{1}{h} \sum_l x_i^{(l)}, i, q = 1, 2, 3.$$

Similarity of the pattern with all the classes of informants was measured for $Z = 1$.

To study the effect of training sets used in learning on a classified set, the above method of speaker identification has been repeated thrice for each of the different sample sizes, namely 1, 3, 5, 7, 10, and 15.

The results of the experiments conducted by Pal and Majumder show an accuracy of vowel sound recognition of about 82 percent when the decision

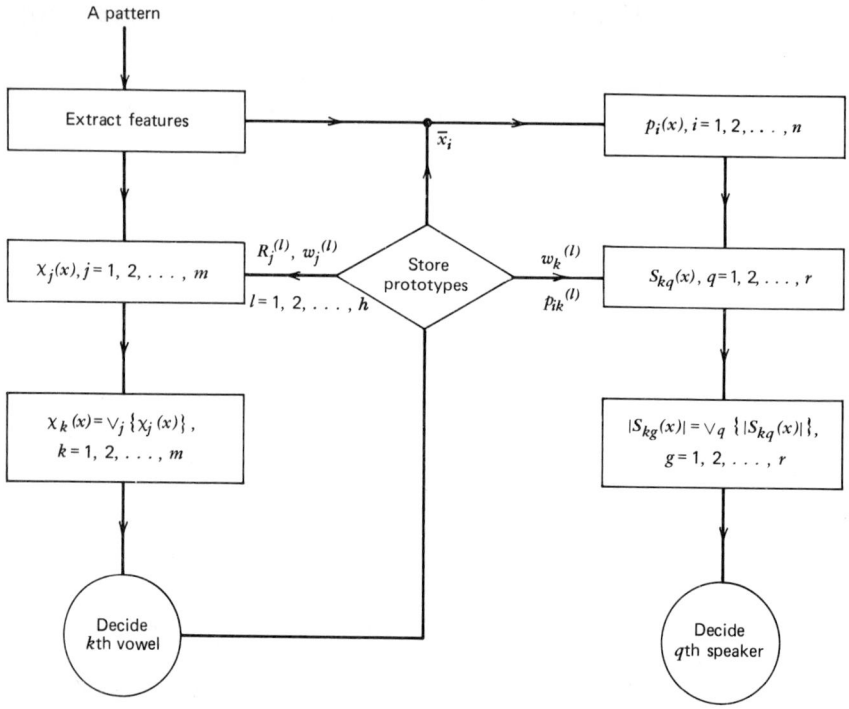

Figure 5.2 Sequential recognition of vowel and its informant.

of the machine was based only on the highest membership values. By incorporating a second choice under the control of a supervisory learning scheme, supposed to be based on linguistic constraints, the above score can be improved by 15 percent. Knowledge of the weighting coefficients and reference vectors used is also available from any size of the training samples containing more than 12–16 utterances, without affecting the overall score.

Although it is well-known that the fundamental voice frequency F_0 and higher formants (F_4 and F_5) are more speaker-dependent, computer decision on the identification of informants in the age group of 28–30 years using property sets and the first three formants only is found to be satisfactory. It may be remarked that informants widely varying in age and sex are expected to result in a large discrimination in F_3, leading to a better identification. At best, for 97 percent of the speech sounds, the machine was seen to render correct recognition of the speaker by comparing the magnitudes of similarity vectors. This is expected to be extended to a large number of informants varying in sex and age. It can be also stated that after an optimum number of learning samples (8–10 only) sufficient to characterize the representative points of a class, the size of the sample space can be extended enormously without disturbing the overall recognition score.

Results of speaker identification were tabulated with the assumption that the vowel in the utterance is correctly recognized. If the immediate knowledge of the vowel is used (for sequential processing), then some but not all of the 18 percent error for vowel recognition will pass to speaker identification, since some samples were found to have error both in identifying the vowel and the speaker. However, our decision algorithm does not preclude the possibility of correct speaker identification from the misrecognized vowel features.

The results reported by Pal and Majumder in [1962] and [1966] do agree with other recent studies [1295], [1611], [1961] showing the conceptual and practical advantages of fuzzy algorithms over the probabilistic approach in demonstrating ill-defined patterns where pattern indeterminacy is due to inherent vagueness rather than randomness of events and especially when the sample size is small. In addition, the method is more significant from the following viewpoints:

1 The burst spectra, an important cue, particularly the antiformants, were not included as recognition features.
2 The CV (consonant-vowel) syllables in the experiment were taken from normally spoken words, and therefore, coarticulation from distant vowels and consonants is likely to affect the transitions.
3 The minimum duration of vowels (250 ms) for arriving at the perfect steady state could not be achieved in these utterances.

The role played by the exponential fuzzy generator is found to be satisfactory in altering the fuzziness within property sets. The fuzzy hedge "slightly," corresponding to the DIL ($\chi_{\text{DIL}(A)}(x) = [\chi_A(x)]^{0.5} \forall x$) operation, as expected, results in better classification than that of the hedge "very" ($F = 2.0$). But successive application of the DIL operation does not ensure an increase in recognition score. In other words, after an optimum value of the exponential fuzzy generator is achieved, the fuzziness in property sets is not much altered. Hence the variation of the score becomes insignificant. A wide variation of about 20–25 percent (excepting the velars) in accuracy rate is achieved with different values of F ranging from 8 to $\frac{1}{16}$.

The reciprocal of the standard deviation provides appropriate phase weights in measuring the importance of the features. This supports the findings in other communications [1603]–[1612]. This characteristic is also found to be significant for the property sets having higher degrees of fuzziness. The recognition score is likely to be further improved by the inclusion of the transitional data of a third formant and the data of the burst spectra. Since their work is purposely restricted to show only the effectiveness of formant transitional data as a recognition parameter and the fuzzy set theoretic approach as a decisional algorithm for automatic consonant recognition, these parameters were excluded by Pal and Majumder.

5.3 FUZZY FILTERS

In the process of character recognition, Wang and Wang [2737] present an effective tool for decision-making that can be employed to solve recognition problems that require a degree of machine intelligence capability, where the theory of fuzzy sets plays a key role in designing the very structure of the system itself.

A printed alphanumeric character represented by different gray levels in 7×5 picture cells can be viewed as a matrix where the numerical value in each element of this 7×5 matrix is a fuzzy variable corresponding to the gray level. Consider the columns and rows of the matrix as two sets R and C, respectively. Let r designate an element of R and c an element of C. The set of ordered pairs (r, c) defines the product set $R \times C$. The fuzzy subset $\mathcal{D}$ such that

$$\forall (r, c) \in R \times C, \qquad \chi_\mathcal{D}(r, c) \in D$$

is a fuzzy graph, where D is the membership set of $R \times C$. An individual entry may be expressed as

$$\chi(r_i, c_j) = \chi_\mathcal{D}(r_i, c_j),$$

where $i = 1, 2, \ldots, 7$ and $j = 1, 2, \ldots, 5$; $\chi(r_i, c_j)$ is the value of the ordered pair (r_i, c_j). An alternate representation of the same character pattern yields

$$\mathcal{D} = \{((r_1, c_1) | \chi_1), ((r_1, c_2) | \chi_2), \ldots, ((r_7, c_5) | \chi_{35})\}.$$

A similar vector representation turns the fuzzy graph $\mathcal{D}$ into a row vector $\chi_\mathcal{D}(d_i)$:

$$\chi_\mathcal{D}(d_i) = \left[\chi_{i_1}^d, \chi_{i_2}^d, \ldots, \chi_{i_{35}}^d\right]$$

$$= [\chi_i(r_1, c_1), \chi_i(r_1, c_2), \ldots, \chi_i(r_1, c_5),$$

$$\times \chi_i(r_2, c_1), \ldots, \chi_i(r_7, c_4), \chi_i(r_7, c_5)].$$

Fuzzy Filters

The value of the ordered pair (r_i, c_j), corresponding to the gray levels, ranges over $[0, 1]$. In practice, the character is scanned by an optical character reader and a gray level for each picture cell is measured. This type of mapping raises no controversy to the issue of membership function determination. It is a linear mapping and the membership function obtained in this manner is certainly an objective one. Hence the highest gray level corresponds to $\chi(r_i, c_j) = 1$ and the lowest gray level corresponds to $\chi(r_i, c_j) = 0$.

Standard masks are prepared for each pattern class such that $\chi^m(v_i, c_j)$ is 0 or 1. Thus $\chi_{\text{mask}}(m_i)$ is the vector membership function of the ith standard mask where $m_i \in M$. Let

$$\chi_{\text{mask}}(m_i) = [\chi_{i_1}^m, \chi_{i_2}^m, \ldots, \chi_{i_{35}}^m]$$

$$= [\chi_i^m(r_1, c_1), \chi_i^m(r_1 c_2), \ldots, \chi_i^m(r_1, c_5),$$

$$\times \chi_i^m(r_2, c_1), \ldots, \chi_i^m(r_7, c_4), \chi_i^m(r_7, c_5)]$$

where $i = 1, 2, \ldots, 37$ and its components are $\chi_i(r_p, c_q)$, where $p = 1, 2, \ldots, 7$ and $q = 1, 2, \ldots, 5$.

On the other hand, $\chi_{\mathcal{D}}(d_j)$, the membership function of the jth data pattern, is a fuzzy vector and $\mathcal{D}$ is the fuzzy data pattern set. Hence we may write

$$\chi_{\mathcal{D}}(d_j) = [\chi_{j_1}^d, \chi_{j_2}^d, \ldots, \chi_{j_{35}}^d]$$

$$= [\chi_j^d(r_1, c_1), \chi_j^d(r_1, c_2), \ldots, \chi_j^d(r_1, c_5),$$

$$\times \chi_j^d(r_2, c_1), \ldots, \chi_j^d(r_7, c_4), \chi_j^d(r_7, c_5)]$$

where $j = 1, 2, 3, \ldots$.

Being a fuzzy set, the vector membership function $\chi_{\mathcal{D}}(d_j)$ is ranged over the closed interval $[0, 1]$.

Define

$$\Theta_M[\chi_{\text{mask}}(m_i), \chi_{\mathcal{D}}(d_j)] = \chi_{\text{mask}}(m_i) \vee \chi_{\mathcal{D}}(d_j)$$

where $\vee$ (max) is performed on the elements of these vectors.

The result of the above operation yields a 35×35 matrix with fuzzy operation $\vee$ (max) applied in a similar manner as matrix multiplication. Note that the transpose of one operation with the other operations is commutative, that is,

$$\Theta_M\left[\chi_{\text{mask}}(m_i), \chi_{\mathcal{D}}(d_j)\right] = \Theta_M^T\left[\chi_{\mathcal{D}}(d_j), \chi_{\text{mask}}(m_i)\right].$$

Similarly, for fuzzy operation $\wedge$ (min), define

$$\Theta_m\left[\chi_{\text{mask}}(m_i), \chi_{\mathcal{D}}(d_j)\right] = \chi_{\text{mask}}(m_i) \wedge \chi_{\mathcal{D}}(d_j).$$

The similar commutative property also holds for Θ_m.

$\Theta_M(\cdot)$ and $\Theta_m(\cdot)$ can be used to form the mixed functions of fuzzy variables, which are particularly useful. Since $\Theta_M[\chi_{\text{mask}}(m_i), \chi_{\text{mask}}(m_j)]$ and $\Theta_m[\chi_{\text{mask}}(m_i), \chi_{\text{mask}}(m_j)]$ are 35×35 matrices, they are expressed as

$$\{M_{i,j}^{m,m}(r,c)\} \quad \text{and} \quad \{m_{i,j}^{m,m}(r,c)\},$$

respectively. The difference of these two matrices is defined as the score matrix

$$\{S_{i,j}^{m,m}(r,c)\},$$

where

$$\{S_{i,j}^{m,m}(r,c)\} = \{M_{i,j}^{m,m}(r,c)\} - \{m_{i,j}^{m,m}(r,c)\}$$

where $r = 1, 2, \ldots, 35$, $c = 1, 2, \ldots, 35$.

Suppose the same relational operation is performed between the vector membership function of incoming data patterns and those of standard masks; this also yields a score matrix

$$\{S_{i,j}^{m,d}(r,c)\} = \{M_{i,j}^{m,d}(r,c)\} - \{m_{i,j}^{m,d}(r,c)\}$$
$$= \Theta_M\left[\chi_{\text{mask}}(m_i), \chi_{\mathcal{D}}(d_j)\right] - \Theta_m\left[\chi_{\text{mask}}(m_i), \chi_{\mathcal{D}}(d_j)\right]$$

where $r = 1, 2, \ldots, 35$ and $c = 1, 2, \ldots, 35$.

To compute the overall weighted distance measure between mask i and data j, the above weights of 1 and $\frac{1}{36}$ are incorporated in the following sums:

$D(i,j) \triangleq$ weighted distance measure between ith mask and jth data

$$= \sum_{r=c} (\text{diagonal distance}) + \frac{1}{36} \sum_{\substack{r \\ r \neq c}} \sum_c (\text{off-diagonal distance}).$$

Fuzzy Filters

The basic idea behind the computation of $S_{i,j}^{m,\,m}(r, c)$ is to treat the vector membership function of the incoming data as the ith mask; hence a reference value of $S_{i,k}^{m,\,m}(r, c)$ can be generated. Also note that both $S_{i,i}^{m,\,m}(r, c)$ and $S_{i,j}^{m,\,m}(r, c)$ are bounded:

$$1 \geqslant S_{i,i}^{m,\,m}(r, c) \geqslant 0$$

$$1 \geqslant S_{i,j}^{m,\,m}(r, c) \geqslant 0.$$

Hence the bounds for $D(i, j)$ can also be established:

$$68.055 \geqslant D(i, j) \geqslant 0.$$

Clearly, the fuzzy filter I detects and selects the jth data vector as the character m_k through the following computation:

$$\max_{\forall i} D(i, j) = D(i, k).$$

Even though the design concept of fuzzy filter II differs from that of fuzzy filter I, the basic mathematical tools used are similar. A square submatrix of our scene, termed a *window*, is selected.

A moving window can be expressed as a Boolean vector depending upon a parameter α, $\alpha = 1, 2, \ldots, 15$, which is used as an index to mark the starting (left-top) position of the window.

Suppose the fuzzy filter receives incoming data as its input, represented by a vector membership function, $\chi_{\mathscr{D}}(d_j)$; then a sequence of fuzzy operations as well as Boolean operations will be acted upon. The key decision-making process for fuzzy filter II is to compute the following distance measure between $\chi_{\mathscr{D}}(d_j)$ and the vector membership function of the ith standard mask, $\chi_{\text{mask}}(m_i)$, that is,

$$D(i, j) = \sum_{\substack{\text{(all components of the vector)} \\ \text{(algebraic sum)}}} \sum_{\alpha=1}^{15} \chi_G(g_\alpha)$$

$$\wedge \left[\left(\chi_{\text{mask}}(m_i) \wedge \chi_{\mathscr{D}}(d_j) \right) \vee \left(\overline{\chi}_{\text{mask}}(m_i) \wedge \overline{\chi}_{\mathscr{D}}(d_j) \right) \right].$$

Once again, the fuzzy filter reaches decision through computing

$$\max_{\substack{i \\ (i=1,2,\ldots,37)}} \{D(i, j)\} = D(k, j)$$

where the jth data pattern is recognized as the kth standard mask.

Like fuzzy filter I, the distance measure has an upper bound and a lower bound:

$$0 < D(i,j) \underset{\forall i}{\leqslant} D(k,j) \leqslant \sum \sum_{\alpha=1}^{15} \chi_G(g_\alpha) = 135$$
(all components of the vector)

where the upper bound is established by direct computation.

The last equation also serves as decision criterion. By searching the maximum distance measure throughout the mask set, fuzzy filter II will choose the kth standard mask as the identified character.

The experiments of character recognition reported by Wang and Wang [2737] illustrate that the type of mathematical manipulation required by fuzzy operators can be implemented much more easily by hardware. Fuzzy filter also enjoys simplicity in programming if a software implementation is desired.

In her recent paper [1032] Hisdal remarks, "Thus inference with the aid of fuzzy set theory, or equivalently with the aid of a possibilistic description, lies somewhere midways between the uncertainty inherent in probabilistic reasoning and the rigid TRUE-FALSE reasoning of Boolean algebra." In the experiments reported in [2737], the findings indicate that fuzzy set theory offers excellent results, making us think that, for some applications, those procedures calling for large statistical experiments should be bypassed and replaced by fuzzy set theory.

By adapting algorithms based upon fuzzy set theory, the intelligent machine has accomplished some degree of fuzzy reasoning. Though more experiments and generalization need to be carried out, a positive justification is provided by the experiments performed by Wang and Wang [2737]. Also, it has been shown that fuzzy algorithms have better performance as the complexity of the problem rises and the degree of uncertainty increases.

5.4 CLUSTERING VIA UNIMODAL FUZZY SETS

It is rather difficult to make meaningful comparisons among the many clustering techniques that have been reported in the literature. The difficulty may be attributed to the fact that many of the algorithms are heuristic in nature and, furthermore, have not been tested on the same standard data sets. In general, it seems that most of the algorithms are not capable of detecting categories that exhibit complicated distributions in the feature space and that a great many are not applicable to large data sets.

In this section we discuss an algorithm that partitions the given data set into "unimodal fuzzy sets." The notion of a unimodal fuzzy set has been chosen by Gitman and Levine [810] to represent the partition of a data set for two reasons. First, it is capable of detecting all the locations in the vector space where there exist highly concentrated clusters of points, since these will appear as modes according to some measure of "cohesiveness." Second, the notion is general enough to represent clusters that exhibit quite general distributions of points.

The general partition is optimal in the sense that the program detects all of the existing unimodal fuzzy sets and realizes the maximum separation among them. The algorithm is economical in memory space and computational time requirements and also detects groups that are fairly generally distributed in the feature space. The algorithm is a systematic procedure (as opposed to an iterative technique) that always terminates and the computation time is reasonable.

An important distinction between this procedure and the methods reported in the literature is that the latter use a distance measure (or certain typical distances) as the only means of clustering. Gitman and Levine have introduced another "dimension," the dimension of the order of "importance" of every point, as an aid in the clustering process. This is accomplished by associating with every point in the set a grade of membership or characteristic value. Thus the order of the points according to their grade of membership, as well as their order according to distance, are used in the algorithm. The latter partitions a sample from a multimodal fuzzy set into unimodal fuzzy sets.

Let X be a finite set of vectors ($|X| = n$) in a metric space and let d be the metric. Let B be a fuzzy set in X with the membership function f such that

$$f(\chi) = \sup_{x \in B} [f(x)]$$

and

$$\Gamma_{x_i} = \{x \mid f(x) \geqslant f(x_i)^2\}$$

$$\Gamma_{x_i,d} = \{x \mid d(\chi, x) \leqslant d(\chi, x_i)\}$$

where x_i is some point in B and d is a metric.

Definition 5.4.1

A fuzzy set B is symmetric if and only if, for every point x_i in B, $\Gamma_{x_i} = \Gamma_{x_i,d}$.

Clearly, if B is symmetric, then for every two points x_i and x_k in B,

$$d(x_i, \chi) \leqslant d(x_k, \chi) \leftrightarrow f(x_i) \geqslant f(x_k).$$

As an example of a symmetric fuzzy set, consider the set B defined as "all the very tall men." B is a symmetric fuzzy set, since the taller the man, the higher the grade of membership he will have in B. Any symmetric (in the ordinary sense) function, or a truncated symmetric function, can represent a characteristic function of a symmetric fuzzy set.

Definition 5.4.2

A fuzzy set B is unimodal if and only if the set Γ_{x_i} is connected for all x_i in B.

In order to consider the problem of clustering data points, it is necessary to define discrete fuzzy sets.

A sample point from B will be a point $x \in X$ with its associated characteristic value, $f(x)$. Further, we denote a sample of N points from B by $S = \{(x_i, f_i)^N\}$, where x_i is a point in X and f_i its corresponding grade of membership. S can be considered as a discrete fuzzy set that includes only those points x_i given by the sample. A large sample S is required; in particular, S is large in comparison to the dimension of the space X, and to the number of local maxima in f.

$\{(S_i, \chi_i)^m\}$ denotes a partition of S into m subsets, where S_i is a discrete fuzzy subset and χ_i the point in S_i at which the maximal grade of membership is attained, such that χ_i is the mode of S_i. A mode is called a local maximum if it is a local maximum of f and it is then denoted by v_j. It is assumed that every local maximal grade of membership is unique.

Definition 5.4.3

Let S be a sample from a fuzzy set B and S_i a proper subset of S. For some point x_k in S_i, we associate a point x_t in $(S - S_i)$ such that

$$d(x_t, x_k) = \min_{x_j \in (S - S_i)} \left[d(x_k, x_j) \right].$$

The point x_k is defined to be an interior point in S_i if and only if the set $\Gamma = \{x \mid d(x_i, x) < d(x_i, x_k)\}$ includes at least one sample point in S.

The clustering procedure decomposes B into unimodal fuzzy sets and realizes the maximum separation among them. The procedure is divided in two main steps: first, local maxima are identified by a systematic search where both the order of the points according to their grade of membership and their order according to distance are used. The second step is the assignment of each point to a cluster. There are as many clusters as local maxima of χ_B. Note that the clusters obtained are not fuzzy sets.

Step 1 Given a sample $S = \{(x_i, f_i)^N\}$ from a multimodal fuzzy set, subject to certain conditions on f and S (see Theorem 5.4.1), step 1 detects all the local maxima of f. It is divided into two parts: in the first part the

Clustering Via Unimodal Fuzzy Sets

sample is partitioned into symmetric subsets and in the second a search for the local maxima in the generated subsets is performed.

Part 1 of Step 1 Let $S = \{(x_i, f_i)^N\}$ be a sample from a fuzzy set (assume, for simplicity, that $f_i \neq f_j$ for $i \neq j$).

1. Initially, it is required to generate the following two sequences.
   (a) $A = (y_1, y_2, \ldots, y_N)$ is a descending sequence of the points in the sample ordered according to their grade of membership; that is, $f_j \geq f_t$ for $j \leq t$, where f_j and f_t are the grades of membership of y_j and y_t, respectively.
   (b) $A_1 = (y_1^1, y_2^1, y_3^1, \ldots, y_N^1)$, where $y_1^1 \equiv y_1$, is the sequence of points ordered according to their distance to y_1^1; that is $d(y_1^1, y_j^1) \leq d(y_1^1, y_t^1)$ for $j \leq t$.

Gitman and Levine [810] refer to A_1 as the sequence of ordered "candidate" points to be assigned into group 1. Thus y_2^1 is the first candidate, and if it is assigned into group 1, then y_3^1 becomes the next candidate, and so on. We can therefore state that the current candidate point for group 1, y_c^1, is the nearest point to its mode $y_1^1 (\equiv y_1 \equiv \chi_1)$ except for points that have already been assigned to group 1. This holds true for any sequence A_i; that is, $y_1^i \equiv \chi_i$ is the mode for group i, and y_c^i is its candidate point.

2. If $y_i \equiv y_i^1$, for $i = 2, 3, \ldots, r - 1$, and $y_r \not\equiv y_r^1$, then the y_i's, $i = 1, 2, \ldots, r - 1$, are assigned into group 1 and a new group is initiated with $y_1^2 \equiv \chi_2 \equiv y_r$ as its mode; that is, the sequence $A_2 = (y_1^2, y_2^2, y_3^2, \ldots, y_p^2)$ is generated. The latter includes, from among the points that have not yet been assigned, those points that are closer to y_r than the shortest distance from y_r to the points that have already been assigned. The points in A_2 are now ordered according to their distance to y_r; that is, $d(y_r, y_j^2) \leq d(y_r, y_t^2)$ for $j \leq t$.

3. Suppose that G groups have been initiated. Thus there exist G sequences A_i, $i = 1, \ldots, G$, each of which displays a candidate point y_c^i. Suppose that y_q in the sequence A is the point currently being considered for assignment; then the following holds:
   (a) If $y_q \equiv y_c^i$ and $y_q \not\equiv y_c^j$, $j = 1, \ldots, G$, $j \neq 1$, then y_q is assigned into group i.
   (b) If $y_q \equiv y_c^i$ for some $i \in I$, where I is a set of integers representing those groups whose candidate points are identical to y_q, then y_q is assigned into that group to which its nearest neighbor with a higher grade of membership has been assigned.
   (c) If $y_q \not\equiv y_c^i$ for $i = 1, \ldots, G$, then a new group is initiated with y_q as its mode.

Part 1 of Step 1 is terminated when the sequence A is exhausted.

Theorem 5.4.1 Let f be a characteristic function of a fuzzy set with K local maxima so that:

1. If v_K is a local maximum of f, then there exists a finite $\varepsilon > 0$ such that the set $\{x \mid d(v_K, x) \leq \varepsilon\}$ is a symmetric fuzzy set.

Let $S = \{(x_i, f_i)^N\}$ be a large sample from f, such that:

2. For every x_i in the domain of f, the set $\{x \mid d(x_i, x) < \varepsilon/2\}$ includes at least one point in S.
3. $\{(v_k, f_k)^K\} \subset S$.

Let $\{(S_i, \chi_i)^m\}$ denote the partition generated by part 1 of Step 1, where S_i denotes the discrete fuzzy set, χ_i its maximal grade of membership (mode), and m the number of groups. Then χ_i is an interior point in S_i if and only if it is a local maximum of f.

Part 2 of Step 1 This part uses the above theorem in order to determine the interior modes among all χ_i's as follows:

Let $\{(S_i, \chi_i)^m\}$ be the partition generated by part 1 of Step 1. For every mode χ_i and set S_i, a point x_{pi} and a distance R_i can be found as follows:

$$R_i = d(\chi_i, x_{pi}) = \min_{x_k \in (S - S_i)} [d(\chi_i, x_k)].$$

R_i is the minimum distance from the mode to a point in S outside the set S_i. Thus χ_i is a local maximum if the set

$$\Gamma_{R_i} = \{x \mid d(x_{pi}, x) < R_i\}$$

includes points in S_i. Otherwise χ_i is not a local maximum since it is a boundary point of S_i.

Step 2 Step 2 partitions a sample from a fuzzy set into unimodal fuzzy sets, providing the local maxima of f are known. Thus this procedure uses the information obtained from the application of Step 1; that is, the number, location, and characteristic values of the local maxima of f. The points are finally assigned in the order in which they appear in the sequence A.

Specifically, let $S = \{(x_i, f_i)^N\}$ be a sample from a fuzzy set, and let $\{(v_i, f_i(v))^K\} \subset S$ be the sample of the K local maxima v_i of f. Assume that $f_i(x_i) \neq f_j(x_j)$ for $i \neq j$, and $f(v_i) > f(v_j)$ for $i \leq j$. Let A be the sequence of the points ordered according to their grade of membership, and suppose that the K local maxima of f are in locations p_i, $i = 1, \ldots, K$ in A. We can infer the following proposition.

Proposition 5.4.1 The point x_j in location j in the sequence A, $p_M \leq j < p_{(M+1)}$, $M \leq K$, can only be assigned into one of the groups $i \in I_M = \{1, 2, \ldots, M\}$.

If $f(x_{p_r})$, $r = M + 1, M + 2, \ldots, K$ is the local maximum of group r, then only points with a lower grade of membership can be assigned into group r. Since all the points that precede location p_r in A have higher grades of membership, none of them can be assigned into group $r, r + M + 1, M + 2, \ldots, K$.

This proposition implies that all the points in A that are found in the locations $p_1 \leq j < p_2$ will automatically be assigned into group 1; the points in locations $p_2 < j < p_3$ will be divided between group 1 and group 2, and so on.

Step 2 uses the following rule: assign the point x_j in location j in the sequence A into the group in which its nearest neighbor with a higher grade of membership (all the points preceding x_j in A) has been assigned. This rule applies to all the points with the exception of the local maxima that initiate new groups. This rule is different from the "nearest neighbor classification rule" because of the particular order in which the points are introduced.

Theorem 5.4.2 Let f be a piecewise continuous characteristic function of a fuzzy set. Let $S = \{(x_i, f_i)^N\}$ be an infinite sample from f, such that, for every x_i in the domain of f and for an $\alpha \geq 0$, the set $\Gamma = \{x \mid d(x_i, x) < \alpha/2\}$ includes at least one sample point S.

If $\alpha \to 0$, then Step 2 partitions the given sample into unimodal fuzzy sets.

Theorem 5.4.3 Let S be a sample from a fuzzy set with a characteristic function f. Let f and S be constrained as in Theorem 5.4.2.

If $\alpha \to 0$, then every final set is a union of the sets S_i generated in part 1 of Step 1.

Utilizing the above results, we can partition a fuzzy set into unimodal fuzzy sets. This is different from clustering a set of elements $\{(x_i)^N\}$ into categories. However, we can associate every element x_i with a grade of membership f_i to represent the importance of the element. There are many possible ways to grade the set of elements. One possibility is to use a clustering model to associate with every point a membership value according to its "contribution" to the desired partition, and the search is for functionals that describe the properties of the desired categories and the structure of the final partition.

Experiments on artificially generated data sets were performed by Gitman and Levine [810] where both the results and errors were discussed in detail.

The algorithm can also be applied effectively in supervised pattern recognition, in particular when the categories are multimodal and this information is not known. This algorithm can be used first to partition every category independently into unimodal fuzzy sets. In this case we associate with every point x_i the distance membership function

$$f_i = \min_{x_j \in (S - C_i)} \left[d(x_i, x_j) \right]$$

where C_i is the set of points in the category in which x_i is a member.

It is suggested that the clustering algorithm reported by Gitman and Levine [810] possesses three advantages over the ones discussed in the literature.

1. It does not require a great amount of fast core memory and therefore can be applied to large data sets. The storage requirement is $(20N + CN + S)^{10}$ bytes, where N is the number of points to be partitioned, $20N$ and CN are required for the fixed portion of the program and the variable length data sequences (A, A_i), respectively, and S is the number of storage locations required for the given set of data points. Obviously, S depends on the particular resolution of the magnitude of the components of the data vectors.

2. The amount of computing time is relatively small.

3. The shape of the distribution of the points in a group (category) can be quite general because of the distributions that the unimodal fuzzy sets include. This can be an advantage, especially in practical problems in which the categories are not distributed in "round" clusters.

The details of the proofs of Theorems 5.4.1–5.4.3 can also be found in [810].

In yet another paper [809], Gitman presents an algorithm for classifying a data set into an initially unknown number of categories. It is composed of a procedure for selecting initial points, a mode estimation procedure, and a classification rule. An integer-valued function is defined on the sample space and a gradient search technique is used for estimating its modes. A procedure for mode estimation in the case of an infinite data set is also proposed. Sufficient conditions for the convergence to the neighborhood of the modes have been stated. The algorithm is used for clustering multicategory artificially generated data sets and is compared with an optimal classification scheme.

Unlike the algorithm that classifies the given data set into unimodal fuzzy sets in [810], the local maximum is not required to be a sample point of the given set. Rather, it can be a point in the space with the highest characteristic value in its neighborhood. Also, no symmetry conditions in its neighborhood are required. These advantages have practical significance since conditions such as 1 and 3 of Theorem 5.4.1 are not necessary for detecting a mode.

This algorithm is composed of the following procedures: (1) a selection of initial points, (2) a gradient search technique, and (3) a classification rule. The technique proceeds as follows. An integer-valued function g is defined on the sample space. A set of initial points is selected and a gradient search technique is applied to each of these, aiming to estimate the local maxima of g. Since a local maximum of g is not necessarily a unique point, we can apply an elimination procedure on the set of final points arrived at. The points not eliminated are considered as local maxima of g. The classification rule of [809] is then used to partition the data set. The conditions under which convergence to the neighborhood of the modes is obtained and the conditions for the procedure of selecting initial points that ensure the detection of all modes are stated in the theorems.

The algorithm is used for clustering data sets drawn from ellipsoidal normal distributions defined in a two-dimensional space. The results are compared with those obtained by using an optimum classification technique [809]. The procedure for selecting initial points was compared with one that uses a uniform distribution over the domain to which the data set is confined. Experiments for determining the sensitivity to change in parameters are also reported in [809].

5.5 FUZZY COVARIANCE MATRIX

When pattern recognition techniques are applied to practical problems, we often find that real-world data contain many missing data points. One method of handling missing data, that seems intuitively reasonable, is via fuzzy expectation.

Let V_i be a vector with K components representing features $X_{i,j}$, $V_i = (X_{i,1}, X_{i,2}, \ldots, X_{i,k})$, and let $d_{UV(i)}$ be the component of distance between vectors U and V along the ith feature, namely,

$$d_{UV(i)} = X_{U,i} - X_{V,i}.$$

A reasonable assumption is that the entire data set includes N vectors. The "typical technique" assumes that the distance to a missing data point is the same as the typical distance between all pairs of vectors along that one feature

$$Q_{i_1} = FEV\{|X_{U,i} - X_{V,i}|\}$$

when V is taken from 1 to $U-1$ and U from 2 to N to describe the proper population.

This typical value of Q_{i_1} can be compared with Q_{i_2} where

$$Q_{i_2} = \frac{2}{N^2 - N} \sum_{U=2}^{N} \sum_{V=1}^{N-1} |X_{U,i} - X_{V,i}|.$$

Thus the distance A from U to V is given by

$$A = \sum_{i=1}^{K} d_{UV(i)}^2$$

where

$$d_{UV(i)} = \begin{cases} Q_{i_1} & \text{if } X_{U,i} \text{ or } X_{V,1} \text{ is missing} \\ X_{U,i} - X_{V,i}, & \text{otherwise.} \end{cases}$$

We can follow this idea with several extensions; from the ideas of Ruspini [2244], Dunn [596], and Bezdek [179], regarding fuzzy partitions, to the concept of fuzzy covariance matrix as a classification tool, introduced by Gustafson and Kessel [921].

The definition of a fuzzy partition used here agrees with that of Ruspini [2244], Dunn [596], and Bezdek [179] and is a natural extension of the conventional partitioning definition. An ordinary, or "hard," partition is a k-tuple of Boolean functions $w(\cdot) = \{w_1, w_2, \ldots, w_k\}$ on the feature space $\Gamma \subset R^n$ that satisfy

$$w_j(x) = 0 \text{ or } 1, \quad \forall x \in \Gamma, 1 \leq j \leq k$$

$$\sum_{j=1}^{k} w_j(x) = 1, \quad \forall x \in \Gamma.$$

If Γ_j represents the jth class, with $\Gamma_i \cap \Gamma_j = \phi, \forall i \neq j$ and $U_{j=1}^{k} \Gamma_j = \Gamma$, then

Fuzzy Covariance Matrix

$x \in \Gamma_m$ and x is a member of precisely one class. A fuzzy partition is a k-tuple of membership functions

$$w(\cdot) = \{w_1(x), w_2(x), \ldots, w_k(x)\}$$

that satisfy

$$0 \leq w_j(x) \leq 1, \quad \forall x \in \Gamma, 1 \leq j \leq k$$

$$\sum_{j=1}^{k} w_j(x) = 1, \quad \forall x \in \Gamma.$$

Denote the distance from a point x to the jth class by

$$d_j(x) = d(x, \theta_j); \quad d_j(x) > 0,$$

where the jth class is parameterized by θ_j. For an indexed set of samples $x_1, x_2, x_3, \ldots, x_N$, denote the distance measure and membership function by

$$d_j(x_i) = d_{ij}, \quad w_j(x_i) = w_{ij}.$$

We are interested in minimizing the following cost:

$$J(w, \theta) = \sum_{i=1}^{N} \sum_{j=1}^{k} w_{ij}^{\alpha} d_{ij}, \quad \alpha \geq 1$$

where $\theta = \{\theta_j\}$, $w = \{w_{ij}\}$, k is the number of classes, and α is a smoothing parameter that controls the "fuzziness" of the clusters. For $\alpha = 1$ the clusters are separated by hard partitions and

$$w_{ij} = 0 \text{ or } 1.$$

As α increases, the partitions become more fuzzy.

Minimizing J with respect to (fuzzy) w, subject to $\alpha > 1$ and the constraints

$$w_{ij} = s_{ij}^2$$

with S_{ij} real, yielding

$$\bar{J}(S, \theta, \lambda) = \sum_{i=1}^{N} \sum_{j=1}^{k} S_{ij}^2 d_{ij} + \sum_{i=1}^{N} \lambda_i \left(\sum_{j=1}^{k} S_{ij}^2 - 1 \right),$$

with the Lagrange multipliers $\{\lambda_i\}$, the first-order necessary conditions for optimality are found by setting the gradients of $\bar{J}$ with respect to S to zero; namely,

$$\frac{\partial \bar{J}}{\partial S_{ij}} = 2\alpha S_{ij}^{2\alpha-1} d_{ij} + 2 S_{ij} \lambda_i.$$

By setting $\partial \bar{J}/\partial S_{ij}$ to zero, we obtain the following first-order necessary conditions:

$$S_{ij}^* \left(\alpha S_{ij}^{*2(\alpha-1)} d_{ij} + \lambda_i^* \right) = 0, \quad \forall i, j$$

$$\sum_{j=1}^{k} S_{ij}^{*2} = 1, \quad \forall i$$

where the asterisk denotes association with optimality.

Assuming that $S_{ij}^* \neq 0$, $\forall i, j$, we get

$$w_{ij}^* = \left(\frac{-\lambda_i^*}{\alpha d_{ij}} \right)^{1/(\alpha-1)}$$

and

$$w_{ij}^* = \frac{1}{\sum_{l=1}^{k} (d_{ij}/d_{il})^{1/(\alpha-1)}}.$$

Thus *for any θ*, the associated extremum of $J(w, \theta)$ is

$$J^*(\theta) = \min_{w} J(w, \theta)$$

$$= \sum_{i=1}^{N} \left[\sum_{j=1}^{k} (d_{ij})^{1/(1-\alpha)} \right]^{1-\alpha}.$$

If $\alpha \to 1$, we get

$$J \to \sum_{i=1}^{M} \sum_{j=1}^{k} w_{ij} d_{ij}$$

Fuzzy Covariance Matrix

and the argument given by Dunn [598] will establish that

$$w_{ik}^* \to \begin{cases} 1, & d_{ik} = \min_j (d_{jk}), \\ 0, & \text{otherwise,} \end{cases} \quad \forall i, k.$$

If

$$d_{ij} = (x_i - \theta_j)^T A (x_i - \theta_j), \quad A > 0,$$

then

$$\sum_{i=1}^N w_{ij}^{*\alpha}(x_i - \theta_j^*) = 0, \quad \forall j,$$

which is equivalent to

$$\theta_j^* = \frac{\sum_{i=1}^N w_{ij}^{*\alpha} x_i}{\sum_{i=1}^N w_{ij}^{*\alpha}}, \quad j = 1, \ldots, k$$

$$\triangleq m_{fj}.$$

Calling m_{fj} the *fuzzy mean* of class j in recognition of its limiting property under hard partitioning, this case comprises fuzzy ISODATA.

As $\alpha \to 1$ and the partitioning becomes hard,

$$w_{ij}^{*\alpha} \to \begin{cases} 1, & j = m \\ 0, & j \neq m \end{cases}$$

where

$$d_{im} = \min_j d_{ij}.$$

That is, under the one-nearest-neighbor rule, $w_{ij}^{*\alpha}|_{\alpha=1} = 1$ for all pattern vectors x_i assigned to class j and is zero otherwise. Thus for hard partitioning

$$\sum_{i=1}^N w_{ij}^* = N_j$$

where N_j is the number of pattern vectors assigned to Γ_j and

$$\theta_j^* \big|_{\alpha \to 1} \to \frac{1}{N_j} \sum_{x_i \in \Gamma_j} x_i$$

$$= \hat{m}_j$$

where $\hat{m}_j$ is the sample mean of Γ_j. This is the hard k-means algorithm; it constitutes the basic idea underlying hard ISODATA.

If we let

$$d_{ij}(\theta_j) = (x_i - v_j)^T M_j (x_i - v_j), \qquad 1 \leq j \leq k$$

with M_j symmetric and positive-definite, such that

$$|M_j| = \rho_j, \qquad \rho_j > 0$$

with ρ_j fixed for each J, then the augmented cost is

$$J(w, \theta, \lambda, \beta) = \sum_{i=1}^{N} \sum_{j=1}^{k} w_{ij}^\alpha d_{ij}(\theta_j)$$

$$+ \sum_{i=1}^{N} \lambda_i \left(\sum_{j=1}^{k} w_{ij} - 1 \right) + \sum_{j=1}^{k} \beta_j (|M_j| - \rho_j)$$

where $\{\beta_j\}$ is a set of Lagrange multipliers.

Thus the necessary conditions are

$$\left. \frac{\partial \bar{J}}{\partial v_j} \right|_* = -2 \sum_{i=1}^{N} w_{ij}^\alpha M_j (x_i - v_j^*) = 0, \qquad j = 1, 2, \ldots, k,$$

which is equivalent to

$$\left. \frac{\partial \bar{J}}{\partial M_j} \right|_* = 0 = \sum_{i=1}^{N} w_{ij}^\alpha (x_i - v_j)(x_i - v_j)^T + \beta_j |M_j^*| M_j^{*-1}$$

where

$$\frac{\partial}{\partial A}(x^T A x) = x x^T, \qquad \frac{\partial}{\partial A}|A| = |A| A^{-1},$$

which hold for a nonsingular matrix A and any compatible vector x.

Using

$$v_j^* = \frac{\sum_{i=1}^{N} w_{ij}^{\alpha} x_i}{\sum_{i=1}^{N} w_{ij}}$$

for the optimal membership functions ($w_{ij} = w_{ij}^*$), v_j^* is the fuzzy mean of Γ_j. Hence for $v_j = v_j^*$ we get

$$M_j^{*-1} = \frac{1}{\beta_j |M_j^*|} \sum_{i=1}^{N} w_{ij}^{\alpha}(x_i - v_j^*)(x_i - v_j^*)^T.$$

Defining the *fuzzy covariance matrix* for Γ_j by

$$P_{fj} = \frac{\sum_{i=1}^{N} w_{ij}^{\alpha}(x_i - m_{fj})(x_i - m_{fj})^T}{\sum_{i=1}^{N} w_{ij}^{\alpha}}, \quad \alpha > 1$$

we get

$$M_j^{*-1} = \frac{\sum_{i=1}^{N} w_{ij}^{\alpha}}{\beta_j |M_j^*|} |P_{fj}|$$

or

$$M_j^{*-1} = \left[\frac{1}{\rho_j P_{fj}}\right]^{1/n} P_{fj}$$

where n is the feature space dimension. The *hard covariance* matrix refers to P_{fj} evaluated at $\alpha = 1$. This matrix is simply the sample class covariance matrix under the cluster assignment rule.

The tests with vectorcardiogram (VCG) data performed by Gustafson and Kessel [921] indicate that the use of fuzzy covariance can enhance clustering performance when compared with the results obtained from a hard covariance matrix. Clearly, further numerical testing is required to verify this behavior in general.

5.6 CLUSTER VALIDITY

The work described in this section is based primarily on three papers by Windham [2834], Bezdek et al. [193], and Backer and Jain [94], who have discussed the problem of clustering performance measures and their relation to fuzziness.

It is clear that fuzzy clustering algorithms have been used extensively to classify information. Windham assumes that the output of a fuzzy clustering algorithm includes a $c \times n$ matrix, $U = [u_{ij}]$, where c is the number of clusters identified, n is the number of data points, and $2 \leq c \leq n$. Each row of the matrix is associated with a particular cluster and each column is the membership function of a particular data point. For each $j = 1, \ldots, n$, he defines $\chi_j = \max_i(\chi_{ij})$ and I_j to be the greatest integer in $1/\chi_j$.

The proportion exponent of U, $P(U)$, is defined [2834] as

$$P(U) = -\log_2 \left\{ \prod_{j=1}^{n} \left[\sum_{k=1}^{I_j} (-1)^{k+1} \binom{c}{k} (1 - k\chi_j)^{c-1} \right] \right\}.$$

The only values used from the matrix U are the n maximum entries from each of its columns. These values indicate the clusters with which the data points have been most closely identified, and the larger these values are, the more clearly the identification has been defined. So, it is natural to focus on them as a *measure of cluster validity*.

It is shown [193] that for $1/c \leq \chi \leq 1$ the proportion of possible c-dimensional membership functions whose maximum entries are at least χ is given by $\sum_{k=1}^{I}(-1)^{k+1}\binom{c}{k}(1-k\chi)^{c-1}$, where I is the greatest integer in $1/\chi$. In other words, for a given value of χ, if the sum is 0.1, then 10% of all possible c-dimensional membership functions will have a maximum entry at least χ, and the interpretation of this result is independent of c.

If a fuzzy clustering algorithm has done at all well, the proportions for each column will be small. If the sample size is very large, the proportion for the matrix is likely to be very small indeed. It is for this reason that the final computation is performed, namely, taking the negative logarithm (base 2). This spreads the values of the functional over a much wider range, particularly for proportions near zero. It also implies that large values for the proportion exponent indicate that the algorithm has worked well.

The proportion exponent was designed primarily to aid in identifying the number of clusters present in the data; however, there is no reason that it cannot be used to evaluate choices of other parameters the investigator sets in using clustering algorithms. For example, many of these algorithms are iterative in nature, and the user starts the iteration process by choosing a membership matrix. The proportion exponent can be used to evaluate the relative merits of different starting points. It cannot be used to evaluate parameters that artificially affect the "fuzziness" of the membership functions. Weight exponents are sometimes used to de-emphasize the effects of noisy or extraneous data. Increasing a weight exponent will cause the membership functions to become fuzzier, that is, the maximums will become smaller and other entries larger. This means that the proportion exponent may decrease even when the user feels that the data classification has improved.

The key building block of the proportion exponent is the proportion of membership functions whose maximum exceeds a given value.

For a given c, the set of possible membership functions is

$$MF_c = \left\{ u \in R^c : \chi_i \geq 0 \text{ for } i = 1,\ldots,c \text{ and } \sum \chi_i = 1 \right\}.$$

This set is the intersection of a $(c-1)$-dimensional plane and the first orthant of R^c. The "size" of a subset of MF_c can be measured by its area, in particular, the area of MF_c itself is $\sqrt{c}/(c-1)!$. So, for a given $A \subset MF_c$, define the proportion of membership functions in A, $p(u \in A)$, by

$$p(u \in A) = \frac{\text{area of } A}{\text{area of } MF_c},$$

since MF_c is a $(c-1)$-dimensional surface in R^c. These areas can be computed by surface integrals, and the surface integrals can be computed using an appropriate parameterization of MF_c. It can be shown that, for $c = 2, 3, \ldots$ and $1/c \leq \chi \leq 1$, if $A_\chi = \{u \in MF_c : \max(\chi_i) \geq \chi\}$, then

$$p(u \in A_\chi) = \sum_{k=1}^{I} (-1)^{k+1} \binom{c}{k} (1 - k\chi)^{c-1}$$

where I is the greatest integer in $1/\chi$.

It should be noted that the proportion exponent may not be defined for a particular membership matrix U. If one of the columns describes a hard cluster, that is, $u_{ij} = 1$ for some i and $u_{kj} = 0$ for $k \neq i$, then the proportion exponent is undefined since the proportion of functions whose maximums exceed one is zero. This rarely occurs in the use of fuzzy clustering algorithms, so it does not present a serious obstacle in the use of the functional.

Let us consider, as an example, the fuzzy ISODATA algorithm, an iterative procedure that, if it converges, does so to a local minimum for the objective function. Then a particular procedure that can be used is based on the following computations:

1 Given U, compute V by

$$v_{ji} = \frac{\sum_k (u_{ik})^m x_{kj}}{\sum_k (u_{ik})^m}$$

where x_{kj} is the jth component of the kth data vector.

2 Given V, compute U by

$$u_{ij} = \frac{1}{\sum_k (d_{ji}/d_{jk})^{2/(m-1)}}.$$

Values for c and m and an initial U matrix are chosen so that 1 and 2 are successively repeated until the distances between cluster centers at two successive stages are less than a preassigned value. Justification for this procedure may be found in [193] and [2834] and convergence properties of this type of algorithm may be found in [181] and [182].

Experiments based on the above and performed by Windham illustrate how the proportion exponent can be used to provide information regarding the clustering validity.

Bezdek et al. [193] have derived the theoretical mean and variance of two related cluster validity functionals. These are the classification entropy and partition coefficients associated with fuzzy c-partitions of finite data sets.

Backer and Jain [94] attempt to put an order on the various partitions of a data set obtained from different clustering algorithms. The goodness of each partition is expressed by means of a performance measure based on a fuzzy set decomposition of the data set under consideration. Several experiments reported show that the proposed performance measure puts an order on different partitions of the same data, which is consistent with the error rate of a classifier designed on the basis of the obtained cluster labelings.

They make reference to the results of five different clustering programs. Four of them exemplify the class of so-called squared-error programs. They differ both in computational details and in the approach taken to minimize the squared error. The fifth clustering program is the induced fuzzy set iterative optimization algorithm, which for a particular choice of the cluster-membership-value assignment is similar to the squared-error approach.

Cluster Validity

The *squared-error cluster programs* are based on a data set given in the form of an object-property table (pattern matrix) denoted by

$$C = \{u_v\}_{v=1}^{N}$$

where

$$u_v = \left(u_v^{(1)}, u_v^{(2)}, \ldots, u_v^{(k)}\right)^T$$

is the vth object.

The number of objects N is assumed to be significantly larger than the number of properties k. A clustering is an m-partition

$$\{C_i\}_{i=1}^{m}$$

of the integers $[1, 2, \ldots, N]$ that assigns each object to a single cluster label.

The objects corresponding to the integers in C_i form the ith cluster, whose center is

$$\chi_i = \left(\chi_i^{(1)}, \chi_i^{(2)}, \ldots, \chi_i^{(k)}\right)^T$$

where

$$\chi_i^{(j)} = \frac{1}{N_i} \sum_{v \in C_i} u_v^{(j)}$$

and N_i is the number of objects in cluster i.

The squared-error for cluster i is

$$e_i^2 = \sum_{v \in C_i} (u_v - \chi_i)^T (u_v - \chi_i)$$

and the squared-error for the clustering is

$$E_m^2 = \sum_{i=1}^{m} e_i^2.$$

All programs try to find a local minimum of E_m^2 and the user hopes that the local minimum coincides with the global minimum.

The four-squared-error programs used by Backer and Jain are the following:

FORGY: Given a set of cluster centers, the cluster label of the closest cluster center is assigned to each object. The cluster centers are then recomputed as sample means of all objects having the same cluster label. A new cluster is created when an object is found that is sufficiently removed from the existing structure.

ISODATA: Clusters are updated as in FORGY. The method is unique in the heuristics employed to create new clusters while trying to achieve the number of clusters requested by the user as described before.

WISH: Cluster centers are updated in a "dispose" box immediately after a cluster label is assigned to each object. (The program processes one object at a time.)

CLUSTER: This program is based on the "hill-climbing" technique and has two phases. Phase 1 creates a sequence of clusterings containing $2, 3, \ldots, m$ clusters where m is specified by the user. Phase 2 merges clusters two at a time to produce a sequence of clusterings containing $m - 1$, $m - 2, \ldots, 2$ clusters. Phase 1 and 2 are alternated until a pass through both decreases the squared-error of none of the clusterings. The best clustering ever achieved for each number of clusters ($\leq m$) is retained.

The *induced fuzzy set iterative optimization program* consists of an algorithm relationship between an inducing object partition $\{C_i\}_{i=1}^m$ on the one hand and a collection of induced fuzzy set $\{(f_i, u), \forall u \in C\}_{i=1}^m$ on the other hand, where

$$f_i: C \to [0, 1], \qquad i = 1, 2, \ldots, m$$

and

$$\sum_{i=1}^m f_i(u) = 1, \qquad \forall u \in C.$$

Usually, the initial inducing object partition $\{C_i\}_{i=1}^m$ stems from a best guess and is subject to iterative optimization. A collection of fuzzy sets is induced by means of a point-to-subset affinity concept on the basis of the structural properties among objects in the representation space. This has been called affinity decomposition and performs in the same way as the Bayes theorem from probability theory. The appropriateness of the induced collection of fuzzy sets is made explicit by applying some performance measure. The collection of induced fuzzy sets may cause a repartition due to a reclassification function. Then a new inducing step follows, so we get an iterative procedure.

More specifically, given an m-partition $\{C_i\}_{i=1}^m$ and some affinity measure $r(u, C)$, the cluster membership value $f_i(u)$ of object $u \in C$ induced by

$C_i \in \{C_j\}_{j=1}^m$ is given by

$$f_i(u) = P_i \frac{r(u, C_i)}{r(u, C)}, \quad \forall u \in C,$$

where

$$r(u, C_i) \geq 0, \quad \forall u \in C$$

and

$$r(u, C) = \sum_{i=1}^m P_i r(u, C_i), \quad \forall u \in C$$

where P_i denotes the relative size of the ith subset given by

$$P_i = \frac{N_i}{N}.$$

The subset affinity is given by

$$r(u, C_i) = 1 - \frac{1}{N_i} \sum_{v \in C_i} h^\beta [\partial(u, v)]$$

where $h^\beta[\partial(u, v)]$ is a nondecreasing distance function on the interval [0, 1], controlled by a certain parameter β, and $\partial(u, v)$ is a distance measure satisfying reflexive and symmetric properties.

Thus

$$r(u, C) = \sum_{i=1}^m \frac{N_i}{N}\left(1 - \frac{1}{N_i}\sum_{v \in C_i} h^\beta[\partial(u, v)]\right) = 1 - \frac{1}{N}\sum_{v \in C} h^\beta[\partial(u, v)]$$

and

$$f_i(u) = \frac{N_i - \sum_{v \in C_i} h^\beta[\partial(u, v)]}{N - \sum_{v \in C} h^\beta[\partial(u, v)]}.$$

Backer and Jain [94] use

$$\partial(u, v) = \left(\sum_k (u^{(k)} - v^{(k)})^2\right)^{1/2}$$

and

$$h^\beta[\partial(u, v)] = \frac{\partial(u, v)^2}{\beta}, \qquad \text{for } \partial(u, v) \leq \beta^{1/2}$$

$$= 1 \qquad \text{for } \partial(u, v) > \beta^{1/2}.$$

Once an m-collection of induced fuzzy sets has been established, we may characterize the partitioning as follows. If the amount of induced fuzziness is low, it means that the m-collection of induced fuzzy sets is reasonably separable and that the inducing partition reflects the real data structure reasonably well. On the other hand, if the amount of induced fuzziness is high, it means that the inter-fuzzy set separability is low and that either the inducing partition does not reflect the real structure well, or that almost no structure is present in the data. Consequently, the performance measure should measure the fuzziness in the gaps between fuzzy sets (along the fuzzy boundaries) and, therefore, should be based on the notion of intersection of fuzzy sets, defined either by

$$f_{i \cap j}(u) = \min\big[f_i(u) \cdot f_j(u)\big], \qquad \forall u \in C$$
$$\small 1$$

or by

$$f_{i \cap j}(u) = f_i(u) \cdot f_j(u), \qquad \forall u \in C.$$
$$\small 2$$

The clustering performance measure can be defined as follows.

Definition 5.6.1 [94]

Given an m-collection of fuzzy sets $\{(f_i, u), \forall u \in C\}_{i=1}^m$ satisfying $\sum_{i=1}^m f_i(u) = 1$, $\forall u \in C$, the clustering performance measure is given by

$$\psi = 1 - \frac{2m}{m-1} \sum_{i=1}^{m-1} \sum_{j=i+1}^{m} \frac{1}{N} \sum_{u \in C} f_i(u) \cdot f_j(u).$$

Clearly, $0 \leq \psi \leq 1$ where $\psi = 0$ corresponds to maximum fuzziness and $\psi = 1$ corresponds to nonfuzziness.

It is also easy to show that

$$\psi = \frac{1}{m-1} \sum_{i=1}^{m-1} \sum_{j=i+1}^{m} \frac{1}{N} \sum_{u \in C} |f_i(u) - f_j(u)|^2,$$

Cluster Validity

which shows that the performance measure can be looked upon as a cluster membership distance measure.

The intuitive appeal of the fuzzy set approach lies in the fact that the data analyst is now equipped with a value of the performance measure as well as the identification of bridging (overlapping) points. This appeal is balanced by the fact that a single performance measure cannot summarize all the information that can be gleaned from a clustering. Unfortunately, no general answer can be given to questions like, "Which of the programs is best?" or, "What is the 'true' structure?"

The method of comparison [94] establishes a ranking of the utility of clustering results obtained from different clustering programs with respect to a certain application domain where utility can be measured uniquely.

The experimental results in [94] indicate at least that the goal-directed comparison of clustering results appears to be promising.

In conclusion, the following questions can be posed when a data set is to be analyzed using clustering methodology:

1 Which clustering algorithm(s) is best suited for a given data set?
2 Once it has been determined which algorithms are suitable for a data set, how do we determine which one of these algorithms gives more valid or useful results?

While no answer is yet available to the first question because we do not have sufficient knowledge regarding the "true" structure of the data set, the paper by Backer and Jain is an attempt to answer the second question. In fact, whenever we are able to define the ultimate goal of the cluster analysis more specifically, a comparison of clustering techniques may become fruitful if an appropriate performance measure is adopted.

It is our belief that a measure of gaps between fuzzy clusters may serve well if classification is the ultimate goal. The empirical evidence just substantiates this belief and indicates that this approach appears to be a promising direction for future investigations in this field.

An important question regarding fuzzy set theory is: "From what kind of data and how can membership function *actually* be derived?" This question is very important for practical applications as well as the check of whether the choices of fuzzy set-theoretic operators have an experimental basis.

The membership function is supposed to be a good model of the way people perceive categories, and experiments made by psychologists have shown a distinction between central members of a category and peripheral members.

Clearly, cluster membership is usually not a yes-or-no matter, but rather a matter of degree. However, Lakoff [1457] pointed out that some people seem to turn relative judgments of category membership into absolute judgments by assigning the member in question to the category in which it has the highest degree of membership. Since category membership is a matter of degree, the question naturally arises as to what determines the degree of membership for each data point or a class member.

Although fuzzy set theory is capable of dealing with degrees of set membership, the membership function is not a primitive concept from a psychological point of view. A membership value is generally not absolutely defined and is a relative and sometimes quite a subjective concept. The choice of continuous set-theoretic operators is consistent with fuzzy knowledge of membership functions: a slight modification of the membership values does not drastically affect the rough shape of the result of a set operation. To take into account the imprecision of membership functions, we may think of using type-2 fuzzy sets, probabilistic sets, tolerance classes of fuzzy sets, or level-2 fuzzy sets [1719]. Estimating the membership function of such higher order fuzzy sets is certainly more difficult than in the case of ordinary fuzzy sets, but the parameters of higher order fuzzy sets tolerate less-precise estimation. On the whole, ordinary membership functions will be sufficient for an approximate quantitative representation of this intrinsically qualitative notion, that is gradual category membership.

The problem of practical estimation of membership functions has not been systematically studied in the literature. Nevertheless, some ideas and methods have been suggested by several authors, independently. Details of these appear in [574] and in other sources cited in the bibliography.

Key References in Fuzzy Pattern Recognition

Albin, M. (1975). Fuzzy sets and their application to medical diagnosis and pattern recognition, Ph.D. Thesis, Univ. of California, Berkeley.

Backer, E., and Jain, A. K. (1979). A clustering performance measure based on fuzzy set decomposition, Delft Univ. of Technology, Delft, Holland, *IEEE Trans. Pattern Anal. Mach. Intell.*, to be published.

Barnes, G. R. (1976). Fuzzy sets and cluster analysis. in *Proc. 3rd Int. Joint Conf. Pattern Recognition.*

Bellacicco, A. (1976). Fuzzy classifications, *Synthese*, **33**, 273-281.

Bellman, R. E., Kalaba, R., and Zadeh, L. A. (1966). Abstraction and pattern classification, *J. Math. Anal. Appl.*, **13**, 1-7.

Bezdek, J. C. (1974a). Numerical taxonomy with fuzzy sets, *J. Math. Biol.*, **1**, 57-71.

Bezdek, J. C. (1974b). Cluster validity with fuzzy sets, *J. Cybern.*, **3** (3), 58-73.

Bezdek, J. C. (1975). Mathematical models for systematics and taxonomy, *Proc. Annu. Int. Conf. Numer. Taxon.*, *8th* (G. Estabrook, Ed.), Freeman Co., San Francisco, pp. 143-164.

Bezdek, J. C. (1976a). A physical interpretation of fuzzy ISODATA, *IEEE Trans. Syst., Man, Cybern.*, **6**, 387-390.

Bezdek, J. C. (1976b). Feature selection for binary data: Medical diagnosis with fuzzy sets, *AFIPS Natl. Comput. Conf. Expo., Conf. Proc.* (S. Winkler, Ed.), pp. 1057-1068.

Bezdek, J. C., and Castelaz, P. F. (1977). Prototype classification and feature selection with fuzzy sets, *IEEE Trans. Syst., Man, Cybern.*, **SMC-7** (2), 87-92.

Bezdek, J. C., and Dunn, J. C. (1975). Optimal fuzzy partitions: A heuristic for estimating the parameters in a mixture of normal distributions, *IEEE Trans. Comput.*, **C-24**, 835-838.

Bezdek, J. C., and Harris, J. D. (1978). Fuzzy partitions and relations: An axiomatic basis for clustering, *Int. J. Fuzzy Sets Syst.*, **1** (2), 111-127.

Bezdek, J. C., Spillman, B., and Spillman, R. (1979). Fuzzy relations spaces for group decision theory—An application, *Int. J. Fuzzy Sets Syst.*, **2** (1), 5-14.

Bezdek, J. C., Windham, M. P., and Ehrlich, R. (1980). Statistical parameters of cluster validity functionals, *Int. J. Comput. Inform. Sci.*, **9** (4), 324-336.

Chang, R. L. P., and Pavlidis, T. (1977). Fuzzy decision-tree algorithms, *IEEE Trans. Syst., Man, Cybern.*, **SMC-7** (1), 28-35.

Conche, B. (1975). La classification dans le cas d'informations incompletes ou non explicites, in Semin., Contribution des Systemes Flous a l'Automatique, Centre Automat. Lille, Lille, France.

De Mori, R., and Laface, P. (1980). Use of fuzzy algorithms for phonetic and phonemic labeling of continuous speech, *IEEE Trans. Pattern Anal. Mach. Intell.*, **2** (2), 136–148.

De Mori, R., and Saitta, L. (1980). Automatic learning of fuzzy naming relations over finite languages, *Inf. Sci.*, **20**.

De Mori, R., and Torasso, P. (1976). Lexical classification in a speech-understanding system using fuzzy relations, in *Proc. IEEE Conf. Speech Process.*, Philadelphia, PA, pp. 565–568.

DePalma, G. F., and Yau, S. S. (1975). Fractionally fuzzy grammars with application to pattern recognition, in *Fuzzy Sets and Their Applications to Cognitive and Decision Processes* (L. A. Zadeh, K. S. Fu, K. Tanaka, and M. Shimura, Eds.), Academic Press, New York, pp. 329–351.

Diday, E. (1972). Optimisation en classification automatique et reconnaissance des formes, *RAIRO-Oper. Res.*, **3**, 61–96.

Diday, E. (1973). The dynamic clusters method and optimization in non hierarchical-clustering, in *Conf. Optim. Tech.*, *5th*, (R. Conti, and A. Ruberti, Eds.), Springer Verlag, Berlin and New York, pp. 241–254.

Dubois, D., and Prade, H. (1980). *Fuzzy Sets and Systems: Theory and Applications*, Academic Press, New York.

Dunn, J. C. (1974a). A fuzzy relative of the ISODATA process and its use in detecting compact well-separated clusters. *J. Cybern.*, **3** (3), 32–57.

Dunn, J. C. (1974b). Well-separated clusters and optimal fuzzy partitions, *J. Cybern.*, **4** (1), 95–104.

Dunn, J. C. (1974c). Some recent investigations of a new fuzzy partitioning algorithm and its application to pattern classification problems, *J. Cybern.*, **4** (2), 1–15.

Dunn, J. C. (1974d). A graph theoretic analysis of pattern classification via Tamura's fuzzy relation, *IEEE Trans. Syst., Man, Cybern.*, **SMC-4**, 310–313.

Dunn, J. C. (1977). Indices of partition fuzziness and detection of clusters in large data sets, in *Fuzzy Automata and Decision Processes* (M. M. Gupta, G. N. Saridis, and B. R. Gaines, Eds.), North-Holland Publishers, Amsterdam, pp. 271–283.

Efstathiou, J., and Tong, R. (1980). Ranking sets using linguistic preference relations, *Proc. 10th Int. Symp. on Multiple—Valued Logic*, Northwestern University, IL, pp. 137–142.

Gitman, I., and Levine, M. D. (1970). An algorithm for detecting unimodal fuzzy sets and its application as a clustering technique, *IEEE Trans. Comput.*, **C-19**, 583–593.

Gustafson, D. E., and Kessel, W. C. (1978). Fuzzy clustering with a fuzzy covariance matrix, Scientific Systems, Inc., Cambridge, MA.

Kandel, A. (1975). Fuzzy hierarchical classifications of dynamic patterns, NATO ISI Pattern Recogn. and Classification, September, France.

Kandel, A., and Lee, S. C. (1979). *Fuzzy Switching and Automatan: Theory and Applications*, Crane Russak and Co., New York, and Edward Arnold, London.

Kandel, A., and Yelowitz, L. (1974). Fuzzy chains, *IEEE Trans. Syst., Man, Cybern.*, **SMC-4**, 472–475.

Kaufmann, A. (1975). *Introduction a la Theorie des Sous-Ensembles Flous. Vol. 3: Applications a la Classification et a la Reconnaissance des Formes, aux Automates et aux Systemes, aux Choix des Criteres*, Masson, Paris.

Kickert, W. J. M., and Koppelaar, H. (1976). Application of fuzzy set theory to syntactic pattern recognition of handwritten capitals, *IEEE Trans. Syst., Man, Cybern.*, **SMC-6** (2), 148-151.

Koczy, L. T., and Hajnal, M. (1977). A new fuzzy calculus and its applications as a pattern recognition technique, in *Modern Trends in Cybernetics and Systems*, Vol. 2 (J. Rose and C. Bilciu, Eds.), Springer-Verlag, Berlin and New York, pp. 103-118.

Kotoh, K., and Hiramatsu, K. (1973). A representation of pattern classes using the fuzzy sets. *Syst. Comput. Controls*, pp. 1-8.

Larsen, L. E., Ruspini, E. H., McNew, J. J., Walter, D. O., and Adey, W. R. (1972). A test of sleep staging systems in the unrestrained chimpanzee, *Brain Res.*, **40**, 318-343.

Lee, E. T. (1972). Proximity measure for the classification of geometric figures, *J. Cybern.*, **2** (4), 43-59.

Lee, E. T. (1973). Application of fuzzy languages to pattern recognition, *Kybernetes*, **6**, 167-173.

Logimov, V. I. (1966). Probability treatment of Zadeh's membership functions and their use in pattern recognition. *Eng. Cybern.*, 68-69.

Majumder, R. (1978). A fuzzy set theoretic approach for recognition of patterns generated from bio-social systems. *ESCSR—PCSR*, **11**, to be published.

Negoita, C. V. (1973). On the application of the fuzzy sets separation theorem for automatic classification in information retrieval systems, *Inf. Sci.*, **5**, 279-286.

Negoita, C. V., and Flondor, P. (1976). On fuzziness in information retrieval. *Int. J. Man-Mach. Stud.*, **8**, 711-716.

Negoita, C. V., and Ralescu, D. A. (1975). *Applications of Fuzzy Sets to Systems Analysis*. Birkhaeuser Verlag, Basel, Switzerland, Chap. 7.

Pal, S. K., and Majumder, D. D. (1977). Fuzzy sets and decision-making approaches in vowel and speaker recognition, *IEEE Trans. Syst., Man, Cybern.*, **SMC-7**, 625-629.

Pal, S. K., and Majumder, D. D. (1980). A self-adaptive fuzzy recognition system for speech sounds, in *Fuzzy Sets Theory and Applications to Policy Analysis and Information Systems* (P. P. Wang and S. K. Chang, Eds.), Plenum, New York, pp. 223-230.

Parrish, E. A., Jr., McDonald, W. E., Aylor, J. H., and Gritton, C. W. K. (1977). Electromagnetic interference source identification through fuzzy clustering, in *Proc. IEEE Conf. Decis. Control*, New Orleans, pp. 1419-1423.

Radecki, T. (1976). Mathematical model of information retrieval system based on the concept of fuzzy thesaurus, *Inf. Process. Manage.*, **12**, 313-318.

Ruspini, E. H. (1969). A new approach to clustering, *Inf. Control*, **15**, 22-32.

Ruspini, E. H. (1970). Numerical methods for fuzzy clustering, *Inf. Sci.*, **2**, 319-350.

Ruspini, E. H. (1973a). New experimental results in fuzzy clustering. *Inf. Sci.*, **6**, 273-284.

Ruspini, E. H. (1973b). A fast method for probabilistic and fuzzy clusters analysis using association measures, in *Proc. 6th Hawaii Int. Conf. Syst. Sci.*, Honolulu, pp. 56-58.

Ruspini, E. H. (1977). A theory of fuzzy clustering, in *Proc. IEEE Conf. Decis. Control*, New Orleans, pp. 1378-1383.

Ruspini, E. H. (1980). Recent developments in fuzzy cluster analysis, *Table Ronde du CNRS sur le Flou*, Lyon, France, June.

Shimura, M. (1975). An approach to pattern recognition and associative memories using fuzzy logic, in *Fuzzy Sets and Their Applications to Cognitive and Decision Processes* (L. A. Zadeh, K. S. Fu, K. Tanaka, and M. Shimura, Eds.), Academic Press, New York, pp. 443-475.

Siy, P. S., and Chen, C. S. (1974). Fuzzy logic for handwritten numerical character recognition, *IEEE Trans. Syst., Man, Cybern.*, **SMC-6**, 570–575.

Sugeno, M. (1973). Constructing fuzzy measure and grading similarity of patterns by fuzzy integrals, *Trans. SICE*, **9**, 361–370 (in Japanese, English summary).

Tamura, S., Higuchi, S., and Tanaka, K. (1971). Pattern classification based on fuzzy relations, *IEEE Trans. Syst., Man, Cybern.*, **SMC-1**, 61–66.

Thomason, M. G. (1973). Finite fuzzy automata, regular fuzzy languages and pattern recognition, *Pattern Recogn.*, **5**, 383–390.

Wang, P. P., and Wang, C. Y. (1979). Fuzzy relations algorithms and fuzzy filters for pattern recognition, 1st Symp. on Policy Anal. and Inf. Syst., Duhram, NC, June.

Windham, M. P. (1981). Cluster validity for fuzzy clustering algorithms. *Fuzzy Sets Syst.*, **5** (2), 177–186.

Woodbury, M. A., and Clive, J. (1974). Clinical pure types as a fuzzy partition, *J. Cybern.*, **4** (3), 111–121.

Yeh, R. T., and Bang, S. Y. (1975). Fuzzy relations, fuzzy graphs and their applications to clustering analysis, in *Fuzzy Sets and Their Applications to Cognitive and Decision Processes* (L. A. Zadeh, K. S. Fu, K. Tanaka, and M. Shimura, Eds.), Academic Press, New York, pp. 125–149.

Zadeh, L. A. (1976). Fuzzy sets and their application to pattern classification and cluster analysis, Memo. UCB/ERL M-607. Univ. of California, Berkeley; also in *Classification and Clustering*, Academic Press, New York, 1977.

BIBLIOGRAPHY

1 Achache, A. (1981). Galois connexion of a fuzzy subset, *Bull. Sous Ensembles Flous Appl.*, **5**, Hiver, (80–81), 4–7.

2 Ackermann, R. (1967). *Introduction to Many-Valued Logics*, Routledge and Kegan Paul, London.

3 Aczel, M. J. (1948). Sur les operations definies pour les nombres reels, *Bull. Soc. Math. Fran.* **76**, 59–64.

4 Aczel, M. J., and Pfanzagl, J. (1966). Remarks on the measurement of subjective probability and information, *Metrika*, **5**, 91–105.

5 Adam, A., and Beran, H. (1978). On the problem of the application of fuzzy-logic concepts, *PCSR* **6**.

6 Adamek, J., and Wechler, W. (1976). Minimization of R-fuzzy automata, in *Studien zur Algebra und Ihre Anwendungen*, Akad.-Verlag, Berlin.

7 Adamo, J. M. (1977). Toward introduction of fuzzy concepts in dynamic modelling. *Conf. Dynamic Modelling and Control of National Economies* (Vienna), North-Holland, Amsterdam.

8 Adamo, J. M. (1980). L.P.L., A fuzzy programming language: 1. Syntactic aspects, *Fuzzy Sets Syst.*, **3** (2), 151–180.

9 Adamo, J. M. (1978). Semantics for a fuzzy programming language. Basic notions and logical expressions. Elementary statements and control structures, *Int. J. Syst. Sci.*

10 Adamo, J. M. (1978). Une implementation de la theorie des sous-ensembles flous—Application a l'analyse de processus de decision, Thèse d'Etat es Sci., Univ. Claude-Bernard, Lyon, France, Mar.

11 Adamo, J. M. (1978). The L.P.L. language—Some applications to combinatorial and non-deterministic programming, *Fuzzy Sets Syst.*

12 Adamo, J. M. (1980). Fuzzy decision trees, *Fuzzy Sets Syst.*, **4** (3), Nov., 207–220.

13 Adamo, J. M. (1980). L.P.L.—A fuzzy programming language. 2. Semantic aspects, *Fuzzy Sets Syst.*, **3** (3), 261–290.

14 Adamo, J. M., and Karsky, M. (1978). Application of fuzzy logic to the design of a behavioral model in an industrial environment, Univ. Claude-Bernard, Lyon, Villeurbanne, France.

15 Adamo, J. M., and Karsky, M. (1978). Application of fuzzy logic to the design of an industrial environment, Univ. Claude-Bernard, Lyon, France, 4th Int. Congr. on Cybern. Syst., Amsterdam.

16 Adams, E. W. (1965). Elements of a theory of inexact measurement, *Philos. Sci.*, **32**, 205–228.

17 Adams, E. W. (1966). Probability and the logic of conditionals, in *Aspects of Inductive Logic* J. Hintikka and P. Suppes, Eds.), North-Holland, Amsterdam, pp. 265–316.

18 Adams, E. W. (1974). The logic of "almost all," *J. Philos. Logic*, **3**, 3–17.

19 Adams, E. W., and Levine, H. P. (1975). On the uncertainties transmitted from premises to conclusions in deductive inferences, *Synthese*, **30**, 429–460.

20 Adavic, P. N., Borisov, A. N., and Golender, V. E. (1968). An adaptive algorithm for recognition of fuzzy patterns, in *Kibernetika I Diagnostika*, vol. 2 (D. S. Kristinkov, J. J. Osis, and L. A. Rastrigin, Eds.). Zinatne, Riga, U.S.S.R., pp. 13–18 (in Russian).

21 Adey, W. R. (1972). Organization of brain tissue—Is the brain a noisy processor?, *Int. J. Neurol.*, **3**, 271–284.

22 Aguilar-Martin, J., and Broit, M. (1978). Reconnaissance d'un solide a l'aide de l'estimation de la probabilite d'appartenance floue de son empreinte sur un capteur plan, Colloq. Int. Theorie et Appl. des Sous-Ensembles Flous, Journ. de Biomath. et d'Inf. Med., Marseille, Sept.

23 Adai, S. (1975). Informatics in "eco-technology," in *Summ. of Pap. on Gen. Fuzzy Problems*, The Working Group on Fuzzy Systems, Tokyo, Nov., pp. 1–4.

24 Aizermann, M. A. (1975). Fuzzy sets, fuzzy proofs and some unsolved problems in the theory of automatic control, Special Interest Disc. Session on Fuzzy Autom. and Decis. Processes, 6th IFAC World Congr., Boston, Ma, Aug.

25 Aizermann, M. A. (1976). Fuzzy sets, proofs and certain unsolved problems in the theory of automatic control, *Automat. Telenehanika*, 171–177.

26 Aizermann, M. A. (1977). Some unsolved problems in the theory of automatic control and fuzzy proofs, *IEEE Trans. Autom. Control*, 116–118.

27 Aizermann, M. A., and Smirnova, I. N. (1978). First Monograph on the theory of fuzzy sets, *Autom. Control*, **10**, 1581–1583.

28 Albert, P. (1977). The algebra of fuzzy logic, *Fuzzy Sets Syst.*, **1**, 203–230.

29 Albin, M. (1975). Fuzzy sets and their application to medical diagnosis, Ph.D. Thesis, Dept. of Math., Univ. of California, Berkeley.

30 Ali, S. T., and Doebner, H. (1976). On the equivalence of non-relativistic quantum mechanics based upon sharp and fuzzy measurements, *J. Math. Phys.*, **17**, 1105–1111.

31 Ali, S. T., and Doebner, H. (1977). Systems of imprimitivity and representation of quantum mechanics of fuzzy phase spaces, *J. Math. Phys.*, **18** (2).

32 Allen, A. D. (1973). A method of evaluating technical journals on the basis of published

comments through fuzzy implications—A survey of the major IEEE Transactions, *IEEE Trans. Syst., Man, Cybern.*, **SMC-3**, 422–425.

33 Allen, A. D. (1974). Measuring the empirical properties of sets, *IEEE Trans. Syst., Man, Cybern.*, **SMC-4**, 66–73.

34 Alley, H., Bacinello, C. P., and Hipil, K. W. (1978). Fuzzy set approaches to planning in the Grand River basin, *Adv. Water Resour.*

35 Almog, J. (1978). Is quantum mechanics a fuzzy system? The vindication of fuzzy logic in microphysics, Hebrew Univ., Jerusalem.

36 Almog, J. (1978). Semantic coniderations on modal counterfactual logic with corollaries on axiomatizatility and recursiveness in fuzzy systems, Hebrew Univ., Jerusalem; *Notre Dame J. Formal Logic*.

37 Alsina, C. (1979). On the sum of normal or gamma fuzzy variables, Report, Dept Math. I Estatistica, Univ. Politechn, Barcelona, Spain.

38 Alsina, C., and Nguyen, H. T. (1979). On the addition of fuzzy variables, *Fuzzy Sets Syst.*, to be published.

39 Alsina, C., Trillas, E., and Valverde, L. (1980). On non-distributive logical connectives for fuzzy sets theory, *Bull. Stud. Exch. Fuzzinesss Appl.*, Ete.

40 Altham, J. E. (1971). *The Logic of Plurality*, Methuen, London.

41 Amagasa, M. and Tazaki, E. (1980). Heuristic structure synthesis in a class of systems using a fuzzy algorithm, in *Summ. of Pap. on Gen. Fuzzy Probl.*, The Working Group on Fuzzy Systems, Tokyo, Dec.

42 Amano, T., and Kunii, T. L. (1974). A fuzzy access to color and texture data in a graphic application system, Conf. on Comput. Graphics and Interactive Techniques, Colorado.

43 Amerongen, V., Lemke, V. N., and Venn, V. (1977). An autopilot for ships designed with fuzzy sets, 5th Conf. Int. IFAC/IFIP, La Haye.

44 Amey, L. R. (1976). Specification of the budget control system, Internal Rep., Fac. of Manage., McGill Univ., Montreal, Canada, Aug.

45 Amey, L. R. (1976). Uncertainty and Imprecision; Fuzzy sets and their application to the control model of the firm, Internal Rep. Fac. of Manage., McGill Univ., Montreal, Canada, Aug.

46 Anthony, J. M., and Sherwood, H. (1979). Fuzzy groups redefined, *J. Math. Anal. Appl.*, **69** (1), May, 124–131.

47 Aoki, Y. (1976). Optimization of Environmental planning under the risk-aversion of non-repairable damage, in *Summ. of Pap. on Gen. Fuzzy Problems*, The Working Group on Fuzzy Systems, Tokyo, pp. 1–13.

48 Arbib, M. A. (1967). Tolerance Automata, *Kybernetika* (Prague), **3**, 223–233.

49 Arbib, M. A. (1977). Book review of *Applications of Fuzzy Sets to Systems Analysis*, *SIAM*, **19**, 753.

50 Arbib, M. A., and Manes, E. G. (1974). Foundations of system theory—Decomposable systems, *Automatica*, 285–302.

51 Arbib, M. A., and Manes, E. G. (1974). Fuzzy morphisms in automata theory, in *Proc. 1st Int. Symp. Category Theory Appl. Comput. Control*.

52 Arbib, M. A., and Manes, E. G. (1975). A categoristic view of automata and systems. *Proc. First Int. Symp. Category Theory Appl. Comput. Control.* **6**, 98–105, Univ. of Massachusetts, Amherst.

53 Arbib, M. A., and Manes, E. G. (1975). A category-theoretic approach to systems in a fuzzy world, *Synthese*, **30**, 381–406.

54 Arbib, M. A., and Manes, E. G. (1975). Fuzzy machines in a category, *Bull. Aust. Math. Soc.*, **13**, 169–210.

55 Arendt, F., and Straube, B. (1977). Zur Auwendung des Theorie Unscharfer Mengen auf Vorhersage und Steuerungsprobleme, in *Unscharfe Modellbildung und Steuerung*, Lecture Notes, Th Karl-Marx-Stadt, Germany, June, pp. 21–29.

56 Arigona, A. O. (1977). Mathematical aspects of language in the calculus of fuzzy restrictions—Working program, *EMCSR—PCSR*, **8**, to be published.

57 Arigona, A. O. (1978). Semantic implication and its utilization in fuzzy set theory, 4th Conf. on Cybern. on Syst. Res., Linz, Austria.

58 Arigona, A. O. (1980). Mathematical developments arising from "semantical implication" and the evaluation of membership characteristic functions, *Fuzzy Sets Syst.*, **4** (2), Sept., 167–184.

59 Arigona, A. O. (1980). Utilization of non-standard analysis in fuzzy measures, Table Ronde du CNRS sur le Flou, Lyon, France, June.

60 Arigona, A. O. (1976). Membership characteristic function of fuzzy elements fundamental theoretical basis, 3rd Eur. Meeting Cybern. Syst. Res., Vienna.

61 Aronson, A. R., Jacobs, B. E., and Minka, J. (1978). On fuzzy resolution CS Tech. Rep. TR-687, Univ. of Maryland, College Park, Aug.

62 Aronson, A. R., Jacobs, B. E., and Minka, J. (1980). A note on fuzzy deduction, *JACM*, **27** (4), Oct., 599–603.

63 Asai, K., and Kitajima, S. (1971). A method for optimizing control of multimodal systems using fuzzy automata, *Inf. Sci.*, **3**, 343–353.

64 Asai, K., and Kitajima, S. (1971). Learning control of multimodal systems by fuzzy automata, in *Pattern Recognition and Machine Learning*, (K. S. Fu, Ed.)., Plenum Press, New York, pp. 195–203.

65 Asai, K., and Kitajima, S. (1972). Optimizing control using fuzzy automata, *Automatica*, **8**, 101–104.

66 Asai, K., and Tanaka, H. (1973). On the fuzzy mathematical programming, *Proc. in 3rd IFAC Symp. Identification Syst. Parameter Estimation*, Pt. II, North-Holland, Amsterdam.

67 Asai, K., and Tanaka, H. (1975). Applications of fuzzy sets theory to decision-making and control, *J. JAACE*, **19**, 235–242.

68 Asai, K., Tanaka, H., and Okuda, T. (1975). Decision making and its goal in a fuzzy environment, in *Fuzzy Sets and Their Applications to Cognitive and Decision Processes*, (L. A. Zadeh, K. S. Fu, K. Tanaka, and M. Shimura, Eds.). Academic Press, New York, pp. 257–277.

69 Asai, K., Tanaka, K., and Okuda, T. (1977). On the discrimination of fuzzy states in probability space, *Kybernetes*, **6**, 185–192.

70 Asai, K., Tanaka, H., Negoita, C. V., and Ralescu, D. A. (1978). *Introduction to Fuzzy Systems Theory*, Ohm-Sha Co. Ltd., Tokyo (in Japanese).

71 Asenjo, F. G. (1966). A calculus for antinomies, *Notre Dame J. Formal Logic*, **7**, 103–105.

72 Assilian, S. (1974). Artificial intelligence in the control of real dynamic systems, Ph.D. Thesis, Queen Mary College, Univ. of London, London.

73. Assilian, S., and Mandani, E. H. (1974). Learning control algorithms in real dynamic systems, in *Proc. 4th Int. IFAC/IFIP Conf. on Digital Comput. Appl. Process Control*, Zurich, Mar.

74. Atkin, R. H. (1974). *Mathematical Structure in Human Affairs*, Heinemann, London.

75. Aubin, J. P. (1974). Coeur et valeur des jeux flous sans paiements lateraux, *C.R. Acad. Sci. Paris*, 891–894.

76. Aubin, J. P. (1974). Fuzzy games, MRC Tech. Summ. Rep. 1480, Math. Res. Center, Univ. of Wisconsin—Madison, Madison,

77. Aubin, J. P. (1974). Coeur et equilibres des jeux flous sans paiements lateraux, *C.R. Acad. Sci. Paris*, Ser. A, **279**, 963–966.

78. Aubin, J. P. (1976). Fuzzy core and equlibria of games defined in strategic form, in *Directions in Large-Scale Systems*, (Y. C. Ho and S. K. Mitter, Eds.). Plenum Press, New York, pp. 371–388.

79. Auray, J. P., and Duru, G. (1976). Contribution a l'etude des structures floues. Doc. Trav. No. 5, Dept. Sci. Econ., Fac. Droit et Sci. Econ. et Politiques, Besancon, France, Jan.

80. Auray, J. P., and Duru, G. (1976). Introduction a la theorie des espaces multiflous. Doc. Trav. de l'IME, No. 16, April.

81. Axinn, A., and Axinn, D. (1976). Notes on the logic of ignorance relations, *Amer. Philos. Quart.*, **13**, 135–143.

82. Ayme, S. (1981). Efficiency of belief functions compared to a classical method of clustering in a computerized system in medical genetics, CORS-TIMS-ORSA Joint Nat. Meeting, Toronto, Canada, May.

83. Ayme, S., Gouvernet, J., Sanchez, E., and Giraud, F. (1980). Aide en mode conversationnel au diagnostic en genetique medicale, Table Ronde du CNRS sur le Flou, Lyon, France, June.

84. Azorin Poch, F. (1979). *Algunes Applicaciones de los Conjuntos Borrosos a la Estadistica*, Publ. Inst. Nac. de Estadistica, Madrid.

85. Baaklini, N., and Mamdani, E. H. (1975). Prescriptive methods for deriving control policy in a fuzzy logic controller *Electron. Lett.*, **11**.

86. Baas, S. M., and Kwakernaak, H. (1977). Rating and ranking of multiple-aspect alternatives using fuzzy sets, *Automatica*, **13** (1), 47–58.

87. Baciu, A., and Pascu, A. (1979). *A Mathematical Model in Fuzzy Systems Theory*, North-Holland, Amsterdam, pp. 1034–1040.

88. Baciu, A., and Pascu, A. (1980). Fuzzy input-output systems and fuzzy Leontief-type languages, 4th Eur. Congr. Oper. Res., July.

89. Backer, E. (1975). A non-statistical type of uncertainty in fuzzy events, in *Colloq. Math. Soc. Janos Bolyai*, Vol. 16, North-Holland, Amsterdam, pp. 53–73.

90. Backer, E. (1978). Cluster analysis by optimal decomposition of induced fuzzy sets, Delftse Univ. Pers.

91. Backer, E. (1978). Cluster analysis formalized as a process of fuzzy identification based on fuzzy relations, Rep. IT-78-15, Delft Univ. of Tech., Delft, Holland.

92. Backer, E. (1979). Agglutintaion spatiale en segmentation d'images avec des relations d'affinites floues, 2nd Congr. AFCET-IRIA, Reconnaissance de Formes et Intell. Artif., Toulouse, France, Sept.

93. Backer, E. (1980). On the minimum distorsion condition in fuzzy relational clustering, Table Ronde du CNRS sur le Flou, Lyon, France, June.

94 Backer, E., and Jain, A. K. (1981). A clustering performance measure based on fuzzy set decomposition. *IEEE Trans. Pattern Anal. Mach. Intell. PAMI* **3** (1), 66-75.

95 Badard, R. (1980). The law of large numbers for fuzzy processes and the estimation problem, Bat. 502, Dept. Inf. Insa-Lyon, Lyon, France.

96 Baldwin, J. F. (1978). A model of fuzzy reasoning and fuzzy logic, Res. Rep. EM/FS10, Eng. Math. Dept., Univ. of Bristol, Bristol, UK.

97 Baldwin, J. F. (1978). A new approach to approximate reasoning using a fuzzy logic, Res. Rep. EM/FS3, Eng. Math. Dept., Univ. of Bristol, Bristol, UK.

98 Baldwin, J. F. (1978). Fuzzy logic and approximate reasoning for mixed input arguments, Res. Rep. EM/FS4, Eng. Math. Dept., Univ. of Bristol, Bristol, UK.

99 Baldwin, J. F. (1979). Fuzzy information structures, Res. Rep. EM/FS23, Eng. Math. Dept., Univ. of Bristol, Bristol, UK.

100 Baldwin, J. F. (1979). Fuzzy logic and fuzzy reasoning, in *Proc. 18th Conf. Dec. Control*, Vol. 2, Fort Lauderdale, FL, Dec., pp. 794-795.

101 Baldwin, J. F. (1979). Fuzzy Logic and its Application to Fuzzy Reasoning, *Advances in Fuzzy Set Theory and Applications*, (M. M. Gupta, R. K. Ragade, and R. R. Yager, Eds.)., North-Holland, Amsterdam.

102 Baldwin, J. F. (1980). A knowledge base using fuzzy logic, in *Proc. Int. Congr. on Appl. Syst. Res. and Cybern.*, Acapulco, Mexico, Dec.

103 Baldwin, J. F. (1980). An efficient automatic fuzzy reasoning method, Table Ronde du CNRS sur le Flou, Lyon, France, June.

104 Baldwin, J. F. (1980). Automated fuzzy reasoning algorithm, Res. Rep. EM/FS27, Eng. Math. Dept., Univ. of Bristol, Bristol, UK.

105 Baldwin, J. F., and Guild, N. C. F. (1978). A model for multi-criterial decision-making using fuzzy logic, in *Proc. Workshop of Fuzzy Reasoning*, Queen Mary College, Univ. of London, London, Sept.; in *Int. J. Man-Machine Stud.*, 1979.

106 Baldwin, J. F., and Guild, N. C. F. (1978). Compaison of fuzzy sets on the same decision space, Res. Rep. EM/FS2, Eng. Math. Dept., Univ. of Bristol, Bristol, UK.

107 Baldwin, J. F., and Guild, N. C. F. (1980). Feasibility algorithms for approximate reasoning using a fuzzy logic, *Fuzzy Sets Syst.*, **3** (3), 225-252.

108 Baldwin, J. F., and Guild, N. C. F. (1978). The resolution of two paradoxes by approximate reasoning using a fuzzy logic, Res. Rep. EM/FS/11, Eng. Math. Dept., Univ. of Bristol, Bristol, UK; submitted for publication.

109 Baldwin, J. F., and Guild, N. C. F. (1979). Comment on the "fuzzy max" operator of Dubois and Prade, *Int. J. Syst. Sci.*, **10** (9), 1063-1064.

110 Baldwin, J. F., and Guild, N. C. F. (1979). FUZLOG, A computer program for fuzzy reasoning, 9th Int. Symp. on Multi-Valued Logic, Beth, U.K.

111 Baldwin, J. F., and Guild, N. C. F. (1979). Management evaluation with fuzzy logic, Working Pap., Dept. Eng. Math., Univ. of Bristol, Bristol, UK.

112 Baldwin, J. F., and Guild, N. C. F. (1980). Modelling controllers using fuzzy relations, *Kybernetes*, **9** (3).

113 Baldwin, J. F., Guild, N. C. F., and Pilsworth, B. W. (1978). A group voting model for fuzzy logic, Dept. Eng. Math., Univ. of Bristol, Bristol, UK.

114 Baldwin, J. F., Guild, N. C. F., and Pilsworth, B. W. (1978). Improved logic for fuzzy controllers, Res. Rep. EM/FS5, Dept. Eng. Math., Univ. of Bristol, Bristol, UK.

115 Baldwin, J. F., Guild, N. C. F., and Pilsworth, B. W. (1978). On the satisfaction of a

fuzzy relation by a set of inputs, Res. Rep. EM/FS7, Eng. Math. Dept., Univ. of Bristol, Bristol, UK.

116 Baldwin, J. F., and Pilsworth, B. (1980). Axiomatic approach to implication for approximate reasoning using a fuzzy logic, in *Int. J. Fuzzy Sets Syst.*, **3** (2), 193–220.

117 Baldwin, J. F., and Pilsworth, B. W. (1978). Fuzzy truth definition of possibility measure for decision classification, Res. Rep. EM/FS12, Eng. Math. Dept., Univ. of Bristol, Bristol, UK.

118 Baldwin, J. F., and Pilsworth, B. W. (1978). Methods of approximate reasoning, Rep. EM/FS6, Eng. Math. Dept., Univ. of Bristol, Bristol, UK.

119 Baldwin, J. F., and Pilsworth, B. W. (1979). A theory of fuzzy probability, 9th Int. Symp. on Multi-Valued Logic, Beth, UK.

120 Baldwin, J. F., and Pilsworth, B. W. (1980). Fuzzy reasoning with probability. Res. Rep. EM/FS29, Dept. Eng. Math., Univ. of Bristol, Bristol, UK.

121 Baldwin, J. K., and Pilsworth, B. W. (1978). A model of fuzzy reasoning through multi-valued logic and set theory, Res. Rep. EM/FS6, Eng. Math. Dept., Univ. of Bristol, Bristol, UK.

122 Ballmer, T. T. (1976). Fuzzy punctuation of the theory of continuing of grammaticality, Memo no. ERL-M590, Electron. Res. Lab., Univ. of California, Berkeley.

123 Banaschewska, B. (1968). Injective hulls in the category of distributive lattices, *J. Reine Angew. Math.*, 102–109.

124 Banaschewska, B., and Bruns, G. (1967). Categorical characterization of the Mac-Neville completion, *Arch. Math.*, 369–377.

125 Bandler, W., and Kohout, L. (1978). Fuzzy relational products and fuzzy implication operators, Int. J. Man-Machine Stud.; Rep. no. FRP.1, Dept. Math, Univ. of Essex; Queen Mary College, Univ. of London, London.

126 Bandler, W., and Kohout, L. (1979). Application of fuzzy logics to computer protection structures, 9th Int. Symp. on Multi-Valued Logic, Beth, UK.

127 Bandler, W., and Kohout, L. (1980). Fuzzy power sets and fuzzy implication operators, *Fuzzy Sets Syst.*, **4** (1), July, 13–30.

128 Bandler, W., and Kohout, L. (1980). Semantics of implication operators and fuzzy relational products, *Int. J. Man-Machine Stud.*, **12** (1).

129 Bandler, W., and Kohout, L. J. (1980). Relational products as a tool for analysis and synthesis of the behavior of complex natural and artificial systems, Fuzzy Sets Theory and Appl. to Policy Anal. and Inf. Syst., Plenum, New York, pp. 341–368.

130 Bang, S. Y., and Yeh, R. T. (1974). Toward a theory of relational data structure, SELTR-1, Univ. of Austin, TX.

131 Banon, G. (1978). Distinctions between several types of fuzzy measures, in *Proc. Int. Colloq. on Theory and Appl. of Fuzzy Sets*, Univ. of Marseille, Marseille, France.

132 Banon, G. (1979). Utilisation de familles de mesures de probabilite comme modele d'information a prior subjective dans l'estimation parametrique. *Busefal* (1), Automne-Hiver (79–80), 36–46; Langages Syst. Inf., Univ. Paul Sabatier, Toulouse, France, Dec.

133 Banon, G. (1979). Use of possibility measures for modelling an a priori knowledge in parametric estimation, *Adv. Control*, **2**.

134 Banon, G. (1981). Distinction between several subsets of fuzzy measures, *Fuzzy Sets Syst.*, **5** (3), 291–306.

135 Baptistella, L. F. B., and Ollero, A. (1980). Fuzzy methodologies for interactive criteria optimization, *IEEE Trans. Syst., Man, Cybern.*, **SMC-10** (7), July, pp. 355–365.

136 Barnes, G. R. (1976). Fuzzy sets and cluster analysis, in *Proc. at 3rd Int. Joint Conf. on Pattern Recognition*, IEEE Cat. no. 76CH1140-3C.
137 Barnev, P., Dimitrov, V., and Stanchev, P. (1974). Fuzzy system approach to decision-making based on public opinion investigation through questionaires, IFAC Symp. Stochastic Control, Budapest, Sept.
138 Barthelmy, J. P., and Borchut, D. (1976). In *Notes of the Univ. Seminar on Fuzzy Sets* (Luong and Massonite, Eds.), Univ. of Besancon, Besancon, France.
139 Bartolin, R., Vovan, L., Gouvernet, J., and Sanchez, E. (1980). Ensembles flous et mesures de possibilities—application a la classification O.M.S. des hyperlipoproteinemies, Table Ronde du CNRS sur le Flou, Lyon, France, June.
140 Basson, D., and Yager, R. (1975). Decision making with fuzzy sets, *Decis. Sci.*, **6**.
141 Battle, N., and Trillas, E. (1979). Entropy and fuzzy integrals, *J. Math. Anal. Appl.*, **69**, 469–474.
142 Becker, J. M. (1973). A structural design process, Ph.D. Thesis, Dept. Civil Eng., Univ. of California, Berkeley.
143 Bel, G., and Dubois, D. (1979). Aide a la decision en planification de reseaux de transport en commun, IFAC/IFORS Symp., Toulouse, France, Mar.
144 Bellacicco, A. (1976). Fuzzy classification, *Synthese*, **33**, 273–281.
145 Bellman, R. E. (1970). Humor and paradox, in *A Celebration of Laughter* (W. M. Mendel, Ed.). Mara Books, Los Angeles, pp. 35–45.
146 Bellman, R. E. (1971). Law and Mathematics, Tech. Rep. 71-34, Univ. of Southern California, Los Angeles, Sept.
147 Bellman, R. E. (1973). Mathematics and the human sciences, in *The Dynamic Programming of Human Systems*, (J. Wilkinson, R. E. Bellman, and R. Garaudy, Eds.). MSS Inf. Corp., New York, pp. 11–18.
148 Bellman, R. E. (1973). Retrospective Futurology—Some introspective comments, in *The Dynamic Programming of Human Systems*, (J. Wilkinson, R. E. Bellman, and R. Garaudy, Eds.). MSS Inf. Corp., New York, pp. 35–37.
149 Bellman, R. E. (1974). Large systems, *IEEE Trans. Autom. Control*, **AC-19**, 465.
150 Bellman, R. E. (1974). Local logics, Tech. Rep. no. USC EE RB 74-9, Univ. of Southern California, Los Angeles.
151 Bellman, R. E. (1975). Communication, ambiguity and understanding, *Math. Biosci.*, **26**, 347–356.
152 Bellman, R. E., and Giertz, M. (1973). On the analytic formalism of the theory of fuzzy sets, *Inf. Sci.*, **5**, 149–156.
153 Bellman, R. E., Kalaba, R., and Zadeh, L. A. (1966). Abstraction and pattern classification, *J. Math. Anal. Appl.*, **13**, 1–7.
154 Bellman, R. E., and Marchi, E. (1973). Games of protocol—The city as a dynamic competitive process, Tech. Rep. RB73-36, Univ. of Southern California, Los Angeles.
155 Bellman, R. E., and Zadeh, L. A. (1970). Decision-making in a fuzzy environment, *Manage. Sci.*, **17**, 141–164.
156 Bellman, R. E., and Zadeh, L. A. (1977). Local and fuzzy logics, in *Modern Uses of Multiple-Valued Logic* (J. C. Dunn and G. Epstein, Eds.). D. Reidel, Dordrecht, Holland, pp. 103–165.
157 Belluce, L. P. (1964). Further results on infinite valued predicate logic, *J. Symb. Logic*, **29**, 69–78.

158 Belluce, L. P., and Chang, C. C. (1963). A weak completeness theorem for infinite valued first-order logic, *J. Symb. Logic*, **28**, 43–50.
159 Belnap, N. D. (1960). Entailment and relevance, *J. Symb. Logic*, **25**, 144–146.
160 Ben Salem, M. (1976). Sur l'extension de la theorie des sous-ensembles flous a l'automatique industrielle, Doct. Thesis, 3rd Cycle, Univ. of Paris, Paris.
161 Benamghar, L., and Martin, J. (1978). Methode de classification simple d'objets definis par des donnees imprecises et recherche sur listing (methode tri-flou), Colloq. Int. theorie et appl. des Sous-Ensembles Flous, journ. de biomath. et d'Inf. Med., Marseille, France, Sept.
162 Benlachen, D., Hirsch, G., and Lamotte, M. (1976). Codage et minimisation des fonctions floues dans une algebre floue, Mem., Lab. Electr. et Autom., Univ. de Nancy, Nancy, France.
163 Benlahcan, D., and Lamotte, M. (1978). Synthese d'un automate flou, Colloq. Int. Theorie et Appl. des sous-ensembles Flous, Journ. de Biomath. et d'Inf. Med., Marseille, France, Sept.
164 Benson, W. H. (1980). An application of fuzzy set theory to data display, in *Proc. Int. Congr. on Appl. Syst. Res. and Cybern.*, Acapulco, Mexico, Dec.
165 Beran, L. (1974). *Grupy a Svazy*, SNTL-Technical Publishers, Prague (in Czech — *Groups and Lattices*).
166 Berndt, R. S., and Caramazza, A. (1977). The development of some adverbial modifiers of dimensional adjectives, in *Proc. 2nd Annu. Boston Univ. Conf. on Lang. Dev.*, Boston, MA.
167 Berndt, R. S., and Caramazza, A. (1978). The development of vague modifiers in the language of pre-school children, *Child Lang.*, to be published.
168 Bertolini, F. (1971). Kripke models and many valued logics, *Symp. Math.*, 113–131.
169 Bertoni, A. (1973). Complexity problems related to the approximation of probabilistic languages and events by deterministic machines, in *Automata, Languages and Programming* (M. Nivat, Ed.). North-Holland, Amsterdam, pp. 507–516.
170 Besson-Berrando, M. (1977). A logical analysis of simple and fuzzy decisions. 6th Eur. Working Group on Fuzzy Sets.
171 Besson, M. (1976). Simple events, fuzzy events, 2nd Eur. Congr. on Oper. Res., Stockholm, Dec.
172 Bezdek, J. C. (1973). Fuzzy-mathematics in pattern classification, Ph.D. Thesis, Cornell Univ., in *Diss. Abstr. Int.*, **34/10-B**, 5066.
173 Bezdek, J. C. (1974). Cluster validity with fuzzy sets, *J. Cybern.*, **3**, 58–73.
174 Bezdek, J. C. (1974). Numerical taxonomy with fuzzy sets, *J. Math. Biol.*, **1**, 57–71.
175 Bezdek, J. C. (1975). Mathematical models for systematics and taxonomy, in *Proc. 8th Annu. Int. Conf. on Numer. Taxon.* (G. Estabrook, Ed.). Freeman, San Francisco.
176 Bezdek, J. C. (1976). A physical interpretation of fuzzy isodata, *IEEE Trans. Syst. Man, Cybern.*, **SMC-6**, 387–389.
177 Bezdek, J. C. (1976). Feature selection for binary data — Medical diagnosis with fuzzy sets, in *Proc. Nat. Comput. Conf.*, AFIPS Press, Montvale, NJ, June.
178 Bezdek, J. C. (1977). Fuzzy preferene rankings, ORSA/TIMS Meet., Atlanta, GA.
179 Bezdek, J. C. (1977). Some current trends in fuzzy clustering, in — *The General Systems Paradigm (Proc. SGSR)*, (J. White, Ed.). pp. 347–352.

Key References in Fuzzy Pattern Recognition

180 Bezdek, J. C. (1979). Particle and grain shape analysis with fuzzy sets, in *Proc. 1st Residential Workshop on Particulate Morphol.*, (K. Beddow, Ed.). CRC Press.

181 Bezdek, J. C. (1980). A convergence theorem for the fuzzy isodata clustering algorithms, *IEEE Trans. Pattern Anal. Mach. Intell.*, **PAMI-2** (1), Jan., 1–8.

182 Bezdek, J. C. (1980). Fuzzy imbeddings that are convex hulls, *J. Cybern.*

183 Bezdek, J. C., and Castelas, P. F. (1977). Prototype classification and feature selection with fuzzy sets, *IEEE Trans. Syst., Man. Cybern.*, **7**, (2), 87–92.

184 Bezdek, J. C., and Dunn, J. C. (1975). Optimal fuzzy partitions—A heuristic for estimating the parameters in a mixture of normal distributions, *IEEE Trans. Comput.*, **C-24**, 835–838.

185 Bezdek, J. C., and Fordon, W. A. (1978). Analysis of hypertensive patients by the use of fuzzy isodata algorithm, *Proc. JACC*, **3**, 349–355.

186 Bezdek, J. C., Gunderson, R., Ehrlich, R., and Meloy, T. (1979). On the extension for detection of linear clusters, in *Proc. 1978 IEEE Conf. on Decis. and Control* (includes the 17th Symp. on Adapt. Processes), 78CH1392-OCS, San Diego, CA, Jan. 1979, pp. 1438–1443.

187 Bezdek, J. C., Gunderson, R., Ehrlich, R., and Meloy, T. (1979). On the extension of fuzzy *K*-means algorithm for detection of linear clusters, in *17th IEEE Conf. on Decis. and Control*, pp. 1438–1443.

188 Bezdek, J. C., and Harris, J. (1977). Convex decompositions of fuzzy partitions, *Inf. Sci.*

189 Bezdek, J. C., and Harris, J. D. (1978). Fuzzy partitions and relations—an axiomatic basis for clustering, *Fuzzy Sets Syst.*, **1**, 111–126.

190 Bezdek, J. C., and Solomon, K. (1980). Simulation of implicit numerical characteristics using small samples, in *Proc. Int. Congr. on Appl. Syst. Res. and Cybern.*, Acapulco, Mexico, Dec.

191 Bezdek, J. C., Spillman, B., and Spillman, R. (1977). Fuzzy measures of preference and consensus in group decision-making, in *Proc. 1977 IEEE Conf. on Decis. and Control* (K. S. Fu, Ed.). IEEE Press, Piscatawy, NJ, pp. 1303–1309.

192 Bezdek, J. C., Spillman, B., and Spillman, R. (1979). Fuzzy relation spaces for group decision theory—An application, *Fuzzy Sets Syst.*, **2** (1), Jan., 5–14.

193 Bezdek, J. C., Windham, M. P., and Ehrlich, R. (1980). Statistical parameters of cluster validity functionals, *Int. J. Comput. Inf. Sci.*, **9** (4), 324–336.

194 Bialnicki-Birula, A. (1957). Remarks on quasi-Boolean algebras, *Bull. Acad. Polonaise Sci., Ser. Math. Astr. Phys.*, **5**, 615–619.

195 Black, M. (1937). Vagueness—An exercise in logical analysis, *Philos. Sci.*, **4**, 427–455.

196 Black, M. (1963). Reasoning with loose concepts, *Dialogue*, **2**, 325–373.

197 Black, M. (1968). *The Labyrinth of Language*, Mentor Books, New York.

198 Black, M. (1970). *Margins of Precision*, Cornell Univ. Press, Ithaca, NY.

199 Blackburn, S. (Ed.) (1975). *Meaning, Reference and Necessity*, Cambridge Univ. Press, Cambridge, UK.

200 Blanchard, N. (1980). Cardinaux flous, Univ. Claude-Bernard, Lyon, France.

201 Blanchard, N. (1980). Fuzzy lip applications—The category "fuzzy-lip," in *Proc. Int. Congr. on Appl. Syst. Res. and Cybern.*, Acapulco, Mexico, Dec.

202 Blanchard, N. (1980). Injectivite, surjectivite, bijectivite floue—Cardinaux flous, Univ. Claude-Bernard, Lyon, France; *Bull. Stud. Exch. Fuzziness Appl.*, Printemps.

203 Blanchard, N. (1980). Parties floues finies d'un referential net. ensembles flous finis, *Bull. Stud. Exch. Fuzziness Appl.*, Ete.

204 Blanchard, N. (1980). Une construction du cardinal flou d'un ensemble flou n'utilisant pas l'axiome du chiox. Une raison de modifier la definition des applications floues?, *Bull. Sous Ensembles Flous Appl.*, Automne.

205 Blin, J. M. (1973). *Patterns and Configurations in Economic Science*, Reidel, Amsterdam.

206 Blin, J. M. et al., (1974). Pattern recognition and macroeconomics, *J. Cybern.*

207 Blin, J. M. (1974). Fuzzy relations in group decision theory, *J. Cybern.*, **4**, 17–22.

208 Blin, J. M. (1975). Fuzzy relations in multiple-criteria decision making, Northwestern Univ., Evanston, IL, Jan.

209 Blin, J. M. (1976). On an isomorphism between aggregation procedures and fuzzy sets, ORSA-TIMS Conf., Miami, FL, Nov.

210 Blin, J. M. (1977). Fuzzy sets in multiple criteria decision making, in *Multiple Criteria Decision Making*, (M. K. Slarr and M. Zeleny, Eds.), Academic Press, New York.

211 Blin, J. M., and Whinston, A. B. (1973). Fuzzy sets and social choice, *J. Cybern.*, **3**, 28–33.

212 Bliss, J. M. (1977). Fuzzy sets in multiple criteria decision making, *Tims Stud. Manage. Sci.*, **6**.

213 Blockley, D. I. (1975). Predicting the likelihood of structural accidents, *Proc. Inst. Civil Eng.*, Part 2, Dec., 659–668.

214 Blockley, D. I. (1978). Analysis of subjective assessments of structural failures, *Int. J. Man-Mach. Stud.*, **10**, 185–195.

215 Blockley, D. I. (1979). The role of fuzzy sets in civil engineering, *Fuzzy Set Syst.*, **2** (4), Oct.

216 Bloore, G. C., and Rutherford, D. A. (1976). The implementation of fuzzy algorithms for control, *Proc. IEEE*, **64** (4), Apr.

217 Blyth, T. S., and Janowitz, M. F. (1972). *Residuation Theory*, Pergamon Press, Oxford, UK.

218 Bocklisch, S. F. (1977). Verfahren des Unscharfen System Identifikation mit Grobmodellen, in *Unscharfe Modellbildung und Steuerung*, Lecture notes, TH Karl-Marx-Stadt, Germany, June, pp. 64–84.

219 Bocklisch, S. F. (1979). Anwendung des Verfahrens der Unscharfen Klassifikation zur Identifikation, Material zur Volesung "Kennwertermittlung und Modellbildung," in Teil II—*Unscharfe Modellbildung und Steuerung*, TH Karl-Marx-Stadt, Germany, Oct.

220 Bocklisch, S. F., Meyer, W., and Straube, B. (1979). Fuzzy clustering and classification for the identification of process types in technical and biomedical systems, Material zur Volesung "Kennwertermittlung und Modellbildung," in Teil II—*Unscharfe Modellbildung und Steuerung*, TH Karl-Marx-Stadt, Germany, Oct.

221 Bocklisch, S. F., and Bilz, F. (1977). Systemidentifikation mit Unscharfen Klassenkonzept, in *Unscharfe Modellbildung und Steuerung*, Lecture notes, TH Karl-Marx-Stadt, June, pp. 64–84.

222 Boicescu, V. (1971). Sur les algebres de lukaciewicz, in *Logique, Automatique, Informatique*, Acad. Rep. Soc. de Roumanie, pp. 91–97.

223 Boichut, D. (1975). Utilisation des sous-ensembles flous pour le diagnostic medical, Semin. Contrib. des Syst. Flous a l'Autom., Centre d'Automatique de Lille, Lille, France, June.

224 Bollman, P., and Konrad, E. (1976). Fuzzy retrieval, 3rd Eur. Meet. on Cybern. and Syst. Res., Vienna, April.

225 Bona, B. (1977). An application of a multicriteria fuzzy similarity relation to the classification of metropolitan districts, in *Proc. of Symp. on Fuzzy Set Theory and Appl., IEEE Conf. on Decis. and Control*, New Orleans.

226 Bonaert, A. P., and Cools, M. J. (1977). Valuation et probabilite dans un processus d'apprentissage, Note interne, Centre Imago, Univ. de Louvain, Louvain-la-Neuve, Belgium, May.

227 Bonissone, P. (1978). A pattern recognition approach to the problem of linguistic approximation in systems analysis, Memo. UCB/ERL M78/57, Univ. of California, Berkeley.

228 Borghi, O. (1972). On a theory of functional probability, *Rev. Union Mat. Argent. Assoc. Fis. Argent.*, **26**, 90–106.

229 Borghi, O., Marchi, E., Zo, F. (1976). A note on embedding of fuzzy sets in a normed space, *Rev. Union Mat. Argent. Assoc. Fis. Argent.*, **28** (1), 36–41.

230 Borisov, A., and Erenstein, R. X. (1970). Comparison of some crisp and fuzzy algorithms of recognition, *Metody i Sredstva Texniceskoi Kibernetiki*, **6**, (in Russian).

231 Borisov, A. N., and Alexeyev, A. V. (1978). Fuzzy situative decision making models, *PCSR*, **6**.

232 Borisov, A. N., and Kokle, E. A. (1970). Recognition of fuzzy patterns by feature analysis, in *Kibernetika i Diagnostika*, Vol. 4 (D. S. Kristinkov, J. J. Osis, and L. A. Rastrigin, Eds.), Zinitne, Riga (in Russian).

233 Borisov, A. N., and Kokle, E. A. (1970). Recognition of fuzzy patterns, in *Kibernetika i Diagnostika*, Vol. 4 (D. S. Kristinkov, J. J. Osis, and L. A. Rastrigin, Eds.), Zinitne, Riga, pp. 135–147 (in Russian).

234 Borisov, A. N., and Osis, J. J. (1970). Methods for experimental estimation of membership functions of fuzzy sets, in *Kibernetika i Diagnostika*, Vol. 4 (D. S. Kristinkov, J. J. Osis, and L. A. Rastrigin, Eds.), Zinitne, Riga, pp. 125–134 (in Russian).

235 Borisov, A. N., and Osis, J. J. (1970). Search for the greatest divisibility of fuzzy sets, in *Kibernetika i Diagnostika*, Vol. 3 (D. S. Kristinkov, J. J. Osis, and L. A. Rastrigin, Eds.), Zinitne, Riga, pp. 79–88 (in Russian).

236 Borisov, A. N., Vulf, G. N., and Osis, J. A. (1972). Applications of the theory of fuzzy sets to state identification of complex systems, in *Kibernetika i Diagnostika*, Vol. 5 (D. S. Kristinkov, J. J. Osis, and L. A. Rastrigin, Eds.), Zinitne, Riga (in Russian).

237 Borisov, A. N., Vulf, G. N., and Osis, J. J. (1972). Prediction of the state of a complex system using the theory of fuzzy sets, in *Kibernetika i Diagnostika*, Vol. 4, (D. S. Kristinkov, J. J. Osis, and L. A. Rastrigin, Eds.), Zinitne, Riga, pp. 79–84 (in Russian).

238 Borisov, A. N., Vulf, G. N., and Osis, J. J. (1972). State Prognosis of complex systems using the theory of fuzzy sets, in *Kybernetika i Diagnostika*, Vol. 5 (D. S. Kristinkov, J. J. Osis, and L. A. Rastrigin, Eds.), Zinitne, Riga (in Russian).

239 Borkowski, L. (1958). On proper quantifiers 1, *Stud. Logica*, **8**, 65–128.

240 Borkowski, L. (Ed.) (1970). *Jan Lukasiewicz, Selected Works*, North-Holland, Amsterdam.

241 Borkowski, L., and Slupecki, J. (1958). The logical works of Lukasiewicz, *Stud. Logica*, **8**, 7–50.

242 Borodkin, L. I. (1979). Aggregated structures of graphs with fuzzy blocks, *Autom.—Remote Control*, **39** (8), 1219–1229.

243 Boruvka, O. (1937). Studies on multiplicative systems (semigroups) Part 1, Publications de la Fac. des Sci. de l'Univ. Masaryk, No. 245 (in English).

244 Boruvka, O. (1938). Studies on multiplicative systems (semigroups), Part 2, Publications de la Fac. des Sci. de l'Univ. Masaryk, No. 265, pp. 1–24 (in English).

245 Boruvka, O. (1939). Theory of groupoids, Publications de la Fac. des Sci. de l'Univ. Masaryk, No. 275 (in Czech).

246 Boruvka, O. (1941). Uber Ketten von Faktoroiden, *Math. Ann.*, **188**, 41–64.

247 Boruvka, O. (1974). *Foundations of the Theory of Groupoids and Groups*, Veb Deutscher Verlag der Wissenschaften, Berlin.

248 Bossel, H., Klaczko, S., and Muller, N. (1976). *Systems Theory in the Social Sciences*, Birkhauser Verlag, Stuttgart, Germany.

249 Bossel, H. H., and Hughes, B. B. (1973). Simulation of value-controlled decision making, Rep. SRC-11, Syst. Res. Center, Case Western Reserve Univ. Cleveland, Ohio.

250 Botta, O., and Delorme, M. (1980). Construction d'un univers flou, *Bull. Stud. Exch. Fuzziness Appl.*, Printemps.

251 Botta, O., and Delorme, M. (1980). Fuzzy Cartesian products and fuzzy mappings, in *Proc. Int. Congr. on Appl. Syst. Res. and Cybern.*, Acapulco, Mexico, Dec.

252 Bouchon, B. (1975). *Une Application Informationnelle de la Theorie des Questionnaires*, Keszthely, Hungary.

253 Bouchon, B. (1977). Application d'une mesure d'information a la construction de questionnaires flous, *Colloq.* CNRS, Les Dev. Recents de la Theorie de l'Inf. et leurs Appl., Cachan, France.

254 Bouchon, B. (1978). Longueur de questionnaire flous, Colloq. Int. sur la Theorie et les Appl. des Sous-Ensembles Flous, Marseille, France, Sept.

255 Bouchon, B. (1980). Information contenue dans un systeme d'evenements flous, Table Ronde du CNRS sur le Flou, Lyon, France, June.

256 Bouchon, B. (1980). Information transmitted by a system of fuzzy events, in *Proc. Int. Congr. on Appl. Syst. Res. and Cybern.*, Acapulco, Mexico, Dec.

257 Bouille, F. (1978). Fuzzy data processing with the hypergraph-based data structure, in *Proc. of the 1978 Int. Conf. on Cybern. and Soc.*, Vol. 2, Univ. Pierredet Marie Curie, France, pp. 1222–1227.

258 Boyd, J. P. (1978). Topai as models of fuzzy sets, in *Proc. 22nd Meet. Soc. for Gen. Syst. Res.* Washington D.C., pp. 443–450.

259 Braae, M., and Rutherford, D. A. (1978). Fuzzy relations in a control setting *Kybernetes*, **7**, 185–188.

260 Braae, M., and Rutherford, D. A. (1979). Theoretical and linguistic aspect of the fuzzy logic controller, *Automatica*, **15**, 553–557.

261 Braae, M., and Rutherford, D. A. (1979). Selection of parameters for a fuzzy logic controller, *Fuzzy Sets Syst.*, **2** (3), July.

262 Bremermann, H. J. (1971). Cybernetic functionals and fuzzy sets, in *IEEE Symp. on Syst., Man, and Cybern.*, 71C4SMC, pp. 248–253.

263 Bremermann, H. J. (1974). Complexity of automata, brains, and behaviour, in *Physics and Mathematics of the Nervous System*, Lecture Notes in Biomath., Vol. 4, (M. Conrad, Guttinger, and M. Dalcin, Eds.), Springer, pp. 304–331.

264 Bremont, J. (1975). Contribution a la reconnaissance automatique de la parole par les sous-ensembles flous, Doct. Thesis, Univ. de Nancy, Nancy, France.

265 Bremont, J., and Lamotte, M. (1974). Reconnaissance de formes, contribution a la reconnaissance automatique de la parole en temps reel par la consideration de sous-ensembles flous, *C.R. Acad. Sci. Paris*, **15**, July.

266 Bremont, J., and Lamotte, M. (1974). Reconnaissance globale de la parole en temps reel par calcul d'un indice de similarite floue, 5th Journ. d'Etud. du Groupe "Commun. Parlee," Orsay, France, May, 15–17.

267 Bremont, J., and Lamotte, M. (1974). Contribution a la reconnaissance automatique de la parole en temps reel par la consideration des sous-ensembles flous, *C.R. Acad. Sci. Paris*, July.

268 Brown, G. S. (1969). *Laws of Form*, Allen and Unwin, London.

269 Brown, J. G. (1969). Fuzzy sets on Boolean lattices, Rep. 1957, Ballistic Res. Lab. Aberdeen, MD, Jan.

270 Brown, J. G. (1971). A Note on fuzzy sets, *Inf. Control*, **18**, 32–39.

271 Brownell, H., and Caramazza, A. (1977). Categorizing, Rep., Johns Hopkins Univ., Baltimore.

272 Brunner, J. (1976). Uberlick zur Theorie und Anwendung von Fuzzy-Mengen, Vortrage aus dem Problemseminar Automaten und Algorithmentheorie, Weissig, Germany, April, pp. 3–15.

273 Brunner J., and Wechler, W. (1976). The behaviour of R-fuzzy automata, in *Lecture Notes in Computer Science*, (A. Mazurkiewicz, Ed.), Vol. 45, Springer-Verlag, Berlin, pp. 210–215.

274 Buckles, B. P. (1982). A fuzzy model for relational data bases. *Fuzzy Sets Syst.*, to be published.

275 Buckles, B. P. (1979). Fuzzy relational databases—A foundational framework and information-theoretic characterization, Intern. Memo., General Research Corp. Inc., Huntsville, AL, July.

276 Buckles, B. P., and Petry, F. E. (1982). A fuzzy representation of data for relational data-bases, *Int. J. Fuzzy Sets Syst.*, Nov. (submitted).

277 Bunge, M. C. (1966). Categories of sets valued functions, Ph.D. Thesis, Dept. Math., Univ. of California.

278 Buoncristiani, J. (1980). Probability on fuzzy sets, Ph.D. Thesis, Boston Univ., Boston, Jan.; in *Diss. Abstr. Int.*, **41/05-B**, 1790.

279 Buroni, E. (1976). Algebres relatives a une loi distributive, *C.R. Acad. Sci. Paris*, **276**, Ser. A, Feb. 5, 443–446.

280 Butnariu, D. (1978). N-Persons fuzzy games, *Fuzzy Sets Syst.*, **1**.

281 Butnariu, D. (1975). Fuzzy automata and fuzzy games, Semin. de Teoria Sistemelor, Dept. Econ. Cybern., Acad. of Economic Studies.

282 Butnariu, D. (1975). L-Fuzzy automata description of a neural model, in *Proc. 3rd Int. Congr. of Cybern. and Syst.*, Bucharest, Aug.

283 Butnariu, D. (1975). L-Fuzzy topologies, *Bull. Math. Soc. Sci. Math. RSR*, **19** (3), 227–236.

284 Butnariu, D. (1976). Fuzzy games and their minimax theorem, *St. Cerc. Math.*, **28** (2), 142–160 (in Romanian).

285 Butnariu, D. (1977). (L', L)-Fuzzy topological spaces, *Anals. Sci. Univ. "Al. I. Cuza"* —Iasi 23 Seria I, Fasc. 1.

286 Butnariu, D. (1977). Fuzzy games—A description of the concept, *Fuzzy Sets Syst.*, **1**, 181–192.

287 Butnariu, D. (1977). Three-person fuzzy games, *St. Cerc. Math.* (2), 1-10, (in Romanian).
288 Butnariu, D. (1978). A fixed point theorem and its applications to fuzzy games, *Rev. Roum. Math. Pures Appl.*
289 Butnariu, D. (1979). Solution concepts for N-persons fuzzy games, in *Advances in Fuzzy Set Theory and Applications*, (M. M. Gupta, R. K. Ragade, and R. R. Yager, Eds.), North-Holland, Amsterdam.
290 Butnariu, D. (1980). Fuzzy topologies—A local point of view, *Fuzzy Sets Syst.*
291 Butnariu, D. (1980). An existence theorem for possible solutions of a 2-persons fuzzy game, *Bull. Math. Soc. Sci. Math. R.S. Roum.*
292 Butnariu, D. (1980). Stability and shapley value for N-persons fuzzy games, *Fuzzy Sets Syst.*, **4** (1), 63-72.
293 Butnariu, D. (1980). Equilibrium points in fuzzy games, *St. Cerc. Math.*, (in Romanian).
294 Cai, M. (1980). Natural number recursion and recursive enumerable fuzzy subsets, News on 1980 Ann. Rep. Meet. of Beijing Working Group on Fuzzy Sets, Beijing, China.
295 Cao, H. (1980). Fuzzy division to processes of climate, News on 1980 Ann. Rep. Meet. of Beijing Working Group on Fuzzy Sets, Beijing, China.
296 Cao, H., and Chen, G. (1980). Fuzzy partition of weather processes, *Kexue Tongbao*, Vol. 25, No. 8, Weather and Climate Inst. Center Meteorol. Bureau, a list of lit. in China, The Univ. of Sci. and Technol. of China, Hofei, Anhwei, China.
297 Cao, Z. (1980). A new method of solution of fuzzy relation equations, News on 1980 Ann. Rep. Meet. of Beijing Working Group on Fuzzy Sets, Beijing, China.
298 Cao, Z. (1980). Convex combination of fuzzy number at interval, News on 1980 Ann. Rep. Meet. of Beijing Working Group on Fuzzy Sets, Beijing, China.
299 Cao, Z. (1980). Discussion on the solutions of fuzzy relation equations, *Collect. Papers Fuzzy Math. (Abstr.)*, Huazhong Inst. of Tech., Huazhong, China, April.
300 Cao, Z. (1980). Semilattice and semilattices's module(I), News on 1980 Ann. Rep. Meet. of Beijing Working Group on Fuzzy Sets, Beijing, China.
301 Capocelli, R. M., and De Luca, A. (1972). Measures of uncertainty in the contex of fuzzy sets theory, in *Atti del ile Congr. Nat. di Cibern. di Casciana Terme*, Pisa, Italy.
302 Capocelli, R. M., and De Luca, A. (1973). Fuzzy sets and decision heory, *Inf. Control*, **23**, 446-473.
303 Cargile, J. (1969). The Sorites paradox, *Brit. J. Philos. Sci.*, **20**, 193-202.
304 Carlsson, C. (1976). An approach to adaptive multigoal control using fuzzy automata, School of Econ., Abo Swedish Univ., Abo; *Rep. 2nd Eur. Cong. on Oper. Res.*, Stockholm, Dec.
305 Carlsson, C. (1978). A system of problems and how to deal with it, Rep., Abo Swedish Univ., Abo.
306 Carlsson, C. (1978). Fuzzy automata as cybernetic control functions, Inst. of Manage. Sci., School of Econ., Abo Swedish Univ., Abo, Finland.
307 Carlsson, C. (1980). Fuzzy sets and management science methodology, *EMCSR—PCSR*, **8**, to be published.
308 Carlsson, C. (1980). Fuzzy systems—An approach to more realistic systems modelling, in *Proc. Int. Congr. on Appl. Syst. Res. and Cybern.*, Acapulco, Mexico, Dec.
309 Carlsson, C. (1980). Some applications of fuzzy sets to economics problems, Table Ronde du CNRS sur le Flou, Lyon, France, June.
310 Carlsson, C. (1980). Tackling a MCDC-problem with the help of some results from fuzzy set theory, 4th Eur. Congr. on Oper. Res. July.

Key References in Fuzzy Pattern Recognition 227

311 Carlstrom, I. F. (1975). Truth and entailment for a vague quantifier, *Synthese*, **30**, 461–495.

312 Carlucci, D., and Donati, F. (1977). Fuzzy Cluster of demand within a regional service system, in *Fuzzy Automata and Decision Processes*, (M. M. Gupta, G. N. Saridis, and B. R. Gaines, Eds.), North-Holland, Amsterdam, pp. 379–386.

313 Carnap, R. (1947). *Meaning and Necessity*, Univ. of Chicago Press, Chicago.

314 Carnap, R. (1950). *Logical Foundations of Probability*, Univ. of Chicago Press, Chicago.

315 Carnap, R. (1963). The philosopher replies, in *The Philosophy of R. Carnap*, The Library of Living Philosophers, Vol. 11, (P. A. Schilpp, Ed.), Open Court, La Salle, IL.

316 Carnap, R. (1964). *The Logical Syntax of Language*, Routledge and Kegan Paul, London (1st ed., 1937).

317 Carson, M. T. (1980). Four fuzzy categories—New norms for typicality and production frequency for children and adults—The effect of those variables on release from PI, Ph.D. Thesis, Rutgers Univ., The State Univ. of New Jersey, New Brunswick, in *Diss. Abstr. Int.*, **41/04-B**, 1534.

318 Carter, G. A., and Hague, M. J. (1973). *Fuzzy control of raw mix permeability at a sinter plant*, in *Discrete Systems and Fuzzy Reasoning*, (E. H. Mamdani, and B. R. Gaines, Eds.), EES-MMS-DSFR-73, Queen Mary College, Univ. of London, London (Workshop Proc.).

319 Castonguay, C. (1972). *Meaning of Existance in Mathematics*, Libirary of Exact Philos., Vol. 9, Springer-Verlag, Vienna.

320 Cattaneo, G. (1980). Fuzzy events and fuzzy logics in classical information systems, *J. Math. Anal. Appl.* **75** (2), 523–548.

321 Cavallo, R. E., and Klir, G. J. (1980). Reconstruction of fuzzy behavior systems, in Proc. *Int. Congr. on Appl. Syst. Res. and Cybern.*, Acapulco, Mexico, Dec.

322 Cayrol, M. (1979). Sous-ensembles flous et intelligence artificielle, *Busefal* (1), Automne–Hiver 79–80, pp. 52–55; Lang. et Syst. Inf., Univ. Paul Sabatier, Toulouse, France.

323 Cayrol, M., Farreny, H., and Prade, H. (1979). Fuzzy production rules directed by fuzzy pattern matching, LSI, Univ. Paul Sabatier, Toulouse, France.

324 Cayrol, M., Farreny, H., and Prade, H. (1979). Vers l'utilisation des sous-ensembles flous en intelligence artificielle, Doc. LSI, Univ. Paul Sabatier, Toulouse, France, Feb.

325 Cayrol, M., Farreny, H., and Prade, H. (1979). Fuzzy instructions—Execution/production in "Towards the use of fuzzy set theory in AI," LSI. Tech Rep., Toulouse, France, Feb.

326 Cayrol, M., Farreny, H., and Prade, H. (1980). Fuzzy reasoning based on multivalent logics in the framework of production-rules systems, in *Proc. 10th IEEE Int. Symp. on Multivalent Logic*, Evanston, IL, June.

327 Cayrol, M., Farreny, H., and Prade, H. (1980). An advanced pattern-matching method taking into account the uncertainty in meaning, Int. Conf. Artif. Intell. and Inf.-Control Syst. of Robots, Czechoslovakia, June–July.

328 Cayrol, M., Farreny, H., and Prade, H. (1980). Possibility and necessity in a pattern-matching process, IXth Int. Cong. Cybern., Namur, Belgium, Sept.

329 Cech, E. (1968). *Topological Spaces in Topological Papers of E. Cech*, Academia, Prague, pp. 436–472 (Translation of 1937 paper in Czech).

330 Cervin, V. B. (1981). Fuzzy manipulations on behavior modification, CORS-TIMS-ORSA Joint Nat. Meet., Toronto, Canada, May.

331 Chamas, S., Pelech, A., and Radosimski, E. (1975). Toward a concept of a R.-D. project fuzzy model, Konf. Prace Naukowe Osronka Badan Prognostycznych Politechniki Wroclawskiej, No. 4, Wroclaw, Poland.

332 Chanas, S., and Kamburowski, J. (1979). The use of fuzzy variables in P.E.R.T., Memoire Kommunikat No. 469, Instytut Organizacz i Zarzadzania Politechniki Wroclawskiej, Wroclaw, Poland.

333 Chanas, S., and Kokalanow, M. (1977). An assignment problem with fuzzy effectiveness estimates, Komunikat 253, Wroclaw, Poland.

334 Chang, C. C. (1958). Algebraic analyses of many valued logics, *Trans. Amer. Math. Soc.*, **88**, 467–490.

335 Chang, C. C. (1958). Proof of an axiom of Lukasiewicz, *Trans. Amer. Math. Soc.*, **87**, 55–56.

336 Chang, C. C. (1959). A new proof of the completeness of the Lukasiewicz axioms, *Trans. Amer. Math. Soc.*, **93**, 74–80.

337 Chang, C. C. (1963). Logic with positive and negative truth values, *Acta Philos. Fenn.*, **16**, 19–39.

338 Chang, C. C. (1963). The axiom of comprehension in infinite valued logic, *Math. Scand.*, **13**, 9–30.

339 Chang, C. C. (1964). Infinite valued ogic as a basis for set theory, in *Proc. 1964 Int. Cong. for Logic Methodol. and Philos. of Sci.* (Y. Bar-Hillel, Ed.), North-Holland, Amsterdam, pp. 93–100.

340 Chang, C. L. (1967). Fuzzy sets and pattern recognition, Ph.D. Thesis, Univ. of California, Berkeley, *Diss. Abstr. Int.*, **29/01-B**, 177.

341 Chang, C. L. (1968). Fuzzy topological spaces, *J. Math. Anal. Appl.*, **24**, 182–190.

342 Chang, C. L. (1971). Fuzzy algebra, fuzzy functions and their application to function approximation, Div. of Comput. Res. and Technol., Nat. Inst. of Health, Bethesda, MD.

343 Chang, C. L. (1975). Interpretation and execution of fuzzy programs, in *Fuzzy Sets and their Applications to Cognitive and Decision Processes* (L. A. Zadeh, K. S. Fu, K. Tanaka, and M. Shimura, Eds.), Academic Press, New York, pp. 191–218.

344 Chang, C. L., and Lee, R. C. T. (1973). *Symbolic Logic and Mechanical Theorem Proving*, Academic Press, New York.

345 Chang, R. L. P. (1976). Application of fuzzy decision techniques to pattern recognition and curve fitting, Ph.D. Thesis, Dept. of EECS, Princeton Univ., in *Diss. Abstr. Int.*, **37/09-8**, 4558.

346 Chang, R. L. P., and Pavlidis, T. (1976). Fuzzy decision trees, *IEEE Conf. on Syst., Man, Cybern.*, Washington, D.C.

347 Chang, R. L. P., and Pavlidis, T. (1977). Fuzzy decision tree algorithms, *IEEE Trans. Syst., Man, Cybern.*, **SMC-7** 28–34.

348 Chang, R. L. P., and Pavlidis, T. (1979). Applications of fuzzy sets, *Curve Fitting, Fuzzy Sets Syst.*, **2** (1), Jan., 67–74.

349 Chang, S. K. (1971). Automated interpretation and editing of fuzzy line drawings, *SJCC*, **38**, 393–399.

350 Chang, S. K. (1971). Fuzzy programs—Theory and applications, in *Proc. of Polytech. Inst. of Brooklyn Symp. on Comput. and Automata*, Brooklyn, NY, p. 147.

351 Chang, S. K. (1971). Picture processing grammar and its applications, *Inf. Sci.*, 121–148.

352 Chang, S. K. (1972). On the execution of fuzzy programs using finite state machines, *IEEE Trans. Comput.*, **C-21**, 241–253.

353 Chang, S. K. (1980). Fuzzy linguistic filters, in *Proc. Int. Cong. on Appl. Syst. Res. and Cybern.*, Acapulco, Mexico, Dec.

354 Chang, S. K., and Ke, J. S. (1976). Database skeleton and its application to fuzzy query translation, Dept. of Inf. Eng., Univ. of Illinois, Chicago.

355 Chang, S. K., and Ke, J. S. (1979). Translation of fuzzy queries for relational database system, *IEEE Trans. Pattern Anal. Mach. Intell.*, **1** (3), 281–294.

356 Chang, S. S. L. (1969). Control systems, Proj. Noaf 9749, Dept. of Elec. Sci., State Univ. of New York, Stony Brook.

357 Chang, S. S. L. (1969). Fuzzy dynamic programming and approximate optimization of partially known systems, in *Proc. 2nd Hawaii Int. Conf. on Syst. Sci.*, Honolulu, p. 123.

358 Chang, S. S. L. (1969). Fuzzy dynamic programming and the decision making process, in *Proc. 3rd Princeton Conf. on Inf. Sci. and Syst.*, Princeton, NJ, pp. 200–203.

359 Chang, S. S. L. (1972). Fuzzy mathematics, man, and his environment, *IEEE Trans. Syst., Man, Cybern.*, **SMC-2**, 93.

360 Chang, S. S. L. (1972). On fuzzy mathematics and control, *IEEE Trans. Syst., Man, Cybern.*, **SMC-2**, 30–34.

361 Chang, S. S. L. (1975). On risk and decision making in a fuzzy environment, in *Fuzzy Sets and their Applications to Cognitive and Decision Processes* (L. A. Zadeh, K. S. Fu, K. Tanaka, and M. Shimura, Eds.), Academic Press, New York, pp. 219–226.

362 Chang, S. S. L. (1977). Application of fuzzy set theory to economics, *Kybernetes*, **6**, 203–207.

363 Chang, S. S. L. (1977). On fuzzy algorithm and mapping, in *Fuzzy Automata and Decision Processes*, (M. M. Gupta, G. N. Saridis, and B. R. Gaines, Eds.), North-Holland, Amsterdam, pp. 191–196.

364 Chang, S. S. L. (1978). Decision implications of fuzzy set theory, *Proc. 1978 JACC*, **3**, 362.

365 Chang, S. S. L. (1978). On a fuzzy algorithm and its implementation, *IEEE Trans. Syst., Man, Cybern.*, **SMC-8** (1), 31.

366 Chang, S. S. L., and Barry, P. E. (1972). Optimal control of systems with uncertain parameters, 5th IFAC Congr., Paris.

367 Chang, S. S. L., and Zadeh, L. A. (1972). On fuzzy mapping and control, *IEEE Trans. Syst., Man, Cybern.*, **SMC-2**, 30–34.

368 Chapin, E. W. (1971). An axiomatization of the set theory of Zadeh, *Not. Amer. Math. Soc.*, **687-02-4**, 753.

369 Chapin, E. W. (1974). Set-valued set theory—Part 1, *Notre Dame J. Formal Logic*, **15**, 619–634.

370 Chapin, E. W. (1975). Set-valued set theory—Part 2, *Notre Dame J. Formal Logic*, **16**, 255–267.

371 Chaudhuri, B. B., and Majumder, D. D. (1978). A generalized fuzzy approach in pattern recognition problems, in *Proc. Forum Interdisciplinary Symp. in Math.*, Rajsthan Univ., Jaipur, India, July.

372 Chaudhuri, B. B., and Majumder, D. D. (1980). Fuzzy sets and possibility theory in reliability studies of man-machine systems, Fuzzy Sets Theory and Appl. to Policy Anal. and Inf. Syst. (P. P. Wang and S. K. Chang, Eds.) Plenum, New York, pp. 267–274.

373 Che, Y. (1980). A tentative exploration for the fuzziness statistics problem, Collect. Pap. on Fuzzy Math. (Abstr.), Huazhong Inst. of Tech., Huazhong, China, Apr.

374 Che, Y. (1980). Fuzzily weighted digraph and its fuzzy stability in the pulse process, Collect. Pap. on Fuzzy Math. (Abstr.), Huazhong Inst. of Tech., Huazhong, China, Apr.

375 Che, Y. (1980). The fuzzy assessment method for the environmental quality, Collect. Pap. on Fuzzy Math. (Abstr.), Huazhong Inst. of Tech., Huazhong, China, Apr.

376 Che, Y. (1980). The fuzzy stability of a fuzzily weighted digraph in the pulse process, in Sel. Pap. on Fuzzy Subsets, Beijing Normal Univ., Beijing, China, Mar.

377 Chen, C. (1974). Realizability of communication nets—An application of the Zadeh criterion, *IEEE Trans. Circuits Syst.*, **CAS-21**, 150–151.

378 Chen, G. (1980). An approach to fuzzy controller algorithms, *Inf. Control*, **9** (3); A List of Lit. in China, The Univ. of Sci. and Tech. of China, Hofei, Anhwei, China.

379 Chen, G. (1980). An example of climate analysis by fuzzy composite evaluation, News on 1980 Ann. Rep. Meet. of Beijing Working Group on Fuzzy Sets, Beijing, China.

380 Chen, G. (1980). Fuzzy sets and its applications, *J. Univ. Sci. Tech. China*, **3**.

381 Chen, G., and Cao, H. (1980). Application of fuzzy cluster to weather forecasting, Collect. Pap. on Fuzzy Math. (Abstr.), Huazhong Inst. of Tech., Huazhong, China, Apr.

382 Chen, G., and Cao, H. (1980). The application of fuzzy sets theory in climatic forecasting, Collect. Pap. on Fuzzy Math. (Abstr.), Huazhong Inst. of Tech., Huazhong, China, Apr.

383 Chen, H. (1980). Comment on determinate membership functions, *Rep. Shanghai Railway Inst.*, A List of Lit. in China, The Univ. of Sci. and Tech. of China, Hofei, Anhwei, China.

384 Chen, H. (1980). Fuzzy mathematics and investigations of dialect, News on 1980 Ann. Rep. Meet. of Beijing Working Group on Fuzzy Sets, Beijing, China.

385 Chen, L. (1980). Fuzzy measure space, Collect. Pap. on Fuzzy Math. (Abstr.), Huazhong Inst. of Tech., Huazhong, China, Apr.

386 Chen, S. C., and Shimura, M. (1974). On-line recognition of hand printed characters using fuzzy logic, Tech. Rep. of Pattern Recogn. and Learning of IECE, PRL74-65.

387 Chen, T., and Wu, X. (1980). Transition logic, fuzzy logic and pansystem logic, *J. Huazhong Inst. Tech.*, **2**, A List of Lit. in China, The Univ. of Sci. and Tech. of China, Hofei, Anhwei, China.

388 Chen, Y. (1980). A new method for analysis of fuzzy controller, *Rep. Beijing Normal Univ.*, **11**, A List of Lit. in China, The Univ. of Sci. andTech. of China, Hofei, Anhwei, China.

389 Chen, Y. (1980). Analysis of controller by using fuzzy network of coordinates, News on 1980 Ann. Rep. Meet. of Beijing Working Group on Fuzzy Sets, Beijing, China.

390 Chen, Y. (1980). Fuzzy categories with additive measure, *Rep. Beijing Normal Univ.*, **11**, A List of Lit. in China, The Univ. of Sci. and Tech. of China, Hofei, Anhwei, China.

391 Cheng, L. (1980). Fuzzy group, *Rep. Shanghai Railway Inst.*, A List of Lit. in China, The Univ. of Sci. and Tech. of China, Hofei, Anhwei, China.

392 Cheng, W. M., Wu, S. C., and Tsuei, T. H. (1980). The intersection of fuzzy subsets and the robustness of fuzzy control, Int. Conf. on Cybern. and Soc., Cambridge, MA, Oct.

393 Chiara, C. S. (1973). *Ontology and the Vicious Circle Principle*, Cornell Univ. Press, Ithaca, New York.

394 Chichinadze, V. (1977). Some problems of organization, in *Proc. of the Symp. on Fuzzy Set Theory and Appl., IEEE Conf. on Decis. and Control*, New Orleans.

395 Chilausky, R., Jacobsen, B., and Michalski, R. S. (1976). An application of variable-valued logic to inductive learning of plant disease diagnostic rules, in *Proc. 6th Int. Symp. Multiple-Valued Logic*, IEEE 76CH1111-4C, pp. 233–240.

396 Chittenden, E. W. (1941). On the reduction of topological functions, in *Lectures in Topology*, (R. L. Wilder, and W. L. Ayres, Eds.), Univ. of Michigan Press, Ann Arbor, pp. 267–285.

397 Chomsky, N., and Halle, M. (1965). Some controversial questions in phonological theory, *J. Linguist.*, **1**, 97–138.

398 Chorayan, O. G. (1979). *Fuzzy Algorithms in Thinking Process*, Rostov Univ. Press, Rostov, U.S.S.R., (in Russian).

399 Chorayan, O. G. (1979). Identifying elements of the probabilistic neuronal ensemble from the standpoint of fuzzy sets theory, *Physiol. J.*, **V**, 65 (in Russian).

400 Chorayan, O. G. (1980). Informational approach to analysis of fuzzy decision-making logic, *Bull. Stud. Exch. Fuzziness Appl.*, Printemps.

401 Choroyan, O. G. (1980). Contribution to the description of the probabilistic neuron ensemble in terms of fuzzy sets theory, *Bull. Stud. Exch. Fuzziness Appl.*, Ete.

402 Christopher, F. T. (1977). Quotient fuzzy topology and local compactness, *J. Math. Anal. Appl.*, **53** (3).

403 Chu, A. T. W., Kalaba, R. E., Spingarn, K. (1979). A comparison of two methods for determining the weights of belonging to fuzzy sets, *J. Optimization Theory Appl.*, **27** (4).

404 Chulin, L., and Hongzu, Y. (1980). Some problems concerning the resolution of fuzzy equation, *J. Huazhong Inst. Tech.*, Special Issue on Fuzzy Math. (2).

405 Chun, Y. (1980). Entropy and amount of information of fuzzy events, Collect. Pap. on Fuzzy Math. (Abstr.), Huazhong Inst. of Tech., Huazhong, China, Apr.

406 Chun, Y. (1980). On computng entropy and amount of information of linguistic variable, Collect. Pap. Fuzzy Math. (Abstr.), Huazhong Inst. of Tech., Huazhong, China, April.

407 Chytil, M. (1969). On constituting of semantical models for Guha-methods, *Cesk. Fysiol.*, **18**, 43–147 (in Czech).

408 Cignoli, R. (1972). Estudio algebraice de logicas polivalentes, Algebra de moisil de orden *N*., Doc. Thesis, Univ. Nac. de Sur-Bahia-Blanco, Bahia Blanco, Argentina.

409 Clark, C. M., Ben-David, and M., Kandel, A. (1981). On the enumeration of distinct fuzzy switching functions, *Fuzzy Sets Syst.*, **5** (1), Jan., 69–82.

410 Clarke, R. D. (1980). Fuzzy preference method for defining group preferences, Thesis, Nav. Postgraduate School, Monterey, CA, Mar.

411 Cleave, J. P. (1970). The notion of vality in logical systems with inexact predicates, *Brit. J. Philos. Sci.*, **21**, 269–274.

412 Cleave, J. P. (1974). The notion of logical consequence in the logic of inexact predicates, *Z. Math. Logik Grundlagen Math.*, **20**, 307–324.

413 Cleave, J. P. (1976). Quasi-Boolean algebras, empirical continuity and three-valued logic, *Z. Math. Logik Grundlagen Math.*

414 Clements, D. (1977). Evaluation of computer security using a fuzzy rating language, ERL Rep. M 77/41, Electron. Res. Lab., Univ. of California, Berkeley.

415 Clements, D. P. (1977). Fuzzy ratings for computer security evaluation, Ph.D. Thesis, Univ. of California, Berkeley, in *Diss. Abstr. Int.*, **39/02-B**, 846.

416 Cohen, L. J. (1975). Probability—The one and the many, *Proc. Brit. Acad.*, **61**, 3–28.

417 Cohen, P. J. (1967). Non-Cantonian set theory, *Sci. Amer.*, Dec., 104–116.

418 Cole, P., and Morgan, J. L. (Eds.) (1975). *Syntax and Semantics*, Vol. 3, Academic Press, New York.
419 Conche, B. (1973). A method of classification based on the use of a fuzzy automaton, Univ. of Paris—Dauphine, Paris.
420 Conche, B. (1973). Application des concepts flous a la psychologie de la perception et des etats de vigilance, Memo., Univ. of Paris—Dauphine, Paris, Dec.
421 Conche, B. (1973). Elements de une methode de classification par utilisation de un automate flou, JEEFLN, Univ. of Paris—Dauphine, Paris.
422 Conche, B. (1974). Apprentissage heuristique en selection de projets de recherche, Une application des concepts flous a la classification automatique, Ph.D. Thesis, Univ. Paris—Dauphine, Paris.
423 Conche, B. (1975). La classification dans le cas d'informations incompletes ou non explicites, Semin. "Contribution des Systemes Flous a l'Autom., Centre d'Autom. de Lille, Lille, France, June.
424 Conche, B., and Courant, P. (1972). Application des concepts flous aux problemes multi-criteres, Semi. Bernard Roy, Univ. of Paris—Dauphine, Paris.
425 Conche, B., Jouault, J. P., and Luan, P. M. (1973). Application des concepts flous a la programmation en languages quasi-naturels, Semin. Bernard Roy, Univ. of Paris—Dauphine, Paris.
426 Conrad, F. (1980). Fuzzy topological concepts, *J. Math. Anal. Appl.* **74** (2).
427 Cools, M. (1973). La semantique dans les processus didactiques, Imago Disc. Pap., Univ. Catholique de Louvain, Louvain, Belgium, Nov.
428 Cools, M. (1975). Vers une evaluation en problem-solving applique a l'E.A.O. par la theorie des sous-ensembles flous, Imago Disc. Pap. No. 102, Univ. Catholique de Louvain, Louvain, Belgium, Sept.
429 Cools, M. (1976). Recherche d'une evaluation de strategie de resolution de problemes formalisables par la theorie des sous-ensembles flous, IDP 110, Centre Imago, Univ. Catholique de Louvain, Louvain, Belgium.
430 Cools, M., and Peteau, M. (1973). Stim 5—Un programme de stimulation inventive utilisant la theorie des sous-ensembles flous, Imago Disc. Pap., Univ. Catholique de Louvain, Louvain, Belgium.
431 Cools, M., and Peteau, M. (1974). Un programme de simulation inventive, Stim 5, Rev. Fr. de RO, AIRO, Nov.
432 Coppo, M., and Saitla, L. (1976). Semantic support for speech understanding based on fuzzy relation, in *Proc. Int. Conf. on Cybern. and Soc.*, Washington, D.C., pp. 520–524.
433 Coulon, J. (1980). Engendrement des sous-espaces vectoriels flous et fermetures de moore floues, *Bull. Stud. Exch. Fuzziness Appl.*, Ete.
434 Coulon, J., and Coulon, J. L. (1979). Structures vectorielles et affines floues, Convexite, Univ. de Lyon, Claude Bernard, Lyon, France.
435 Coulon, J. J. L. (1978). Notion de topologie floue, Semin. "Math. Floue," Univ. de Lyon, Lyon, France.
436 Coulon, J. J. L. (1978). Notions de limite et d'adherence dans les H-topologies floues. Journ. de Biomath. et d'Appl. d'Inf. Med., Marseille, France.
437 Coulon, J. J. L. (1979). Fuzzy Boolean algebras, in *Proc. of the Int. Semin. on Fuzzy Set Theory*, Johannes Kepler Univ., Linz, Austria.
438 Coulon, J. J. L. (1979). Structures vectorielles floues. Semin. "Math. Floue," Univ. de Lyon, Lyon, France.

439 Coulon, J. J. L. (1979). The cuts of a fuzzy subset, and their use, in *Proc. of the Int. Semin. on Fuzzy Set Theory*, Johannes Kepler Univ., Linz, Austria.

440 Creswell, M. J. (1973). *Logics and Languages*, Methuen, London.

441 Crocq, L., Sanchez, E., Fondaria, J., and Gouvernet, J. (1980). Psychiatric diagnosis assistance, in *Proc. Int. Congr. on Appl. Syst. Res. and Cybern.*, Acapulco, Mexico, Dec.

442 Curry, H. B. (1942). The inconsistency of certain formal logics, *J. Symb. Logic*, **7**, 115–117.

443 Dalcin, M. (1975). Fuzzy-state automata, their stability and fault-tolerance, *Int. J. Comput. Inf. Sci.*, **4**, 63–80.

444 Dalcin, M. (1975). Modification tolerance of fuzzy-state automata, *Int. J. Comput. Inf. Sci.*, **4**, 81–93.

445 Dale, A. I. (1980). Probability, vague statements, and fuzzy sets. *Philos. of Sci.*, **47**.

446 Damerau, F. J. (1975). On fuzzy adjectives, RC5340, IBM Res. Lab., Yorktown Heights, NY.

447 Danes, F. (1966). The relation of centre and periphery as a language universal, *Trav. Linguist. Prague*, **2**, 9–21.

448 Danes, F., and Vachek, J. (1964). Prague studies in structural grammar today, *Trav. Linguist. Prague*, **1**, 21–31.

449 Dang, J. (1980). Study for fuzzy decision-making systems, *J. Huazhong Inst. Tech.*, **2**, A List of Lit. in China, The Univ. of Sci. and Tech. of China, Hofei, Anhwei, China.

450 Danielsson, S. (1967). Modal logic based on probability theory, *Theoria*, **33**, 189–197.

451 Daumerie, J. (1979). La semantique dans l'enseignement special, Rap. IDP 300, Lab. Prof. A Jones, Univ. Catholique de Louvain, Louvain, Belgium, Feb.

452 Davio, M., and Thayse, A. (1973). Representation of fuzzy functions, Philips Res. Rep., **28**, 93–106.

453 Defays, D. (1975). K-Recouvrements et sous-ensembles flous, *Bull. Soc. Roy. Sci. Liege*, **44**.

454 Defays, D. (1978). Les methodes d'analyse hierarchique, Un essai de classification, *Rev. Belg. Stat. Inf. RO*, **18** (1), Mar., 19–45.

455 Defays, D. (1975). Recherche des ultrametriques a distance minimale d'une similarite donnee, *Bull. Soc. Roy. Sci. Liege*, **44E**, 330–343.

456 Defays, D. (1975). Relations floues et analyse hierarchique des questionnaires, *Math. Sci. Hum.* (55), 45–60.

457 Defays, D. (1975). Ultrametrique et relations floues, *Bull. Soc. Roy. Sci. Liege*, **44E** (1–2), 104–118.

458 Defays, D. (1977). Analyse hierarchique et theorie des sous-ensembles flous, Doc. Thesis, Univ. of Liege, Liege, France.

459 Defays, D. (1978). Analyse hierarchique des preferences et generalisation de la transitivite, *Math. Sci. Hum.*, **61**, Nov.

460 Defays, D. (1978). Application de la theorie et applications des sous-ensembles flous a l'analyse des donnees, Colloq. Int. Theorie et Appl. des Sous-Ensembles Flous, Journ. Biomath. et Inf. Med., Marseille, France, Sept.

461 Defays, D. (1978). Ultrametriques et relations floues, Doc. Inst. de Psych. des Sci. de l'Educ., Liege, France.

462 De Finetti, B. (1972). *Probability, Induction and Statistics*, John Wiley, London.

463 Deherder, R. (1975). Ensembles flous, *Rev. Belg. Stat., Inf. RO*, **15** (4), Dec., 1–4.

464 De Kerf, J. (1974). Vage Verzamelingen, *Ingenieurstijdingen*, **23E**, Jaargang, 581–589.
465 De Kerf, J. (1974). Vage Verzamelingen, *Omega* (*Ver. voor Wis-en Natuur-Kundigen Lovanienses*), **2**, 2–18.
466 De Kerf, J. (1975). A bibliography on fuzzy sets, *J. Comput. Appl. Math.*, **1**, 205–212.
467 Delgado, M. (1977). On a definition of fuzzy-function, Doc. Dept. de Estat. Math., Univ. de Granada, Granada, Spain.
468 Deloche, R. (1975). Theorie des sous-ensembles floues et classification en analyse enonomique spatiale, Doc. Inst. de Math. Econ., Fac. Sci. Econ., et de Gestion, Dijon, France, July.
469 Deloche, R. (1977). Theorie des sous-ensembles flous et classification en analyse spatiale, Doc. de Trav. No. 11, Ime, Rev. Econ., Polit., No. 3.
470 Deluca, A. (1971). Fuzzy sets e theoria degli algoritmi, Atti del Congr. de Cibern., Casciano Terme, Italy, Oct.
471 Deluca, A., and Capocelli, R. M. (1973). Fuzzy sets and decision theory, *Inf. Control*, **23**, 446–473.
472 Deluca, A., and Termina, S. (1977). Measures of ambiguity in the analysis of complex systems, in *Mathematical Foundations of Computer Science*, Springer-Verlag, New York, 382—389.
473 Deluca, A., and Termini, S. (1971). Algorithmic aspects in complex systems analysis, *Scientia*, **106**, 659–671.
474 Deluca, A., and Termini, S. (1972). A definition of a nonprobabilistic entropy in the setting of fuzzy sets, *Inf. Control*, **20**, 301–312.
475 Deluca, A., and Termini, S. (1972). Algebraic properties of fuzzy sets, *J. Math. Anal. Appl.*, **40**, 373–386.
476 Deluca, A., and Termini, S. (1974). Entropy of *L*-fuzzy sets, *Inf. Control*, **24**, 55–73.
477 Deluca, A., and Termini, S. (1976). Una condizione necesseria et sufficiente per la convergenza dell'entropia logarithmica di un "fuzzy sets," IV Congr. Nax. di Cibern. e Biol., Siena, Italy, Oct.
478 Deluca, A., and Termini, S. (1977). On the convergence of entropy measures of a fuzzy set, *Kybernetes*, **6**, 219–227.
479 Deluca, A., and Termini, S. (1979). Entropy and energy measures of a fuzzy set, *Advances in Fuzzy Set Theory and Applications*, (M. M. Gupta, R. K. Ragade, and R. R. Yager, Eds.), North-Holland, Amsterdam.
480 Demant, B. (1971). Fuzzy Retrieval Strukturen, *Angew. Inf., Appl. Inf.*, **13**, 500–502.
481 De Mori, R., and Laface, P. (1980). Use of fuzzy algorithms for phonetic and phonemic labeling of continuous speech, *IEEE Trans. Pattern Anal. Mach. Intell.* **PAMI-2** (2), Mar.
482 De Mori, R., and Saitta, L. (1978). Automatic learning of fuzzy relations, *Inf. Sci.*
483 De Mori, R., and Saitta, L. (1979). Scheduling of processes in a speach understanding system based on approximate reasoning, in *Int. Joint Conf. on Artif. Intell.*, pp. 204–207.
484 De Mori, R., and Saitta, L. (1980). Automatic learning of fuzzy naming relations over finite languages, *Inf. Sci.*, **20**.
485 De Mori, R., and Torasso, P. (1976). Lexical classification in a speech understanding system using fuzzy relations, in *Proc. IEEE ASSP Conf.*, Philadelphia, PA, pp. 565–568.
486 Deng, Z. (1980). Convergence and sequential compactness in fuzzy topological spaces, Collect. Pap. on Fuzzy Math. (Abstr.), Huazhong Inst. of Tech., Huazhong, China, Apr.

487 Deng, Z. (1980). Fuzzy metric spaces, Collect. Pap. on Fuzzy Math. (Abstr.), Huazhong Inst. of Tech., Huazhong, China, Apr.

488 Deng, Z. (1980). Neighborhoods in fuzzy topological spaces, Collect. Pap. on Fuzzy Math. (Abstr.), Huazhong Inst. of Tech., Huazhong, China, Apr.

489 Depalma, G. F., and Yau, S. S. (1975). Fractionally fuzzy grammars with application to pattern recognition, in *Fuzzy Sets and their Applications to Cognitive and Decision Processes* (L. A. Zadeh, K. S. Fu, K. Tanaka, and M. Shimura, Eds.), Academic Press, New York, pp. 329–351.

490 Dhar, S. B. (1979). Power system long-range decision analysis under fuzzy environment, *IEEE Trans. Power Appar. Syst.*, **PAS-98** (2), Mar.-Apr.

491 Diamond, P. (1975). Fuzzy chaos, Dept. of Math., Univ. of Queensland, Brisbane, Australia.

492 Diarra, N. (1975). A propos des ensembles flous, Ph.D. Thesis, Centre Pedagogique Superieur de l'Ecole Normale Superieur, Bamako, Tunisia, Oct.

493 Diday, E. (1972). Optimisation en classification automatique et reconnaissance des formes, Notes Sci. IRIA, Bull. No. 12, May-June.

494 Diday, E. (1973). The dynamic clusters method and optimization in non-hierarchical clustering, in *5th Conf. on Optim. Techniques*, Pt. I. (R. Conti and A. Ruberti, Eds.), Springer-Verlag, Berlin.

495 Dienes, Z. P. (1949). On an implication function in many-valued systems of logic, *J. Symb. Logic*, **14**, 95–97.

496 Diffenbach, J., Renda, A., and Fiksel, J. R. (1979). Feasibility and usefulness of possibility analysis in a policy assessment of tme-of-day electricity rate structures, 18th IEEE Conf. on Decis. and Control, Fort Lauderdale, FL, Dec.

497 Dijkman, J. G., and Lowen, R. (1976). Fuzzy relations on countable sets, Tech. Highschool, Delft, Holland and Vrige Univ., Brussels, Belgium.

498 Dill, A. M., and Emptoz, H. (1979). Entropy and indetermination measure in the setting of fuzzy sets theory, in *Proc. IEEE Int. Symp. on Inf. Theory*, Grignano, Italy, June.

499 Dill, A. M., Emptoz, J., and Fages, R. (1980). Dispersion, Mesures floues et classification automatique, Table Ronde du CNRS sur le Flou, Lyon, France, June.

500 Dillman, I. (1973). *Introduction and Deduction*, Basil Blackwell, Oxford, UK.

501 Dimitrov, V. (1976). Learning decision-making with a fuzzy automata, in *Computer-Oriented Learning Processes* (J. C. Simon, Ed.), Noordhoff, pp. 149–154.

502 Dimitrov, V., and Cuntchev, O. (1977). Efficient fuzzy governing humanistic systems by fuzzy instructions, in *Modern Trends in Cybernetic and Systems*, (J. Rose and C. Bilcin, Eds.), Vol. 2, pp. 125–130, Springer-Verlag, Berlin.

503 Dimitrov, V., Wechler, W., and Barnev, P. (1974). Optimal fuzzy control of humanistic systems, Inst. of Math. and Mech., Bulgaria Acad. of Sci., Sofia.

504 Dimitrov, V. D. (1970). G.M.D.H. algorithms on fuzzy sets of Zadeh, *Sov. Autom. Control* (3), 40–45.

505 Dimitrov, V. D. (1975). Efficient governing humanistic systems by fuzzy instructions, 3rd Int. Congr. of Gen. Syst. and Cybern., Bucharest.

506 Dimitrov, V. D. (1977). Social choice and self-organization under fuzzy management, *Kybernetes*, **6**, 153.

507 Dimitrov, V. D., and Driankova, L. D. (1977). Programme system for social choice under fuzzy managing, in *Information Processing 77* (B. Gilchrist, Ed.), North-Holland, Amsterdam.

508 Dimitrov, V. D., Wechler, W., Drjankov, D., and Petrov, A. (1975). Computer Execution of fuzzy algorithms, in *Proc. Conf. on Appls. Math. Models and Comput. in Linguist.*, Varna, Bulgaria, May (in Russian).

509 Di Nola, A., and Ventre, A. G. S. (1978). On some convergent sequences of (N-$)-valued fuzzy sets, Doc. Inst. Math., Fac. di Archit., Univ. di Napoli, Naples, Italy.

510 Di Nola, A., and Ventre, A. G. S. (1979). A model for a synthesis process, Doc. Inst. Math., Fac. di Archit., Univ. di Napoli, Naples, Italy, Oct.

511 Di Nola, A., and Ventre, A. G. S. (1979). On some chains of fuzzy sets, Doc. Inst. Math., Fac. Di Archit., Univ. di Napoli, Naples, Italy, Oct.

512 Di Nola, A., and Ventre, A. G. S. (1979). On some sequences on fuzzy sets, *RAIRO, INF.* (Paris), **13**, 199-204.

513 Di Nola, A., and Ventre, A. G. S. (1980). Synthesis and decision by defuzzification, *Econ. Comput. ECSR* (Rumania), **4**, 99-106.

514 Dishkant, H. (1981). About membership functions estimation, *Fuzzy Sets Syst.*, **5** (2), March, 141-148.

515 Dockery, J. (1977). A treatment of effectiveness measures as fuzzy sets, unpublished report.

516 Dockery, J. J. (1977). A fuzzy definition of effective measures from a study of military force structure, Working Pap., U.S. Army Concept Anal. Agency, MD.

517 Dockery, J. J. (1977). The use of fuzzy sets in the analysis of military commands, in *Proc. 39th Mil. Oper. Res. Soc.* (*MORS*).

518 Dockery, J. T. (1980). Fuzzy design of military information systems, ORSA-TIMS MIL.

519 Dodson, C. T. J. (1974). Hazy spaces and fuzzy spaces, *Bull. London Math. Soc.*, **6**, 191-197.

520 Dodson, C. T. J. (1975). Tangent structure for hazy spaces, *J. London Math. Soc.*, **2** (11), 465-473.

521 Dodson, C. T. J. (1976). Interactions and hazy directions, Dept. of Math., Univ. of Lancaster, Bailrigg, Lancaster, UK.

522 Dodson, C. T. J. (1977). Hazy spaces, tangent spaces, manifolds and groups, Int. Rep. Int. Atomic Energy Agency and Unesco, Int. Centre for Theoretical Phys., Miramare, Trieste, Italy, Mar.

523 Dombi, J. (1980). Basic concepts for a theory of evaluation, The 10th Meet. of the Eur. Working Group on Fuzzy Sets, in connection with the Eur. IV Congr., July.

524 Dombi, J., and Gyimothy, T. (1978). Syntactic pattern recognition with acceptor fuzzy automata, *EMCSR—PCSR*, **11**, to be published.

525 Dorris, A. L., and Sadosky, T. L. (1973). A fuzzy set theoretic approach to decision making, 44th Nat. Meet. of ORSA, San Diego, CA, Nov.

526 Dowker, C. H., and Papert, D. (1966). Quotient frames and subspaces, *Proc. London Math. Soc.*, **16**, 275-296.

527 Dowker, C. H., and Papert, D. (1967). On Ursohns lemma. General topology and its relations ot modern analysis and algebra 2, in *Proc. 2nd Prague Topol. Symp. 1966*, Academia, Prague and Academic Press, New York, pp. 111-114.

528 Dravecky, J., and Riecan, B. (1975). Measurability of functions with values in partially ordered spaces, *Cas. Pestovani Math.*, **100**, 27-35.

529 Dreyfess, G. R., Kochen, M., Robinson, J., and Badre, A. N. (1975). On the psycholinguistic reality of fuzzy sets, in *Functionalism* (R. E. Grossman, L. J. San, and T. J. Vance, Eds.), Univ. of Chicago Press, Chicago, pp. 135-149.

530 Driankov, D. (1978). Dialectics, fuzzy sets and models of human actions, *EMCSR— PCSR*, **8**, to be published.
531 Drosselmeyer, E., and Wonneberger, R. (1975). Studies on a fuzzy system in the parachial field, Spec. Interest Disc. Sess. on Fuzzy Automata and Decis. Processes, 6th IFAC World Congr., Boston, MA, Aug.
532 Dubois, D. (1977). Quelques outils methodologiques pour la conception de reseaux de transport, Doc. Thesis, ENSEA, Toulouse, France, Oct.
533 Dubois, D. (1978). A fuzzy heuristic, interactive approach of the optimal network problem, Internal Memo., School of Elec. Eng., Purdue Univ., West-Lafayette, IN, Apr.
534 Dubois, D. (1978). An Application of fuzzy sets to bus transportation network modification, in *Proc. 1978 JACC*, Vol. 3, pp. 53–60.
535 Dubois, D. (1979). New results about properties and semantics of fuzzy-set-theoretic operations, 1st Int. Symp. on Policy Anal. and Inf. Syst., June.
536 Dubois, D. (1979). Notes sur l'interet des sous-ensembles flous en analyse de l'attraction des points de vente, Note Ime Np. 35, Inst. Math. Econom., Univ. de Dijon, Dijon, France, Feb.
537 Dubois, D. (1979). Quelques classes d'operateurs remarquables pour combiner les sous-ensembles flous, *Busefal* (1), Automne-Hiver 1979–1980, 29–35; Lang. Syst. Inf., Univ. Paul Sabatier, Toulouse, France, Dec.
538 Dubois, D. (1979). Selection automatique par specifications floues, Rap. Interne Adepa, Mortrouge, France.
539 Dubois, D. (1980). Ensembles flous et conception assistee par ordinateur, Imag Grenoble Rap. de Rech. No. 199, to be published.
540 Dubois, D. (1980). Sur les liens entre les notions de probabilite et de possibilite, Table Ronde du CNRS sur le Flou, Lyon, France, June.
541 Dubois, D. (1980). Theorie des nombres flous, Table Ronde du CNRS sur le Flou, Lyon, France, June.
542 Dubois, D. (1980). Un apercu de la theorie des sous ensembles flous et quelques applications 2, *Bull. Sous Ensembles Flous Appl.*, Automne.
543 Dubois, D. (1980). Un apercu de la theorie des sous-ensembles flous et quelques applications 1, *Bull. Sous Ensembles Flous Appl.*, Ete.
544 Dubois, D. (1980). Un economiste precurseur de la theorie des possibilites, G. L. S. Shackle, *Bull. Stud. Exch. Fuzziness Appl.*, Printemps.
545 Dubois, D. (1980). Vers l'application de la theorie des ensembles flous a la conception aidee par ordinateur, *Bull. Stud. Exch. Fuzziness Appl.*, Printemps.
546 Dubois, D. (1981). Proprietes de la cardinalite floue d'un ensemble flou fini, *Bull. Sous Ensembles Flous Appl.*, **5**, Hiver (80–81), 11–12.
547 Dubois, D. (1981). Un apercu de la theorie des sous-ensembles flous et quelques applications, Partie III, *Bull. Sous Ensembles Flous Appl.*, **5**, Hiver (80–81), 69–75.
548 Dubois, D., and Prade, H. (1976). Le flou, Kouacksewa \ Note Interne Cert., Dept. d'Etudes et Rech. en Autom. Complexe Aerosp., Toulouse, France, Oct.
549 Dubois D., and Prade, H. (1977). Agorithmes de plus courts chemins pour graiter des donnees floues, Doc. ONERA/CERT/DERA, Toulouse, France, Dec.
550 Dubois, D., and Prade, H. (1977). Flou et catastrophes dans les systemes, Rap. DERA/CERT, Toulouse, France, Oct.
551 Dubois, D., and Prade, H. (1977). Le flou, Mercedonksa NPTE Interne DERA-CERT, Dept. d'Etudes et de Rech. an Autom., Toulouse, France, Apr.

552 Dubois, D., and Prade, H. (1977). Nombres flous a plateau, choix de la valeur moyenne, Memo. ONERA/CERT/ENSEA, Toulouse, France.

553 Dubois, D., and Prade, H. (1978). A procedure for multiple-aspect decision making, *Int. J. Syst. Sci.*, **9** (3) 357-360.

554 Dubois, D., and Prade, H. (1978). A summary comment on theory of fuzzy sets, fuzzy sets as a basis for a theory of possibility, a theory of approximate reasoning, and Pruf—A meaning representation language for natural languages, Rep. No. TR-EE 78-13, School of Elec. Eng., Purdue Univ., West Lafayette, IN, pp. E1–E20.

555 Dubois, D., and Prade, H. (1978). Algebre et analyse floue. Leurs applications potielles, Colloq. Int. on Theorie et Appl. des Sous-Ensembles Flous. Journ. de Biomath. et d'Inf. Med., Marseille, France, Sept.

556 Dubois, D., and Prade, H. (1978). Algorithmes de plus courts chemins pour traiter des donnes floues, *R.A.I.R.O. Op. Res.*, **12** (2), 213–227.

557 Dubois, D., and Prade, H. (1978). An alternative fuzzy logic, Rep. No. TR-EE 78-13, School of Elec. Eng., Purdue Univ., West Lafayette, IN, pp. F1–F12.

558 Dubois, D., and Prade, H. (1978). Comment on tolerance analysis using fuzzy sets and a procedure for multiple aspect decision-making, *Int. J. Syst. Sci.*, **9**, 357–360.

559 Dubois, D., and Prade, H. (1978). Decision making with fuzziness, 2nd Lawrence Symp. on Syst. and Decis. Sci., Berkeley, CA.

560 Dubois, D., and Prade, H. (1978). Fuzzy algebra, analysis and logic, Tech. Rep. 78-13, School of Elec. Eng., Purdue Univ., West Lafayette, IN.

561 Dubois, D., and Prade, H. (1978). Fuzzy real algebra—Some results, Rep. No. TR-EE 78-13, School of Elec. Eng., Purdue Univ., West Lafayette, IN, pp. A1–A37.

562 Dubois, D., and Prade, H. (1978). Graphes flous et plus courts chemins flous, *Rev. RAIRO, AFCET*, **12** (2), May.

563 Dubois, D., and Prade, H. (1978). Operations in a fuzzy-valued logic, Rep. No. TR-EE 78-13, School of Elec. Eng., Purdue Univ., West Lafayette, IN, pp. D1–D21.

564 Dubois, D., and Prade, H. (1978). Operations on fuzzy numbers, *Int. J. Syst. Sci.*, **9**, 613–626.

565 Dubois, D., and Prade, H. (1978). Summary of technical memos of Zadeh, Rep. Stanford Artif. Intell. Lab., Stanford, CA.

566 Dubois, D., and Prade, H. (1978), Systems of linear fuzzy constraints, Rep. No. TR-EE 78-13, School of Elec. Eng., Purdue Univ., West Lafayette, IN, pp. B1–B21.

567 Dubois, D., and Prade, H. (1978). Towards fuzzy analysis—integration and derivation of fuzzy functions, Rep. No. TR-EE 78-13, School of Elec. Eng., Purdue Univ., West Lafayette, IN, pp. C1–C51.

568 Dubois, D., and Prade, H. (1979). Decision-making under fuzziness, in *Advances in Fuzzy Set Theory and Applications*, (M. M. Gupta, R. K. Ragade, and R. R. Yager, Eds.), North-Holland, Amsterdam.

569 Dubois, D., and Prade, H. (1979). Outline of fuzzy set theory, an introduction, in *Advances in Fuzzy Set Theory and Applications*, (M. M. Gupta, R. K. Ragade, and R. R. Yager, Eds.), North-Holland, Amsterdam.

570 Dubois, D., and Prade, H. (1979). The advantages of fuzzy approach in OR/MS demonstrated on two-examples of resources allocation problems, Eur. Working Group on Fuzzy Sets.

571 Dubois, D., and Prade, H. (1979). Various kinds of interactive addition of fuzzy

numbers application to decision analysis in presence of linguistics probabilities, in *Proc. 18th IEEE Conf. on Decis. and Control*, Dec.

572 Dubois, D., and Prade, H. (1979). Criteria aggregation and ranking of alternatives in the framework of fuzzy set theory, unpublished.

573 Dubois, D., and Prade, H. (1980). Fuzzy logics and fuzzy control, *Int. J. Man-Mach. Stud.*

574 Dubois, D., and Prade, H. (1980). *Fuzzy Sets and Systems: Theory and Applications*, Academic Press, New York.

575 Dubois, D., and Prade, H. (1980). A unifying view of comparison indices in a fuzzy set theoretic framework, in *Proc. Int. Congr. on Appl. Syst. Res. and Cybern.*, Acapulco, Mexico, Dec.

576 Dubois, D., and Prade, H. (1980). Analysis and synthesis of fuzzy mappings, in *Proc. Int. Congr. on Appl. Syst. Res. and Cybern.*, Acapulco, Mexico, Dec.

577 Dubois, D., and Prade, H. (1980). Criteria aggregation and ranking of alternatives in the framework of fuzzy set theory. *Bull. Stud. Exch. Fuzziness Appl.*, Printemps.

578 Dubois, T. (1974). Line methode d'evaluation par les sous-ensembles flous appliquee a la simulation, Imago Centre, IDP 13, Univ. Catholique de Louvain, Louvain, Belgium.

579 Dubois, T. (1975). Les sous-ensembles flous dans les systemes didactiques multimedia, Semin. "Contrib. des Syst. Flous a l'Automat.," Centre d'Autom. de Lille, Lille, France, June.

580 Dubois, T. (1975). Methodes d'evaluation et de gestion dans les systemes didactiques multimedia, Ph.D. Thesis, Sci. Appl., Univ. Catholique de Louvain, Louvain, Belgium.

581 Dubois, T. (1977). A teaching system using fuzzy subsets and multi criteria analysis, *Int. J. Math. Educ. Sci. Tech.*, **8** (2), 203–217.

582 Dubois, T., Jones, A., Peteau, M., and Huynen, A. N. (1977). Toward continuous learning management in CAI, *Int. J. Math. Educ. Sci. Tech.*, **8** (3), 335–350.

583 Dubreil, P., and Dubreil-Jacotin, L. (1973). Proprietes des relations d'equivalence, *C.R. Acad. Sci. Paris*, **205**, 704–706.

584 Dubuisson, A. (1979). Conception et realisation du compilateur et de l'execution du langage L.P.L., Thesis, Univ. Claude Bernard, Lyon, France, Oct.

585 Duchaussoy, A. (1979). Transitivite max—De relations floues definies a partir de l'information transmise entre deux variables aleatoires, Doc. Lab. d'Inf., Univ. Claude Bernard, Lyon, France, Oct.

586 Dufour, J., Gilles, G., and Foulard, C. (1976). Applications de la theorie des sous-ensembles flous a la partition dessystemes dynamiques complexes, *C.R. Acad. Sci. Paris*, Ser. A., **282**, 491–494.

587 Dugundju, J. (1940). Note on a property of matrices for Lewis and Langfords calculi of propositions, *J. Symb. Logic*, **5**, 150–151.

588 Dukman, J. G., and Lowen, R. (1976). Fuzzy relations on countable sets, Tech. Highschool, Delft, Holland, and Vrije Univ., Brussels, Belgium.

589 Dumitrescu, D. (1977). A definition of an informational energy in fuzzy sets theory, *Stud. Univ. Babes-Bolyai, Math.*, **2**, 57–59.

590 Dummett, M. A. E. (1959). A propositional calculus with denumerable matrix, *J. Symb. Logic*, **24**, 97–106.

591 Dummett, M. A. E. (1973). The justification of deduction, *Proc. Brit. Acad.*, **59**, 3–34.

592 Dummett, M. A. E. (1975). Wangs paradox, *Synthese*, **30**, 301–324.

593 Dunn, J. C. (1973). A fuzzy relative of the Isodata process and its use in detecting compact well-separated clusters, *J. Cybern.*, **3**, 32–57.
594 Dunn, J. C. (1974). A graph theoretic analysis of pattern classification via Tamura's fuzzy relation, *IEEE Trans. Syst., Man, Cybern.*, **SMC-3**, 310–313.
595 Dunn, J. C. (1974). Some recent investigations of a new fuzzy partitioning algorithm and its application to pattern classification problems, *J. Cybern.*, **4**, 1–15.
596 Dunn, J. C. (1974). Well-separated clusters and optimal fuzzy partitions, *J. Cybern.*, **4**, 95–104.
597 Dunn, J. C. (1975). Canonical forms of Tamura's fuzzy relation matrix—A scheme for visualizing cluster hierarchies, in *Proc. Comput. Graphics, Pattern Recogn. and Data Struct. Conf.*, Beverly Hills, CA, May.
598 Dunn, J. C. (1977). Indices of partition fuzziness and the detection of clusters in large data sets, in *Fuzzy Automata and Decision Processes*, (M. M. Gupta, G. N. Saridis, B. R. Gaines, Eds.), North-Holland, Amsterdam, pp. 271–284.
599 Dunn, M., and Epstein, G. (1977). *Modern Use of Multi-Valued Logic*, Reidel, Dordrecht, Holland.
600 Dunst, A. J. (1971). Application of the fuzzy set theory, Jan.
601 Dussauchoy, A. (1979). Transitivite max delta de relations floues, 2nd Int. Symp. on Data Anal. and Inf., Versailles, France, Oct.
602 Dussauchoy, A. (1980). Information generalisee sur les ensembles ordonnes, relation de similarite et classification automatique, Table Ronde CNRS sur le Flou, Lyon, France, June.
603 Dussauchoy, A. (1980). Resultate recents en theorie de l'information. Application a l'analyse structurale, *Bull. Sous Ensembles Flous Appl.*, Automne.
604 Dykman, J. G., Van Haeringen, H., and Lange, S. J. (1980). Fuzzy numbers, in *Proc. Int. Congr. on Appl. Syst. Res. and Cybern.*, Acapulco, Mexico, Dec.
605 Dyson, R. G. (1980). Maximin programming, fuzzy linear programming and multi-criteria decision making, *J. Opl. Res. Soc.*, **31**.
606 Economakos, E. (1979). Application of fuzzy concepts to power demand forecasting, *IEEE Trans. Syst., Man, Cybern.*, **SMC-9** (10), 651–657.
607 Economakos, E. (1979). Application of fuzzy concepts to power demand forecasting, *IEEE Trans. Syst., Man, Cybern.*, **SMC-9** (10), Oct.
608 Edwards, W. (1962). Subjective probabilities inferred from decisions, *Psychol. Rev.*, **69**, 109–135.
609 Edwards, W., Phillips, L. D., Hayes, W. L., and Goodman, B. C. (1968). Probabilistic information processing systems—Design and evaluation, *IEEE Trans. Syst. Man, Cybern.*, **SMC-4**, 248–265.
610 Efstathiou, J. (1979). A practical development of multi-attribute decision making using fuzzy set theory, Ph.D. Thesis, Dept. of Comput., Univ. of Durham, Australia.
611 Efstathiou, J., and Rajkovic, V. (1978). Multi-attribute decision making and fuzzy set theory, in *Proc. Workshop on Fuzzy Reasoning*, Queen Mary College, Univ. of London, London.
612 Efstathiou, J., and Rajkovic, V. (1979). Multi-attribute decision making using a fuzzy heuristic approach, *IEEE Trans. Syst., Man, Cybern.*, **SMC-9** (6), 326–333.
613 Efstathiou, J., and Tong, R. (1980). Ranking sets using linguistic preference relations, Int. Symp. Multiple-Valued Logic.

614 Egea, M. (1980). Jeux flous cooperatifs, Table Ronde du CNRS sur le Flou, Lyons, France, June.

615 Ehrenfeucht, A., and Orlowska, E. (1967). Mechanical proof procedure for propositional calculus, *Bull. Acad. Polonaise Sci.* (*Ser. Math., Phys.*), **15**, 25–30.

616 Eikmeyer, H. J., and Rieser, H. (1978). Vagheitstheorie, Project Vagheitstheorie, Univ. Bielefeld, Bielefeld, Germany.

617 Elder, R. C. (1976). Fuzzy systems theory and medical decision making, M.S. Thesis, School of Ind. Syst. Eng., The Georgia Inst. of Tech., Atlanta, GA.

618 El-Fattah, Y. M. (1976). Control of complex systems by fuzzy learning automata, in *Discrete Systems and Fuzzy Reasoning*, (E. H. Mamdani; and B. R. Gaines, Eds.), EES-MMS-DSFR-76, Queen Mary College, Univ. of London, London (Workshop Proc.).

619 Elliot, J. L. (1976). Fuzzy Kiviat Graphs, Eur. Comput. Congr. (Eurocomp 76), Online, London, Sept.

620 Ellis, C. A. (1971). Probabilistic tree automata, *Inf. Control*, **19**, 401–416.

621 Emptos, E., and Fages, R. (1980). Dispersion, Mesures flous et classification automatique, Table Ronde du CNRS sur le Flou, Lyon, France, June.

622 Emptoz, H. (1974). Indices de flou et indices de dispersion, Colloq. Int. sur la Theorie des Sous-Ensembles Flous, Marseille, France, Sept.

623 Emptoz, H. (1978). Les indices de flou, Rap. de Rech. Santi, Univ. de Lyon, Lyon, France.

624 Emptoz, H. (1981). Nonprobabilistic entropies and indetermination measures in the setting of fuzzy sets theory, *Fuzzy Sets Syst.*, **5** (3), 307–318.

625 Emptoz, H., Terrenoire, M., and Tounissoux, D. (1978). Methodes arborescentes pour la reconnaissance de variables du type continu, Congr. AFCET-IRIA, Reconnaissance de Variables du Type Continu, Congr. AFCET-IRIA, Reconnaissance des Formes et Traitement des Images, Feb.

626 Emptoz, H., and Tounissoux, D. (1977). Utilisation de pseudo-questionnairres dans une approache metrique des problemes de reconnaissance des formes, Colloq. Int. CNRS, Cachan, France, July.

627 Emptoz, H., and Tounissoux, D. (1978). Une methode sequentielle pour la reconnaissance d'une forme floue, Colloq. Int. Theorie et Appl. des Sous-Ensembles Flous, Journ. de Biomath. et Inf. Med., Marseille, France, Sept.

628 Emptoz, H., Tounissoux, D., and Terrenoire, H. (1978). Indetermination measure for a sequential identification process, 4th Int. Joint Congr. on Pattern Recogn., Kyoto, Japan, Nov.

629 Emsellem, B., and Rochfeld, A. (1975). Bases de donnees et recherche documentaire par voisinage, *Rev. Metra*, **XIV** (2), 321–334.

630 Endo, Y., and Tsukamoto, Y. (1973). Apportion models of tourists by fuzzy integrals, Ann. Conf. Rec. of SICE, Japan.

631 Engel, A. B., and Buonomano, V. (1973). Towards a general theory of fuzzy sets 1, Inst. of Math., Univ. Estaduel de Campinas, Brazil.

632 Engel, A. B., and Buonomano, V. (1973). Towards a general theory of fuzzy sets 2, Inst. of Math., Univ. Estaduel de Campinas, Campinas, Brazil.

633 Enta, Y. (1976). Fuzzy choice models, Rep., Dept. of Bus., Hosea Univ.

634 Enta, Y. (1978). A measure for the discrimination effect of information, in *Proc. 1978 JACC*, Vol. 3, pp. 69–80.

635 Enta, Y. (1978). Discriminative effect of information, the 3rd measure of information, Memo No. UCB/ERL M78-62, Electron. Res. Lab., Univ. of California, Berkeley.
636 Enta, Y. (1980). Fuzzy decision theory, in *Proc. Int. Congr. on Appl. Syst. Res. Cybern.*, Acapulco, Mexico, Dec.
637 Epstein, G. (1972). Multiple-valued signal processing with limiting, Symp. on Multiple-Valued Logic Design, Buffalo, NY.
638 Epstein, G., Frieder, G., and Rine, D. C. (1974). The development of multiple-valued logic as related to computer science, *Computer*, **7**, 20–32.
639 Epstein, G., and Horn, A. (1974). P-algebras, an abstraction from post algebras, *Alg. Univers.*, **4**, 195–206.
640 Epstein, G., and Horn, A. (1975). Chain based lattices, *Pacific J. Math.*, **55**, 65–84.
641 Epstein, G., and Horn, A. (1975). Logics which are characterized by subresiduated lattices, Tech. Rep. 24, Comput. Sci. Dept., Indiana Univ., Bloomington.
642 Epstein, G., and Shapiro, S. C. (1975). The development of language and reasoning in the child as connected with mathematical linguistics and logic, Tech. Rep. 41, Oct.
643 Erceg, M. A. (1979). Metric spaces in fuzzy set theory, *J. Math. Anal. Appl.*, **69** (1), May, 205–231.
644 Erceg, M. A. (1980). Functions, equivalence relations, quotient spaces and subsets in fuzzy set theory, *Fuzzy Sets Syst.*, **3** (1), Jan.
645 Ernst, C. (1980). An approach to management expert systems using fuzzy logic, Cybernetique des Enterprises Reconnaissance des Formes Intelligence Artificielle, Dec.; in *Recent Developments in Fuzzy Set and Possibility Theory*.
646 Ernst, C. J. (1980). Fuzzy logic and management expert systems, in *Proc. Int. Congr. on Appl. Syst. Res. and Cybern.*, Acapulco, Mexico, Dec.
647 Erson, A. R., and Belnap, N. D. (1962). The pure calculus of entailment, *J. Symb. Logic*, **27**, 19–25.
648 Erson, A. R., and Belnap, N. D. (1975). *Entailment*, Princeton Univ. Press, Princeton, NJ.
649 Esculier, C. (1979). Bases de donnees floues et convivialite, *Busefal* (1), Automne-Hiver 1979–1980, 47–51; Lang. Syst. Inf., Univ. Paul Sabatier, Toulouse, France, Dec.
650 Eshragh, F. (1979). Conversational programs for decision making using fuzzy set theory, Ph.D. Thesis, Dept. of Elec. and Electron. Eng., Queen Mary College, Univ. of London, London.
651 Eshragh, F., and Mamdani, E. H. (1979). A general approach to linguistic approximation, *Int. J. Man-Mach. Stud.*, **11**, 501–519.
652 Eshragh, E., Mandic, N. J., and Mamdani, E. H. (1981). Multi-criteria decision making using fuzzy sets. *EMCSR—PCSR*, **8**, to be published.
653 Esogbue, A. O. (1975). On the application of fuzzy allocation theory to the modelling of cancer research appropriation process, in *Proc. 3rd Int. Congr. of Cybern. and Syst.*, Bucharest, Aug.
654 Esogbue, A. O. (1980). Some novel applications of fuzzy sets in water quality pollution control planning, in *Proc. Int. Congr. on Appl. Syst. Res. and Cybern.*, Acapulco, Mexico, Dec.
655 Esogbue, A. O., and Elder, R. C. (1977). Fuzzy sets and modelling physician decision processes, Part 1, Initial interview information session, Ind. and Syst. Eng. Rep. No. J-77-6, Georgia Inst. of Tech., Atlanta.

656 Esogbue, A. O., and Elder, R. C. (1977). Fuzzy sets and the modeling of physician decision processes, Part 2, Fuzzy diagnosis decision, Ind. and Syst. Eng. Rep. No. J-77-6, Georgia Inst. of Tech., Atlanta.

657 Esogbue, A. O., and Ramesh, V. (1970). Dynamic programming and fuzzy allocation processes, Tech. Memo. 202, Oper. Res. Dept., Case Western Res. Univ., Cleveland, Ohio.

658 Eto, H. (1975). Multivariate analysis of ambiguous opinions on opening the sports facilities of firms to the public, in *Summ. of Pap. on Gen. Fuzzy Problems*, The Working Group on Fuzzy Syst., Tokyo, Nov., pp. 5–9.

659 Eto, H. (1976). Fuzzy operational approach to analysis of Delphi technology forecasting, in *Summ. of Pap. On Gen. Fuzzy Problems*, The Working Group on Fuzzy Syst., Tokyo, pp. 25–34.

660 Eto, H. (1977). Generalized domination and fuzzy domination, in *Summ. of Pap. on Gen. Fuzzy Problems*, The Working Group on Fuzzy Syst., Tokyo, pp. 1–10.

661 Eto. H. (1978). Various concepts of ambiguity and their relationship with the existing formulation of fuzziness, in The Working Group on Fuzzy Syst., Tokyo, Dec., pp. 5–9. Rep. No. 4.

662 Etschmaier, M. M. (1980). Fuzzy controls for maintenance scheduling in transportation systems, *Automatica*, **16**, 255–264.

663 Evans, G., and McDowell, J. (Eds.) (1976). *Truth and Meaning*, Clarendon Press, Oxford, UK.

664 Evenden, J. (1974). Generalized logic, *Notre Dame J. Formal Logic*, **15**, 35–44.

665 Eytan, M. (1977). Semantique preordonnee des ensembles flous, in *Congr. AFCET Model. Maitrise Syst.*, Versailles, France, pp. 601–608.

666 Eytan, M. (1978). Fuzzy sets, a Topos-logical point of view, Univ. Rene Descartes, Paris.

667 Ezhkova, I. V., and Pospelov, D. A. (1978). Decision making on fuzzy premises, II, Deduction schemes, engineering cybernetics, *Sov. J. Comput. Syst.*, **16** (2), 1–6.

668 Ezoe, T. (1975). Cause picture method introduced into categorical analysis of multi-variable's data, in *Summ. of Pap. on Gen. Fuzzy Problems*, The Working Group on Fuzzy Syst., Tokyo, Nov., pp. 10–13.

669 Fadina, A., Lettieri, A., and Liguoir, F. (1980). Classification of some fuzzy preference relations, Part 1, *Bull. Sous Ensembles Flous Appl.*, Automne.

670 Fadini, A., Lettieri, A., and Liguori, F. (1981). Classifications of some fuzzy preference relations, Part 2, *Bull. Sous Ensembles Flous Appl.*, **5**, Hiver (80–81), 30–41.

671 Farreny, H., and Prade, H. (1980). Fuzzy monitoring of an assembly line, The 10th Meet. of The Eur. Working Group on Fuzzy Sets, in Connection with the Eur. IV Congr., July.

672 Farreny, H., and Prade, H. (1980). Towards fuzzy monitoring of an assembly line, *Bull. Sous Ensembles Flous Appl.*, Automne.

673 Feagans, T., and Biller, W. F. (1979). Fuzzy concepts in the analysis of public health risks, 1st Int. Symp. on Policy Anal. and Inf. Syst., Duke Univ., NC., June.

674 Fee, W. G., and Fu, K. S. (1969). A formulation of fuzzy automata and its applications as a model of learning systems, *IEEE Trans. Solid State Circuits*, **SSC-5** (3), 215–223.

675 Fellinger, W. L. (1974). Specifications for a fuzzy systems modelling language, Ph.D. Thesis, Oregon State Univ., Corallis, in *Diss. Abstr. Int.*, **35/07-B**, 327.

676 Fenstad, J. E. (1964). On the consistency of the axiom of comprehension in the Lukasiewicz infinite valued logic, *Math. Scand.*, **14**, 65–74.

677 Fenstad, J. E. (1967). Representations of probabilities defined on first order languages, in *Sets Models and Recursion Theory*, (J. N. Crossley, Ed.), North-Holland, Amsterdam, pp. 156–172.

678 Feron, R. (1976). Economie d'exchange aleatiore flou, *C. R. Acad. Sci. Paris*, **282**, 1379–1382.

679 Feron, R. (1976). Ensembles aleatoires flous, *C. R. Acad. Sci. Paris*, **282** (4), 903–906.

680 Feron, R. (1976). Ensembles flous attaches a un ensemble aleatoire flou, *Publ. Econ.*, **IX**, Fasc. 2.

681 Feron, R. (1976). Ensembles flous, ensembles aleatiores flous et economie aleatiore floue, *Publ. Econ.*, **IX**, Fasc. 1.

682 Feron, R. (1976). Economic d'exchange alteatoire floue, *C. R. Acad. Sci. Paris*, **282** (9), 1379–1382.

683 Feron, R. (1977). Les theories de champs aleatoires et des ensembles aleatoires flou pervent-elle etre utilisees en geographie, Actes Colloq. Math. Appl. a la Geogr., Oct.

684 Feron, R. (1978). Ensembles flous dont le referentiel est un espace de frechet, *Semin. Math. Floues*, Dept. Math., Univ de Lyon, Lyon, France, Sept., pp. 59–71.

685 Feron, R. (1978). Random fuzzy correspondence, C. R. Symp. on Stochast. Geometry and Directional Stat., Erevan, U.S.S.R.; *J. Appl. Prob.*

686 Feron, R. (1978). On the notions of distance and deviation in a fuzzy structure, *EMCSR—PCSR*, **8**, to be published.

687 Feron, R. (1979). Ensemble aleatoire, ensemble flou, ensemble aleatoire flou, processus aleatoire flous et leur application a l'etude du profil fiscal, Doc. Univ. degli Studi di Pavona, Fac. di Econ. e Commercio, Verona, Italy.

688 Feron, R. (1979). Ensembles aleatoires flous dont la fonction d'appartenance prend ses valeurs dans un treillis distributif ferme, *Publ. Econ.*, **XII**, Fasc. I.

689 Feron, R. (1979). Sur les notions de distance et d'ecart dans une structure floue et leurs applications aux ensembles aleatoires flous. Cas ou le referentiel n'est pas metrique, *C. R. Acad. Sci.*, **T 289**, Ser. A (35), July.

690 Feron, R., and Kuzmin, V. B. (1980). Probabilistic and statistical study of random fuzzy sets whose referential is $(R)N$th, in *Proc. Int. Congr. on Appl. Syst. Res. and Cybern.*, Acapulco, Mexico, Dec.

691 Ferrier, D. (1976). Approche methodogique d'un traitement mathematique applique au phonomene criminel, Mem. Fac. de Droit et Sci. Politique, Univ. Aix-Marseille, France, Sept.

692 Ferrier, D. (1978). Applications de la theorie des sous-ensembles flous a la recherche criminologique, Colloq. Int., Theorie et Appl. des sous-ensembles flous, Journ. de Biomath. et d'inf. Med., Marseille, France, Sept.

693 Fevrier, P. (1976). On the representation of measurements results by fuzzy sets, 3rd Eur. Meet. Cybern. Syst. Res., Vienna.

694 Fevrier, P. (1978). Expressing measurements by means of fuzzy sets, *EMCSR—PCSR*, **8**, to be published.

695 Fiksel, J. (1980). Applications of fuzzy systems in management science, soft methods for hard problems, in *Proc. Int. Congr. on Appl. Syst. Res. and Cybern.*, Acapulco, Mexico, Dec.

696 Fiksel, J. (1980). Fuzzy logics with applications, in *Proc. JACC*, San Francisco, Aug.
697 Fiksel, J., Diffenbach, J., and Renda, A. (1978). A methodology employing possibility theory for a tech. assess. of time of day El. Rts., ORSA/TIMS Meet., Los Angeles.
698 Fiksel, J., Diffenbach, J. and Renda, A. (1979). Time-of-use pricing of electricity a policy assessment methodology, *Advances in Fuzzy Set Theory and Application*, (M. M. Gupta, R. K. Ragade, and R. R. Yager, Eds.), North-Holland, Amsterdam.
699 Fiksel, J. R. (1979). Stress and stability, fuzzy concepts for risk management, Doc. Arthur D. Little, Boston, MA, Dec.
700 Fillmore, C. J., and Jangendoen, D. T. (Eds.) (1971). *Studies in Linguistic Semantics*, Holt, Rinehart and Winston, New York.
701 Findler, N. V. (1979). Fuzzy retrieval of fuzzy information, Colloq. de St Maximin Iria-Lish, Sept.
702 Fine, K. (1975). Vagueness, truth and logic, *Synthese*, **30**, 265–300.
703 Fine, T. L. (1973). *Theories of Probability*, Academic Press, New York.
704 Fishburn, P. C. (1975). A theory of subjective expected utility with vague preference, *Theory Decis.* **6**, 287–310.
705 Flachs, J., and Pollatschek, M. A. (1977). Duality theorems for certain programs involving minimum or maximum operations, Doc. Technion, Israel Inst. of Tech., June.
706 Flachs, J., and Pollatschek, M. A. (1977). Some properties of optimization in $L(K)$ norm, Tech. Rep. 113 Comput. Sci. Dept., Technion, Israel Inst. of Tech., Dec.
707 Flachs, J., and Pollatschek, M. A. (1978). Further results on fuzzy-mathematical programming, *Inf. Control*, **38**, 241–257.
708 Flonder, P. (1975). On C-sets, Working Group on Fuzzy Syst., Acad. of Econ. Stud., Bucharest, unpublished.
709 Flonder, P. (1975). Models for property assignment, Semin. on Fuzzy Syst., Dept. of Cybern., ASE, Bucharest.
710 Flonder, P. (1977). An example of a fuzzy system, *Kybernetes*, **6**, 229–230.
711 Floyd, R. W. (1967). Non-deterministic algorithms, *J. Assoc. Comput. Mach.*, **14**, 636–644.
712 Foradori, E. (1933). Stetigkait und Kontinuitat als Teilbarkeitseigenschaften, *Monatsckefte Math. Phys.*, **40**, 161–180.
713 Fordon, W. A., and Bezdek, J. C. (1979). The application of fuzzy set theory to medical diagnosis, in *Advances in Fuzzy Set Theory and Applications*, (M. M. Gupta, R. K. Ragade, and R. R. Yager, Eds.), North-Holland, Amsterdam.
714 Fortet, R., and Kambouzia, M. (1976). Ensembles aleatoires et ensembles flous, *Publ. Econ.*, **IK**, Fasc. 1.
715 Foster, D. H. (1979). Fuzzy topological Groups, *J. Math. Anal. Appl.*, **67** (2), Feb., 549–565.
716 Foster, M. H., and Martin, M. L. (Eds.) (1966). *Probability, Confirmation and Simplicity*, Odyssey Press, New York.
717 Fox, J. (1978). Fuzzy logic and natural inference—A reply to Haack, *EMCSR—PCSR*, **9**, to be published.
718 Franksen, O. I. (1979). On fuzzy sets, subjective measurements and utility, *Int. J. Man-Mach. Stud.*, **11**, 521–545.
719 Fraser, B. (1975). Hedged performatives, in *Syntax and Semantics*, Vol. 3 (P. Cole, and J. L. Morgan, Eds.), Academic Press, New York, pp. 187–210.

720 Freeling, A. N. S. (1979). Decision analysis and fuzzy sets, Masters Thesis, Cambridge Univ., Cambridge, UK.
721 Freeling, A. N. S. (1980). Fuzzy sets and decision analysis, *IEEE Trans. Syst., Man, Cybern.*, **SMC-10** (7).
722 Freksa, C. (1980). L-Fuzzy—An AI language with linguistic modification of patterns, AISB Conf., Amsterdam, July.
723 Frink, O. (1938). New Algebras of logic, *Amer. Math. Mon.*, **45**, 210–219.
724 Fu, K. S. (Ed.) (1971). A critical review of learning contro research, in *Pattern Recognition and Machine Learning*, Plenum Press, New York.
725 Fu, K. S. (1974). Pattern recognition and some socio-economic problems, Purdue Univ., West Lafayette, IN.
726 Fu, K. S., and Li, T. J. (1968). On the behavior of learning automata and its applications, Report No. TR-EE68-20, Lafayette School of Elec. Eng., Purdue Univ., West Lafayette, IN.
727 Fu, K. S., and Li, T. J. (1969). Formulation of learning automata and games, *Inf. Sci.*, **1**, 237–256.
728 Fu, K. S., and Wee, W. G. (1967). On general adaptive algorithms and applications of the fuzzy sets concept to pattern class?, Rep. TR-EE67-7, Lafayette School of Elec. Eng., Purdue Univ., West Lafayette, IN.
729 Fu, K. S., and Yao, J. T. P. (1980). Application of fuzzy sets in earthquake engineering research, in *Proc. Int. Congr. on Appl. Syst. Res. and Cybern.*, Acapulco, Mexico, Dec.
730 Fujisake, H. (1971). Fuzziness in medical sciences and its processing, in *Proc. Symp. on Fuzziness in Syst. and its Process*. Prof. Group of Syst. Eng. of SICE.
731 Fukami, S., Mizumoto, M., and Tanaka, K. (1980). Some considerations on fuzzy conditional inference, *Fuzzy Sets Syst.*, **4**.
732 Fukami, F., Umano, M., Mizumoto, M., and Tanaka, K. (1979). Fuzzy database retrieval and manipulation languages, *Tech. Rep. of IECE of Jap.*, **78** (233) (On Autom. and Lang.), 65–72, AL78-85 (in Japanese).
733 Fukuda, T. (1980). A fuzzy synthesis of complex systems, Int. Conf. on Cybern. and Soc., Oct., Cambridge, MA.
734 Fung, L. W., and Fu, K. S. (1973). Decision making in a fuzzy environment, TR-EE73-22, School of Elec. Eng., Purdue Univ., Lafayette, IN.
735 Fung, L. W., and Fu, K. S. (1974). The K'th optimal policy algorithm for decision making in fuzzy environments, in *Identification and System Parameter Estimation*, (P. Eykhoff, Ed.), North-Holland, Amsterdam, pp. 1025–1059.
736 Fung, L. W., and Fu, K. S. (1975). An axiomatic approach to rational decision making in a fuzzy environment, in *Fuzzy Sets and their Applications to Cognitive and Decision Processes*, (L. A. Zadeh, K. S. Fu, K. Tanaka, and M. Shimura, Eds.), Academic Press, New York, 227–256.
737 Fung, L. W., and Fu, K. S. (1977). Characterization of a class of fuzzy optimal control problems, in *Fuzzy Automata and Decision Processes*, (M. M. Gupta, G. N. Saridis, and B. R. Gaines, Eds.), North-Holland, Amsterdam, pp. 209–220.
738 Furami, S., Mizumoto, M., and Tanaka, K. (1980). Some considerations on fuzzy conditional inference, *Fuzzy Sets Syst.*, **4** (3), 243–274.
739 Furukawa, M., Nakumura, K., and Oda, M. (1972). Fuzzy models of human decision-making process, Annu. Conf. Rec. of JAACE.

740 Furukawa, M., Nakamura, K., and Oda, M. (1973). Fuzzy variant process of memories, Annu. Conf. Rec. of SICE, Japan.

741 Fustier, B. (1975). L'attraction des points de vente dans les espaces precis et imprecis, Doc. Inst. de Math. Econ., Fac. Sci. Econ. et de Gestion, Dijon, France, July.

742 Fustier, B. (1977). Theorie des sous-ensembles flous et gravitation economique, Doc. IXth Colloq. Ann. IME., Univ of Dijon, Dijon, France.

743 Fustier, B. (1978). Contribution a l'analyse spatiale de l'attraction imprecise. Theorie et applications des sous-ensembles flous, Journ. de Biomath. et d'Inf. Med., Marseille, France, Sept.

744 Fustier, B. (1980). Ensemble de pateto ordinaire et decision floue, *Bull. Stud. Exch. Fuzziness Appl.*, Ete.

745 Fustier, B. (1980). Introduction a l'etude de series statistiques flous, *Bull. Stud. Exch. Fuzziness Appl.*, Printemps.

746 Gadreau, M. (1975). Economie formalisee d'un systeme de sante et sous-ensembles flous. Note Interne, Inst. Math. Econ., Univ. de Dijon, Dijon, France, pp. 1–27.

747 Gaffney, J. E. (1972). Navigation in space, the future and aritificial intelligence history of navigation, Orlando, FL.

748 Gaifman, H. (1964). Concerning measures in first order calculi, *Isr. J. Math.*, **2**, 1–18.

749 Gaines, B. R. (1975). A calculus of possibility, eventuality, and probability, in EES-MMS-FUZI-75, Dept. of Elec. Eng. Sci., Univ. of Essex, Colchester, UK.

750 Gaines, B. R. (1975). Approximate identification of automata, *Electron. Lett.*, **11**, 444–445.

751 Gaines, B. R. (1975). Control engineering and artificial intelligence, Lecture Notes of BCS AISB Summer School, Cambridge, UK, July, pp. 52–60.

752 Gaines, B. R. (1975). Multivalued logics and fuzzy reasoning, Lecture Notes of BCS AISB Summer School, Cambridge, UK, July, pp. 100–112.

753 Gaines, B. R. (1975). Stochastic and fuzzy logics, *Electron. Lett.*, **11**, 188–189.

754 Gaines, B. R. (1976). Behaviour-structure transformations under uncertainty, *Int. J. Man-Mach. Stud.*, **8**, 337–365.

755 Gaines, B. R. (1976). Fuzzy and stochastic probability logics, EES-MMS-FUZ-76, Dept. of Elec. Eng. Sci., Univ. of Essex, Colchester, UK.

756 Gaines, B. R. (1976). Fuzzy reasoning and the logics of uncertainty, in *Proc. 6th Int. Symp. Multiple-Valued Logic*, IEEE 76CH1111-4C, pp. 179–188.

757 Gaines, B. R. (1976). General fuzzy logics, 3rd Eur. Meet. Cybern, Syst. Res., Vienna.

758 Gaines, B. R. (1976). Research notes on fuzzy reasoning, in *Discrete Systems and Fuzzy Reasoning*, EES-MMS-DSFR-76 (E. H. Mamdani and B. R. Gaines, Eds.), Queen Mary College, Univ. of London, London (Workshop Proc.).

759 Gaines, B. R. (1976). Survey of fuzzy reasoning, Workshop on Discrete Syst. and Reasoning, Dept. of Elec. Eng., Queen Mary College, Univ. of London, London.

760 Gaines, B. R. (1976). System identification, approximation and complexity, *Int. J. Gen. Syst.* (3).

761 Gaines, B. R. (1976). Understanding uncertainty, in *Proc. Workshop on Discrete Systems and Fuzzy Reasoning*, Queen Mary College, Univ of London, London.

762 Gaines, B. R. (1976). V-fuzzy Q-analysis, EES-MMS-QFUZ-76, Dept. of Elec. Eng. Sci., University of Essex, Colchester, UK.

763 Gaines, B. R. (1976). Why fuzzy reasoning?, in *Discrete Systems and Fuzzy Reasoning*, EES-MMS-DSFR-76 (E. H. Mamdani and B. R. Gaines, Eds.), Queen Mary College, Univ. of London, London (Workshop Proc.).

764 Gaines, B. R. (1977). Foundations of fuzzy reasoning, in *Fuzzy Automata and Decision Processes*, (M. M. Gupta, G. N. Saridis, and B. R. Gaines, Eds.), North-Holland, Amsterdam, pp. 19–76.

765 Gaines, B. R. (1977). Sequential fuzzy system identification, in *Proc. 1977 IEEE Conf. on Decis. and Control*, New Orleans, pp. 1309–1314.

766 Gaines, B. R. (1978). Fuzzy and probability uncertainty logics, *Inf. Control*, **38**, 154–169.

767 Gaines, B. R. (1978). *General Systems Research—Recent Developments and Trends*, NATO Conf. Series, Series II Syst. Sci., Vol. 5, pp. 91–105.

768 Gaines, B. R., and Kohout, L. J. (1975). Possible automata, in *Proc. 1975 Int. Symp. Multiple-Valued Logic*, IEEE 75CH0959-7C, pp. 183–196.

769 Gaines, B. R., and Kohout, L. J. (1975). The logic of automata, *Int. J. Gen. Syst.*, **2**, 191–208.

770 Gaines, B. R., and Kohout, L. J. (1977). The fuzzy decade—A bibliography of fuzzy systems and closely related topics, *Int. J. Man-Mach. Stud.*, **9**, 1–68.

771 Gale, S. (1972). Inexactness, fuzzy sets and the foundations of behavioral geography, *Geograph. Anal.*, **4**, 337–349.

772 Gale, S. (1974). A prolegomenon to an interrogative theory of scientific enquiry, WP9 Res. on METRO. Change and Conflict Resolut., Peach Sci. Dept., Univ. of Pennsylvania.

773 Gale, S. (1974). A resolution of the regionalization problem and its implications for political geography and social justice, WP3 Res. on Metro. Change and Conflict Resolut., Peace Sci. Dept., Univ. of Pennsylvania, Philadelphia.

774 Gale, S. (1975). Boundaries, tolerance spaces and criteria for conflict resolution, *J. Peace Sci.*

775 Gale, S. (1975). Conjectures on many-valued logic, regions, and criteria for conflict resolution, in *Proc. 1975 Int. Symp. Multiple-Valued Logic*, IEEE 75CH0959-7C, pp. 212–225.

776 Gale, S., and Atkinson, M. (1977). Fuzzy regions and social justice, ORSA/TIMS Meet., Atlanta, GA.

777 Gallin, D. (1975). *Intensional and Higher Order Modal Logic*, North-Holland, Amsterdam.

778 Gang, C. (1981). The applications of fuzzy mathematics to some kind of information process, A List of Lit. in China, The Univ. of Sci. and Tech. of China.

779 Ganter, T. E., Steinlage, R. C., and Warren, R. H. (1975). Compactness in fuzzy topological spaces, Dept. of Math., Univ. of Dayton, Dayton, OH.

780 Ganter, T. E., Steinlage, R. C., and Warren, R. H. (1978). Errata to "Compactness in fuzzy topological spaces," *J. Math. Anal. Appl.*, **65**, 503.

781 Gao, S. (1981). The applications of fuzzy mathematics to agriculture meteorology, Rep. Weather and Climate Inst., Central Meteorol. Bureau, A List of Lit. in China, The Univ. of Sci. and Tech. of China.

782 Gearing, C. E. (1975). Generalized Bayesian posterior analysis with ambiguous information, 45th ORSA/TIMS Joint Nat. Meet., Boston, MA, Apr.

783 Gearing, C. E. (1976). A fuzzy-set-theoretic generalization of Bayes theorem, Joint Nat. Meeting of ORSA/TIMS.

784 Gehring, H., and Zimmermann, H. J. (1975). Fuzzy information profiles for information selection, Aachen Working Pap. No. 75/04, Aachen Univ., Aachen, Germany.

785 Gelman, I. (1970). Organizational data, A model and computation algorithm that uses the notion of fuzzy systems., Thesis, McGill Univ., Montreal, Canada.

786 Gen. C., and Shen, J. (1980). The applications of fuzzy mathematics to some kind of information processing, Collect. Pap. on Fuzzy Math. (Abstr.), Huazhong Inst. of Tech., Huazhong, China, Apr.

787 Gentilhomme, Y. (1968). Les ensembles flous en linguistique, *Notes Theor. Appl. Linguist.*, **5**, Bucharest, Rumania.

788 Georges, J. (1980). Ensembles Flous. (Reaction apres l'article—Le flou devient mathematique—Henri Prade), *Le Monde-Dimanche*, June.

789 Georgescu, G. (1971). Algebras de Lukasiewicz de orden theta II, *Rev. Roum. Math. Pures Appl.*, **16**, 363–369.

790 Georgescu, G. (1971). Les algebres de Lukaciewicz theta-valentes, in *Logique, Automatique, Informatique* (Acad. Rep. Soc. Roum., Eds.), pp. 99–176.

791 Georgescu, G. (1971). N-valued complete Lukasiewicz algebras, *Rev. Roum. Math. Pures Appl.*, **16**, 41–50.

792 Georgescu, G. (1971). The theta-valued Lukasiewicz algebras III, *Rev. Roum. Math. Pures Appl.*, **16**, 1365–1390.

793 Georgescu, G. (1971). The theta-valued Lukasiewicz algebras 1, *Rev. Roum. Math. Pures Appl.*, **16**, 195–209.

794 Georgescu, G., and Vraciu, C. (1970). On the characterization of central Lukasiewicz algebras, *J. Algebra*, **16**, 486–495.

795 Gerhardts, M. D. (1965). Zur Charakterisierung Distributiver Scheifverbande, *Math. Ann.*, **161**, 231–240.

796 Gerhardts, M. D. (1969). Schragverbande und Quasiordnungen, *Math. Ann.*, **181**, 65–73.

797 Gero, J. S., and Oguntade, O. (1977). Fuzzy sets evaluation in architecture and building analysis, Doc. Dept. Archit. Sci., Univ. of Sydney, Sydney, Australia.

798 Gero, J. S., and Oguntade, O. (1978). Fuzzy sets evaluation in architecture; Prelim. Stud., Doc. Comput. Center CR 29, Dept. of Archit., Univ. of Sydney, Sydney Australia.

799 Gershman, A. (1976). Fuzzy sets methods for understanding vague hints for maze running, M. S. Thesis, Comput. Sci. Dept., Univ. of California, Los Angeles.

800 Giles, R. (1974). A nonclassical logic for physics, *Stud. Logica*, **33**.

801 Giles, R. (1974). A pragmatic approach to the formalization of empirical theories, in *Proc. Conf. on Formal Methods in the Methodol. of Empirical Sci.*, Warsaw, June.

802 Giles, R. (1974). Formal languages and the foundations of physics, Proc. Int. Res. Semin. on Abstr. Represent., in *Mathematical Physics*, D. Reidel, London, Ontario, Dec.

803 Giles, R. (1976). A logic for subjective belief, in *Foundations of Probability Theory, Statistical Inference, and Statistical Theories of Science*, Vol. 1, (W. Harper and C. A. Hooker, Eds.), Reidel, Dordrecht, Holland, pp. 41–72.

804 Giles, R. (1976). Formal languages and the foundations of physics and quantum mechanics, in *The Logico-Algebraic Approach to Quantum Mechanics*, Vol. 2, (C. A. Hooker, Ed.), D. Reidel, Dordrecht, Holland.

805 Giles, R. (1976). Lukasiewicz logic and fuzzy set theory, *Int. J. Man-Mach. Stud.*, **8**, 313–327.
806 Giles, R. (1979). A formal system for fuzzy reasoning, *Fuzzy Sets Syst.*, **2** (3), July, North-Holland, Amsterdam.
807 Giles, R. (1980). A computer program for fuzzy reasoning, *Fuzzy Sets Syst.*, **4** (3), Nov., 221–234.
808 Giles, R. (1981). Semantics for fuzzy reasoning, CORS-TIMS-ORSA Joint Nat. Meet. Toronto, Canada, May.
809 Gitman, I. (1970). Organization of data—A model and computational algorithm that uses the notion of fuzzy sets, Ph.D. Thesis, McGill Univ., Montreal, Canada, in *Diss. Abstr. Int.*, **31/10-B**, 6098.
810 Gitman, I., and Levine, M. D. (1970). An algorithm for detecting unimodal fuzzy sets and its application as a clustering technique, *IEEE Trans. Comput.*, **C-19**, 583–593.
811 Gluss, B. (1973). Fuzzy multistage decision making, *Int. J. Control*, **17**, 177–192.
812 Godal, R. C., and Goodman, T. J. (1980). Fuzzy sets and borel, *IEEE Trans. Syst., Man, Cybern.*, **SMC-10** (10), Oct.
813 Goddard, G., and Routley, R. (1973). *The Logic of Significance and Context*, Scottish Academic Press, Edinburgh.
814 Goetcherian, V. (1980). Image processing using fuzzy logic concept, *Pattern Recogn.*, **12**.
815 Goguen, J. A. (1980). Fuzzy sets and the social nature of truth, in *Proc. Int. Congr. on Appl. Syst. Res. and Cybern.*, Acapulco, Mexico, Dec.
816 Goguen, J. A. (1966). Fuzzy sets, Proj. NRP 49-170, NR 314–103, Dept. of Math., Univ. of California, Berkeley.
817 Goguen, J. A. (1967). *L*-fuzzy sets, *J. Math. Anal. Appl.*, **18**, 145–174.
818 Goguen, J. A. (1968). Categories of fuzzy sets—Applications of non-Cantonian set theory, Ph.D. Thesis, Dept. of Math., Univ. of California, Berkeley, in *Diss. Abstr. Int.*, **29/09-B**, 3393.
819 Goguen, J. A. (1969). Categories of *L*-fuzzy sets, *Bull. Amer. Soc.*, **75** (3), May, 622–624.
820 Goguen, J. A. (1969). Representing inexact concepts, ICR Quart. Rep. No. 20, Inst. for Comput. Res., Univ. of Chicago, Chicago.
821 Goguen, J. A. (1969). The logic of inexact concepts, *Synthese* **19**, 325–373.
822 Goguen, J. A. (1970). Mathematical representation of heirarchically organized system, in *Global System Dynamics*, (E. O. Attinger, Ed.), S. Karger, Berlin, pp. 111–129.
823 Goguen, J. A. (1972). Hierarchical inexact data structures in artificial intelligence problems, in *Proc. 5th Hawaii Int. Conf. Syst. Sci.*, Honolulu, p. 345.
824 Goguen, J. A. (1973). Axioms, extensions and applications for fuzzy sets, IBM Res. Rep., IBM Res. Center, Yorktown Hts., NY.
825 Goguen, J. A. (1973). Some comments on applying mathematical systems theory, Doc. Comput. Sci. Dept., Univ. of California, Los Angeles.
826 Goguen, J. A. (1973). Systems theory concepts in computer science, in *Proc. 6th Hawaii Int. Conf. on Syst. Sci.*, Honolulu, pp. 77–80.
827 Goguen, J. A. (1974). Concept representation in natural and artificial languages—Axioms extensions and applications for fuzzy sets, *Int. J. Man-Mach. Stud.*, **6**, 513–561.
828 Goguen, J. A. (1974). The fuzzy Tychonoff theorem, *J. Math. Anal. Appl.*, **43**, 734–742.
829 Goguen, J. A. (1975). Objects, *Int. J. Gen. Syst.*, **1**, 237–243.

Key References in Fuzzy Pattern Recognition

830 Goguen, J. A. (1975). On fuzzy robot planning, in *Fuzzy Sets and their Applications to Cognitive and Decision Processes*, (L. A. Zadeh, K. S. Fu, K. Tanaka, and M. Shimura, Eds.), Academic Press, New York, pp. 429-447.

831 Goguen, J. A. (1976). Robust programming languages and the principle of maximal meaningfulness, *Milwaukee Symp. on Autom. Comput. and Control*, Milwaukee, WI, pp. 87-90.

832 Goguen, J. A. (1979). Fuzzy sets and the social nature of truth, in *Advances in Fuzzy Set Theory and Applications*, (M. M. Gupta, R. K. Ragade, and R. R. Yager, Eds.), North-Holland, Amsterdam.

833 Goguen, J. A. (1979). Review of C. V. Negoita and D. A. Ralescu, Application of fuzzy sets to systems analysis, *J. Symb. Logic*.

834 Goguen, J. A., Thatcher, J., Wagner, E., and Wright, J. (1973). A junction between computer science and category theory, basic concepts and examples, II., Universal constructions, IBM Rep. Res.

835 Good, I. J. (1962). Subjective probability as the measure of a non-measurable set, in *Logic, Methodology and Philosophy of Science*, (E. Nagel, P. Suppes, and A. Tarski, Eds.), Stanford Univ. Press, Stanford, CA, pp. 319-329.

836 Goodman, I. R. (1976). Some relations between fuzzy sets and random sets, Apr., Unpublished.

837 Goodman, I. R. (1980). Fuzzy sets as equivalence classes of random sets, in *Proc. Int. Congr. on Appl. Syst. Res. and Cybern.*, Acapulco, Mexico, Dec.

838 Goodman, I. R. (1980). Identification of fuzzy sets with a class of canonically induced random sets, in *Proc. Int. Congr. on Appl. Syst. Res. and Cybern.*, Acapulco, Mexico, Dec.

839 Goodman, J. S. (1974). From multiple balayage to fuzzy sets, Inst. of Math., Univ. of Florence, Italy.

840 Gorvernet, J., Ayme, S., and Sanchez, E. (1980). Approximate reasoning in medicine genetics, in *Proc. Int. Congr. on Appl. Syst. Res. and Cybern.*, Acapulco, Mexico, Dec.

841 Gottinger, H. W. (1973). Competitive processes—Application to urban structure, *Cybernetica*, 16, 177-197.

842 Gottinger, H. W. (1973). Towards a fuzzy reasoning in the behavioural science, *Cybernetica*, 16, 113-135.

843 Gottinger, H. W. (1975). A fuzzy algorithmic approach to the definition of complex or imprecise concepts, Conf. on Syst. Theory, Univ. of Bielefeld, Bielefeld, Germany, Apr.

844 Gottinger, H. W. (1976). Some basic issues connected with fuzzy analysis, in *Systems Theory in the Social Sciences*, (H. Bossel, S. Klaczko, and N. Muller, Eds.), Birkhauser Verlag, Basel, Switzerland, 323-325.

845 Gottinger, H. W. (1976). Toward an algebraic theory of complexity and catastrophe, 3rd Eur. Meet. Cybern. Syst. Res., Vienna.

846 Gottinger, H. W. (1977). Complexity and social decision rules, in *Proc. 1977 IEEE Conf. on Decis. and Control*, New Orleans.

847 Gottinger, H. W., Hirota, K., and Iiyama, Y. (1977). Complexity and social decision, in *Proc. of the Symp. on Fuzzy Set Theory and Appl., IEEE Conf. on Decis. and Control*, New Orleans.

848 Gottwald, S. (1969). Konstruktion von Zahlbereichen und die Grundlagen der Inhaltstheorie in einer Mehrwertigen Mengenlehre, Ph.D. Thesis, Univ. of Leipzig, Leipzig, Germany.

849 Gottwald, S. (1971). Elementare inhals—Und Masstheorie in einer Mehrwertigen Mengenlehre, *Math. Nachr.*, **50**, 27-68.

850 Gottwald, S. (1971). Zahlbereichskonstruktionen in einer Mehrwertigen Mengenlehre, *Z. Math. Logik Grundlagen Math.*, **17**, 145-188.

851 Gottwald, S. (1973). Uber Einbettungen in Zahlenbereiche einer Mehrwertigen Mengenlehre, *Math. Nachr.*, **56**, 43-46.

852 Gottwald, S. (1974). Fuzzy topology, product and quotient theorems, *J. Math. Anal. Appl.*, **45**, 512-521.

853 Gottwald, S. (1974). Mehrwertige Anordnungsrelationen in Klassischen Mengen, *Math. Nachr.*, **63**, 205-212.

854 Gottwald, S. (1975). A cumulative system of fuzzy sets, in *Proc. 2nd Colloq. Set Theory and Hierarchy Theory*, Bierutovice, Poland, Sept.

855 Gottwald, S. (1975). Ein Kumulatives System Mehrwertiger Mengen, Habilitationeschrift, Univ. of Leipzig, Leipzig, Germany.

856 Gottwald, S. (1976). Untersuchen zur Mehrwertigen Mengenlehre, *Math. Nachr.*, **72**, 297-303; **74**, 329-336.

857 Gottwald, S. (1976). Fuzzy propositional logics, Sekt. Math., Karl-Marx-Univ., Leipzig, Germany.

858 Gottwald, S. (1977). Mengentheoretische Eigenschaften Unscharfer Begriffe, *Math. Nachr.*, **91**.

859 Gottwald, S. (1978). Theoretische Betrachtungen uber Fuzzy Logik, Schriftenreihe Weiterbildungszentrum Math. Kybern. Rechentechnik, Tu Dresden, Heft, pp. 3-22.

860 Gottwald, S. (1978). Universes of fuzzy sets closed under fuzzification, Sekt. Math., Karl-Marx-Univ., Leipzig, Germany.

861 Gottwald, S. (1979). A note on measures of fuzziness, *Elek. Informationsverarb. Kybern.*, **15** (4), 221-223.

862 Gottwald, S. (1979). Eine Anwendungsvariante der Mehrwertigen Logik, Doc., Sekt. Marxistisch-Leninistiche Philos., Karl-Marx-Univ., Leipzig, Germany.

863 Gottwald, S. (1980). Fuzzy uniqueness of fuzzy mappings, Sekt. Math., Karl-Marx-Univ., Leipzig, Germany, pp. 1-41.

864 Gottwald, S. (1980). A note on fuzzy cardinals, in *Kybernetika*, **16** (2), Prag.

865 Gottwald, S. (1981). Fuzzy points and local properties of fuzzy topological spaces, *FSS*, 199-202.

866 Gottwald, S. (1980). Set theory for fuzzy sets of higher level, Sekt. Math., Karl-Marx-Univ., Leipzig, Germany.

867 Govind, R. (1978). Synthesis of fuzzy controllers for process plants, Carnegie-Mellon Inst. of Res., Pittsburgh, PA, *Proc. of the 1978 Int. Conf. on Cybern. and Soc.*, Vol. 2, pp. 1228-1230.

868 Goyet, G., and Lepoint, P. (1978). A family of quasi-distances generating the topology of convergence in fuzzy measure, Colloq. Int., Theorie et Appl. des Sous-Ensembles Flous, Journ. de Biomath. et d'Inf. Med., Marseille, France, Sept.

869 Gozalczany, M. B., Kiszka, J. B., and Stachowicz, M. S. (1980). Some problems of studying adequacy of fuzzy models, in *Proc. Int. Congr. on Appl. Syst. Res. and Cybern.*, Acapulco, Mexico, Dec.

870 Grainier, E. (1977). Recherche inventive sur le 4th age en utilisant la theorie des sous-ensembles flous, Doc. Thesis in Med., Fac. de Med. de Montpellier, France, Feb.

Key References in Fuzzy Pattern Recognition

871 Gratten-Guiness, I. (1976). Fuzzy membership mapped onto interval and many-valued quantities, *Z. Math. Logik Grundlagen Math.*, **22**, 149–160.

872 Gratten-Guinness, I. (1979). Forays into the meta-theory of fuzzy set theory, *Logique et Anal.* (Louvain), No. 87.

873 Grigolia, R. (1975). On the algebras corresponding to the N-valued Lukasiewicz-Tarski logical systems, in *Proc. 1975 Int. Symp. Multiple-Valued Logic*, IEEE 75CH0959-7C, 234–239.

874 Groen, M. (1977). On the effect of smoothing density membership values in a fuzzy clustering algorithm. Delft Univ. of Tech., Delft, Holland.

875 Grofman, B., and Hyman, G. (1973). Probability and logic in belief systems, *Theory Decis.*, **4**, 179–195.

876 Grudzewski, W. M., and Pelech, A. (1976). Modeling a structure of the project in the fuzzy sets language, Rep. No. 189, Inst. of Organ. and Manage., Tech. Univ. of Wroclaw, Wroclaw, Poland.

877 Gu, X. (1980). Applications of fuzzy sets theory to the problems dealing with data of geophysical and geochemical prospecting, News on 1980 Ann. Rep. Meet. of Beijing Working Group on Fuzzy Sets, Beijing, China.

878 Gunderson, R. (1978). Application of fuzzy Isodata algorithms to star-tracker pointing systems, 7th Trienn. IFAC World Congr., Helsinki.

879 Gunderson, R. (1979). Application of fuzzy C-variates and entropy to earth sun experiment sampling rates, 1st Int. Symp. on Policy Anal. and Inf. Syst., Durham, NC, June.

880 Gunderson, R. B., and Watson, J. D. (1979). Sampling and interpretation of atmospheric science experimental data, 1st Int. Symp. on Policy Anal. and Inf. Syst., June.

881 Guo, R., Xie, M., Chen, W., Ma, B. (1980). Fuzzy set model for computerized diagnosis system in traditional Chinese medicine, News on 1980 Ann. Rep. Meet. of Beijing Working Group on Fuzzy Sets, Beijing, China.

882 Guo-Quan, C. (1981). Manual control strategies described by fuzzy mathematics and its mechanization, Univ. of Sci. and Tech. of China, Hofei, Anhwei, China.

883 Guo-Quan, C. (1978). Fuzzy sets theory and applications, A review, No. 1, Articles 1–3 were read out at 1978 Ann. Meet. of Autom. Inst. of China.

884 Guo-Quan, C. (1979). Advance from accuracy to fuzzy, Central People's Broadcasting Station, Apr.

885 Guo-Quan, C. (1980). An approach to fuzzy controller algorithms, *Inf. Control*, **9** (3).

886 Guo-Quan, C. (1981). Fuzzy mathematics description of manual control strategies and its mechanization, in *Proc. 1st Nat. Conf. on Mach. Intell. and Pattern Recogn.*

887 Guo-Quan, C. (1981). *Fuzzy Sets, Fuzzy Linguistic Variable and Fuzzy Logic*, Translation and edition supported by Chinese Academic Press, New York.

888 Gupta, M. M. (1974). Fuzzy automata and control, in *Proc. Comput., Electron., and Control*—An Int. Trade Show and Symp., Calgary, May, VI.3.1–3.8.

889 Gupta, M. M. (1974). Introduction to fuzzy control *Proc. Comput., Electron., and Control*—An Int. Trade Show and Symp., Calgary, May.

890 Gupta, M. M. (1975). Fuzzy automata and decision processes—A decade, 6th Trien. IFAC World Congr., Boston, MA, Aug.

891 Gupta, M. M. (1975). On the estimation and control in a fuzzy environment—Report on the round table discussion, IFAC *J. Automatica*, **11**, Mar., 209–212.

892 Gupta, M. M. (1976). Fuzziness and human behavior in decision making, in *Joint Nat. Meet. of Oper. Res. Soc. of Amer. (ORSA) and the Inst. of Manage. Sci. (TIMS)*, Miami, FL, Nov., Pap. No. FA 10.6, pp. 1–13.

893 Gupta, M. M. (1976). Fuzziness and human conciousness in decision-making, ORSA/TIMS Conf., Miami, FL, Nov.

894 Gupta, M. M. (1976). Theory of fuzzy sets and its applications to operations research and management science, ORSA/TIMS 1976, Joint Nat. Meet., Philadelphia, PA, Mar. 31–Apr. 2.

895 Gupta, M. M. (Ed.) (1977). Fuzzy set theory and applications, in *Proc. 1977 IEEE Conf. on Decis. and Control*, New Orleans, Dec. 8, pp. 1301–1450.

896 Gupta, M. M. (1977). Fuzzy-ism, the first decade, in *Fuzzy Automata and Decision Processes*, (M. M. Gupta, G. N. Saridis, and B. R. Gaines, Eds.), North-Holland, Amsterdam, pp. 5–10.

897 Gupta, M. M. (1977). The notion of fuzziness, a perspective, in *Proc. Symp. on Fuzzy Sets*, New Orleans, p. 1302.

898 Gupta, M. M. (Ed.) (1978). Special issue of fuzzy sets and applications—An editorial, *Int. J. Fuzzy Sets Syst.*, Jan.

899 Gupta, M. M. (1979). A survey of process control applications of fuzzy set theory, Univ. of Saskatchewan, in *Proc. 1978 IEEE Conf. on Decis. and Control*, includes the 17th Symp. on Adaptive Processes, 78CH1392-OCS, Jan. 1979, San Diego, CA, pp. 1454–1461.

900 Gupta, M. M. (1978). Fuzzy concepts and its future, ORSA/TIMS, New York, May.

901 Gupta, M. M. (1978). IFAC Report on the 3rd round table discussion session (RT-15) on fuzzy decision making and applications, in *Proc. 7th Trienn. IFAC World Congr.*, Helsinki, June.

902 Gupta, M. M. (1978). The manifold uses of graded membership, RT-15, in *Proc. 7th Trienn. IFAC World Congr.*, Helsinki, June.

903 Gupta, M. M. (1979). Guest editorial to the Special Issue on "Fuzzy Sets and Applications," *Fuzzy Sets Syst.*, **2** (1), Jan., 1–4.

904 Gupta, M. M. (1979). The field of fuzzy sets, Some remarks, 1st Int. Symp. on Policy Anal. and Inf. Syst., June.

905 Gupta, M. M. (1980). Fuzzy logic controllers, in *Proc. JACC*, San Francisco, Aug.

906 Gupta, M. M. (1980). Stochastic control versus fuzzy logic control in feedback control systems, in *Proc. Int. Congr. on Appl. Syst. Res. and Cybern.*, Acapulco, Mexico, Dec.

907 Gupta, M. M., and Mamdani, E. H. (1976). 2nd IFAC round table on fuzzy automata and decision processes, *Automatica*, **12**, 291–296.

908 Gupta, M. M., and Nikiforuk, P. N. (1978). Feedback control of industrial processes via fuzzy set theory, Some remarks, in *Proc. 1978 JACC*, Vol. 3, pp. 37–52.

909 Gupta, M. M., and Nikiforuk, P. N. (1979). On the characterization of fuzzy processes, some observations, in *Proc. 18th IEEE Conf. on Decis. and Control*, Dec.

910 Gupta, M. M., Nikiforuk, P. N., and Kanai, K. (1973). Decision and control in a fuzzy environment—A rationale, in *Proc. 3rd IFAC Symp. Identification and Syst. Parameter Estimation*, The Hague, June, pp. 1048–1049.

911 Gupta, M. M., and Ragade, R. K. (1977). Fuzzy set theory and its applications—Survey (Invited Pap.), in *Proc. IFAC Symp. on Multivariable Syst. (MVTS)*, Fredericton, Canada, July, pp. 247–259.

Key References in Fuzzy Pattern Recognition

912 Gupta, M. M., Ragade, R. K., and Yager, R. R. (1979). *Advances in Fuzzy Set Theory and Applications*, North-Holland, Amsterdam.

913 Gupta, M. M., Ragade, R. K., and Yager, R. R. (1979). Report on the IEEE Symp. on Fuzzy Set Theory and Applications (Short communication), *Fuzzy Sets Syst.*, **2** (1), Jan., 105-111.

914 Gupta, M. M., Ragade, R. K., and Yager, R. R. (Eds.) (1979). *Advances in Fuzzy Set Theory and Applications*, North-Holland, Amsterdam.

915 Gupta, M. M., Saridis, G. N., and Gaines, B. R. (1977). *Fuzzy Automata and Decision Process*, North-Holland, New York.

916 Gupta, M. M., and Tsukamoto, Y. (1980). Fuzzy logic controllers—A perspective, JACE, Univ. of Saskatchewan, Saskatoon, Canada.

917 Gupta, M. M., Tsukamoto, Y., and Ikiforuk, P. N. (1980). Truth qualifications and numerical truth values, in *Proc. JACC*, San Francisco, Aug.

918 Gurbutt, P. A. (1975). A model of problem solving in physics, IDP 106, Centre Imago, Univ. Catholique de Louvain, Louvain, Belgium.

919 Gusev, L. A., and Smirnova, I. M. (1973). Fuzzy sets—Theory and applications (A survey), *Autom. Remote Control* (5), May, 66-85.

920 Gusev, L. A., and Smirnova, I. M. (1975). Simulation of behavior and intelligence, *Aumatika Telemek.*, Moskow (5), 65-85 (in Russian).

921 Gustafson, D., and Kessel, W. (1979). Fuzzy clustering with Covar, in *Matrix, 17th IEEE Conf. on Decis. and Control*, pp. 761-766.

922 Gyorfi, L, and Koczy, L. T. (1979). An algorithm for non parametric decision motivated by fuzzy approach, *Probl. Control. Inf. Theory*, **8** (3).

923 Haack, S. (1974). *Deviant Logic*, Cambridge Univ. Press, Cambridge, UK.

924 Haack, S. (1975). "Alternative" in "alternative logic," in *Meaning, Reference and Necessity* (S. Blackburn, Ed.), Cambridge Univ. Press, Cambridge, UK, pp. 32-55.

925 Haack, S. (1976). The justification of deduction, *Mind*, **85**, 112-119.

926 Haack, S. (1979). Do we need "FUZZY LOGIC"?, *Int. J. Man-Mach. Stud.*, **11**, 437-445.

927 Haake, J. S. (1978). Maze running using fuzzy logic, M.Sc. Thesis, Comput. Sci. Dept., Univ. of California, Los Angeles.

928 Haar, R. L. (1977). A fuzzy relational data base system, TR-586, Dept. of Comput. Sci. Annu. Rep., Univ. of Maryland, College Pk.

929 Hacking, I. (1963). What is strict implication?, *J. Symb. Logic*, **28**, 51-71.

930 Hacking, I. (1975). All kinds of possibility, *Philos. Rev.*, **84**, 319-337.

931 Hacking, I. (1975). *The Emergence of Probability*, Cambridge Univ. Press, Cambridge, UK.

932 Hackstaff, H. H. (1966). *Systems of Formal Logic*, D. Reidel, Dordrecht, Holland.

933 Hagg, C. (1977). Possibility and cost in decision analysis, *Fuzzy Sets Syst.*, **1**, 81-86.

934 Hajek, P. (1967). Sets, semisets, models, in *Axiomatic Set Theory*, *Proc. Symp. on Pure Math.*, Vol. 13, Amer. Math. Soc., RI, pp. 67-81.

935 Hajek, P. (1968). Problem obecneho pojeti metody guha, *Kybernetika* (Prague), **6**, 505-515 (in Czech—The question of the general concept of Guha-methods).

936 Hajek, P. (1973). Automatic listing of important observational statements 2, *Kybernetika* (Prague), **9**, 251-271.

937 Hajek, P. (1973). Automatic listing of important observational statements I, *Kybernetika* (Prague), **9**, 187–206.
938 Hajek, P. (1973). Some logical problems of automated research, in *Proc. Symp. Math. Found. Comput. Sci.*, High Tatras, Czechoslovakia.
939 Hajek, P. (1974). Automatic listing of important observational statements III, *Kybernetika* (Prague), **10**, 95–124.
940 Hajek, P. (1974). Generalized quantifiers and finite sets, in *Proc. Autumn School in Set Theory and Hierarchy Theory*, Wroclaw, Poland.
941 Hajek, P. (1975). On logics of discovery, in *Mathematical Foundations of Computer Science 1975*, Lecture Notes in Comput. Sci., Vol. 32 (J. Becvar, Ed.) Springer-Verlag, Berlin, pp. 30–45.
942 Hajek, P., Bendova, K., and Renc, Z. (1971). The Guha Method and the three valued logic, *Kybernetika* (Prague), **7**, 421–435.
943 Hajek, P., and Harmancova, D. (1973). On generalized credence functions, *Kybernetika* (Prague), **9**, 343–356.
944 Hajek, P., Havel, I., and Chytil, M. (1966). The Guha method of automatic hypotheses determination, *Computing*, **1**, 293–308.
945 Hajnal, M., and Koczy, L. T. (1980). Texture analysis by vector valued fuzzy sets, *Bull. Sous Ensembles Flous Appl.*, Automne.
946 Halmos, P. R. (1962). *Algebraic Logic*, Chelsea Publ. Co., New York.
947 Halpern, J. (1975). Set adjacency measures in fuzzy graphs, *J. Cybern.*, **5** (4), 77–87.
948 Hamacher, H. (1976). On logical connectives of fuzzy statements and their affiliated truth-functions, 3rd Eur. Meet. Cybern. Syst. Res., Vienna.
949 Hamacher, H. (1978). *Uber Logische Aggregationen Nicht-Binar Explizierter Entscheidungskriterien, Ein Axiomatischer Beitrag zur Normativen Entscheidungstheorie*, R. G. Fischer Verlag, Frankfort/Main, Germany.
950 Hamacher, H. (1979). A representation theorem for the negation in fuzzy logics, 1st Int. Symp. on Policy Anal. and Inf. Syst., June, Duke Univ., NC.
951 Hamacher, H. (1979). On "and" and "or" connectives—An axiomatic approach, *Fuzzy Sets Syst.*
952 Hamacher, H., Leberling, H., and Zimmermann, H. J. (1978). Sensitivity Analysis in fuzzy linear programming, *Fuzzy Sets Syst.*, **1**, 269–281.
953 Hamblin, C. L. (1959). The modal "probably," *Mind*, **68**, 234–240.
954 Hammerbach, I. M., and Yager, R. R. (1981). The personalization of security selection—An application of fuzzy set theory, *Fuzzy Sets Syst.*, **5** (1), Jan., 1–10.
955 Hanakata, K. (1974). A methodology for interactive systems, in *Learning and Intelligent Robots*, (K. S. Fu and J. T. Tou, Eds.), Plenum Press, New York, pp. 317–324.
956 Hannan, E. L. (1979). On the efficiency of the product operator in fuzzy programming with multiple objectives, *Fuzzy Sets Syst.*, **2** (3), 259–362, July.
957 Hao-Xuan, Z. (1980). Relations between topological spaces and fuzzy topological spaces. Rep. prepared by the Group of Topology in Sichuan Univ., Sichuan, China, Nov.
958 Hara, F. (1975). A dynamic model of collective human flow from big fires, in *Summ. of Pap. on Gen. Fuzzy Problems*, The Working Group on Fuzzy Syst., Tokyo, Nov., pp. 14–18.
959 Hara, F. (1980). Fuzzy control in computer simulations of civil evaluation from

large-scale fires, in *Summ. of Pap. on Gen. Fuzzy Problems*, The Working Group on Fuzzy Syst., Tokyo.

960 Hara, F. and Kitagawa, S. (1977). A new face graph and its application to system fault diagnosis, in *Summ. of Pap. on Gen. Fuzzy Problems*, The Working Group on Fuzzy Syst., Tokyo.

961 Haroche, C. (1975). Grammar, implicitness and ambiguity—Foundations of inherent ambiguity of discourse, *Found. Lang.*, **13**, 215–236 (in French).

962 Harris, J. I. (1974). Fuzzy implication—Comments on a paper by Zadeh, DOAE Res. Working Pap., Minist. of Def., Byfleet, Surrey, UK.

963 Harris, J. I. (1974). Fuzzy sets—How to be imprecise precisely, DOAE Res. Working Pap., Minist. of Def., Byfleet, Surrey, UK.

964 Hart, W. D. (1972). Probability as a degree of possibility, *Notre Dame J. Formal Logic*, **13**, 286–288.

965 Hatten, M. L., Whinston, A. B., and Fu, K. S. (1975). Fuzzy set and automata theory applied to economics, Reprint Ser. No. 533, H. C. Krannert Graduate School, Purdue Univ., Lafayette, IN.

966 Havranek, T. (1971). The statistical modification and interpretation of the Guha method, *Kybernetika* (Prague), **7**, 13–21.

967 Havranek, T. (1974). Some aspects of automatic systems of statistical inference, in *Proc. Eur. Meet. of Stat.*, Prague.

968 Havranek, T. (1975). Statistical quantifiers in observational calculi—An application in Guha-methods, *Theory Decis.*, **6**, 213–230.

969 Havranek, T. (1975). The approximation problem in computational statistics, in *Mathematical Foundations of Computer Science 1975*, Lecture Notes in Comput. Sci., Vol. 32, (J. Becvar, Ed.), Springer-Verlag, Berlin, pp. 260–265.

970 Hay, L. S. (1963). Axiomatization of the infinite-valued predicate calculus, *J. Symb. Logic*, **28**, 77–86.

971 He, J. (1980). Fuzzy Events and Fuzzy Probability, Collect. Pap. on Fuzzy Math. (Abstr.), Huazhong Inst. of Tech., Huazhong, China, Apr.

972 He, J. (1980). Fuzzy measure theory (1)—Classes of fuzzy sets, Collect. Pap. on Fuzzy Math. (Abstr.), Huazhong Inst. of Tech., Huazhong, China, Apr.

973 He, J. (1980). Fuzzy measure theory (2)—Fuzzy measurable sets and fuzzy measures, Collect. Pap. on Fuzzy Math. (Abstr.), Huazhong Inst. of Tech., Huazhong, China, Apr.

974 He, J. (1980). Fuzzy measure theory (3)—Fuzzy measurable transformation, Collect. Pap. on Fuzzy Math. (Abstr.), Huazhong Inst. of Tech., Huazhong, China, Apr.

975 He. J. (1980). Fuzzy measure theory (4)—Fuzzy integral, Collect. Pap. on Fuzzy Math. (Abstr.), Huazhong Inst. of Tech., Huazhong, China, Apr.

976 He, J. (1980). Fuzzy measure theory (5)—Fuzzy space, Collect. Pap. on Fuzzy Math. (Abstr.), Huazhong Inst. of Tech., Huazhong, China, Apr.

977 He, J. (1980). On fuzzy measure, Rep. Dept. of Math., Sichuan Normal Univ. 5, A List of Lit. in China, The Univ. of Sci. and Tech. of China, Hofei, Anhwei, China.

978 He, T. (1980). A general theorem of lower extreme solution of fuzzy relation equation and concrete solution of ordinary solutions, News. on 1980 Ann. Rep. Meet. of Beijing Working Group on Fuzzy Sets, Beijing, China.

979 He, T. (1980). Fuzzy relational equations, Rep. Beijing Normal Univ. 11, A List of Lit. in China, The Univ. of Sci. and Tech. of China, Hofei, Anhwei, China.

980 He, Z. (1980). Fuzzy mathematics—Mathematical model on kindred and genetics relationships, Collect. Pap. on Fuzzy Math. (Abstr.), Huazhong Inst. of Tech., Huazhong, China, Apr.

981 He, Z. (1980). Fuzzy mathematics and computer, Collect. Pap. on Fuzzy Math. (Abstr.), Huazhong Inst. of Tech., Huazhong, China, Apr.

982 He, Z. (1980). Fuzzy optimal strategy in games with perfect informations, News on 1980 Ann. Rep. Meet. of Beijing Working Group on Fuzzy Sets, Beijing, China.

983 He, Z. (1981). The mathematical model of blood relation and heredity and some considerations of the realization of this mathematics model on computer, Beijing Ind. Inst., A List of Lit. in China, The Univ. of Sci. and Tech. of China, Hofei, Anhwei, China.

984 He, Z., and Ming, S. (1980). Lysis to the study of semantics in poetry, News on 1980 Ann. Rep. Meet. of Beijing Working Group on Fuzzy Sets, Beijing, China.

985 Hejek, P. (1973). Automatic listing of important observational statements, *Kybernetics*, **9**, 187–206.

986 Hejek, P. (1973). Why semisets, *Commentat. Math. Univ. Carolinae*, **14**, 397–420.

987 Hempel, C. G. (1937). A purely topological form of non-Aristotelian Logic, *J. Symb. Logic*, **2**, 97–112.

988 Hendry, W. L. (1972). Fuzzy sets and Russell's paradox, Los Alamos Sci. Lab., Univ. of California, Los Alamos, NM.

989 Henkin, L. (1963). A class of non-normal models for classical sentential logic, *J. Symb. Logic*, **28**, 300.

990 Henry-Labordere, A., and De Backer, P. (1979). Intelligence creative, automatisasation de processus d'association, in *Praktische Anwendung der Morphologischen Methode*, Fritz Swicky Stiftung, Glarus, Switzerland, May, pp. 20–29.

991 Hentschel, B. (1977). Nachbildung der Empirie des Experten Zur Progressteuerung, in *Unscharfe Modellbildung und Steuerung*, Lecture Notes, TH Karl-Marx-Stadt, pp. 84–98, June.

992 Hentschel, B. (1979). Ein Verfahren zur Experimentellen Bestimmung und Nutzung Unscharfer Prozessinverser und ihr Einsatz zur Losung des Folgesteuerungsproblems Nicht-Linear-Dynamischer Mehrgrossen-Systeme, Material zur Volesung "Kennwertermittlung und Modellbildung", Teil II—"Unscharfe Modellbildung und Steuerung" der TH Karl-Marx-Stadt, Oct.

993 Herbert, J. H. (1974). Light-shattering from fuzzy spheres—A study of the effects of light structure and a model for inhomogeneous and nonspherical (aerosol) particles, Ph.D. Thesis, Univ. of Washington, in *Diss. Abstr. Int.*, **35/08-B**, 4094.

994 Hersch, H. M. (1977). A fuzzy model of human reasoning, The ORSA/TIMS Meet., Atlanta, GA.

995 Hersch, H. M. (1977). A fuzzy set-theoretic analysis of age terms, Unpublished.

996 Hersh, H., and Caramazza, A. (1975). Integrating verbal quantitative information, *Bull. Psychonomic Soc.* **6** (6), 589–591.

997 Hersh, H. M. (1976). Fuzzy reasoning—The integration of vague information, Ph.D. Thesis, The John Hopkins Univ., Baltimore, MD, in *Diss. Abstr. Int.*, **37/04-B**, 1941.

998 Hersh, H. M., and Caramazza, A. (1975). The quantification of vague concepts, Psychometric Soc. Meet., Iowa City, IA, Apr.

999 Hersh, H. M., and Caramazza, A. (1976). A fuzzy set approach to modifiers and vagueness in natural language, *J. Exp. Psychol.*, **105**, 254–276.

Key References in Fuzzy Pattern Recognition

1000 Hersh, H. M., Caramazza, A., and Brownell, H. H. (1979). Effects of context of fuzzy membership functions, in *Proc. 18th IEEE Conf. on Decis. and Control*, Fort Lauderdale, FL, Dec.

1001 Hersh, H. M., and Spiering, J. (1976). How old is old, Eastern Psychol. Assoc. Meet., New York, Apr.

1002 Hinde, C. J. (1977). Algorithms embedded in fuzzy sets, in *Comput. Symp.* (Morlet and Ribbens, Eds.), North-Holland. Amsterdam, pp. 361–367.

1003 Hintikka, J., Moravscsik, J., and Suppes, P. (1973). *Approach to Natural Languages*, D. Reidel, Dordrecht, Holland.

1004 Hintikka, J., and Suppes, P. (Eds.) (1970). *Information and Inference*, D. Reidel, Dordrecht, Holland.

1005 Hirai, H., Asai, K., and Kitajima, S. (1968). Fuzzy automata and its application to learning control systems, *Mem. Fac. Eng. Osaka City Univ.*, **10**, 67–73.

1006 Hiramatsu, K., Kabasawa, K., and Kaibara, S. (1974). Fuzzy logic applied to the medical diagnosis, *Med. Electron. Biosci.*, **12**, 34–41.

1007 Hirokawa, S., and Miyano, S. (1978). A note on the regularity of fuzzy languages, *Mem. Fac. Sci. Kyushu Univ.*, Ser. A, **32** (1), 61–66.

1008 Hirota, K. (1977). "Kakuritsu-Shugoron" fundamental research works of fuzzy system theory and artificial intelligence, S.51 Mombusho Kaken-Hi Hokoku, pp. 193–213.

1009 Hirota, K. (1977). Concepts of probabilistic sets, in *Proc. 1977 IEEE Conf. on Decis. and Control*, New Orleans.

1010 Hirota, K. (1977). Probabilistic sets—Expansion of fuzzy concepts based on probability theory, *Summ. of Pap. on Gen. Fuzzy Problems*, The Working Group on Fuzzy Syst., Tokyo, pp. 135–155.

1011 Hirota, K. (1978). Extended fuzzy expression of probability sets, vagueness functions and monitor, Rep. No. 4, Working Group on Fuzzy Syst., Tokyo, Dec., pp. 26–30.

1012 Hirota, K. (1978). Extended fuzzy expression of probabilistic sets—Analytical expression of ambiguity and subjectivity in pattern recognition, Semin. on Appl. Funct. Anal., July.

1013 Hirota, K. (1980). On the number of states in questionnaires, Table Ronde du CNRS sur le Flou, Lyon, France, June.

1014 Hirota, K., and Iijima, T. (1976). A decision making model using the probabilistic set theory, PRL76-21, Tech. Rep. on Pattern Recogn. and Learning of IECE.

1015 Hirota, K., and Iijima, T. (1976). Probabilistic set theory, PRL76-36, Tech. Rep. on Pattern Recogn. and Learning of IECE.

1016 Hirota, K., and Iijima, T. (1978). A decision making model—A new approach based on the concept of probability sets, IEEE Int. Conf. on Cybern. and Soc., Tokyo, Japan.

1017 Hirota, K., and Iijima, T. (1978). The bounded variation quantity and its applications to feature extractions, 4th Int. Joint Conf. on Pattern Recogn., Kyoto, Japan.

1018 Hirsch, G., Lamotte, M., Mas, M. T., and Vigneron, H. J. (1978). Classification des phonemes au moyen d'une relation de dissimilitude floue. Colloq. Int. Theorie et Appl. des Sous-Ensembles Flous, Journ. de Biomath. et d'Inf. Med., Marseille, France, Sept.

1019 Hirsch, G., Lamotte, M., Mas, M. T., and Virneron, M. J. (1981). Phonemic classification using a fuzzy dissimilitude relation, *Fuzzy Sets and Syst.*, **5** (3), 267–276.

1020 Hisdal, E. (1978). Conditional and joint possibility of type 2, particularization, Inst. of Inf., Univ. of Oslo, Oslo.

1021 Hisdal, E. (1978). Conditional and joint possibilities of higher order, ISBN 82-90230-37-0, Res. Rep. No. 42, Inst. of Inf., Univ. of Oslo, Oslo.

1022 Hisdal, E. (1978). Conditional possibilities independence and noninteraction, *Fuzzy Sets Syst.*, **1**, 283–297.

1023 Hisdal, E. (1978). Particularization—The theory of fuzzy sets versus classical theories, Inst. of Inf., Univ. of Oslo, Oslo.

1024 Hisdal, E. (1978). Possibilities and grades of membership concrete and mathematical sets, Inst. of Inf., Univ. of Oslo, Oslo.

1025 Hisdal, E. (1978). Generalized fuzzy set systems and particularization, Inst. of Inf., Univ. of Oslo, Oslo.

1026 Hisdal, E. (1979). Concrete and mathematical sets—Measures of 2nd order possibilities, ISBN 82-90239-38-9, Res. Rep. No. 43, Inst. of Inf., Univ. of Oslo, Oslo.

1027 Hisdal, E. (1979). Developments in the wake of the theory of possibility, 1st Int. Symp. on Policy Anal. and Inf. Syst., Duke Univ., Durham, NC, June.

1028 Hisdal, E. (1979). Possibilistically dependent variables and a general theory of fuzzy sets, in *Advances in Fuzzy Set Theory and Application* (M. M. Gupta, R. K. Ragade, and R. R. Yager, Eds.), North-Holland, Amsterdam.

1029 Hisdal, E. (1979). Possibilistically dependent variables, 1st Symp. on Policy Anal. and Inf. Syst., Durham, NC, June.

1030 Hisdal, E. (1980). Fuzzy sets of higher type, interval valued fuzzy sets and a new if then else relation with guaranteed corrected inference, Table Ronde du CNRS sur le Flou, Lyon, France, June.

1031 Hisdal, E. (1980). Fuzzy sets of higher type, interval-valued fuzzy sets and a new if then else relation with guaranteed correct influence, in *Proc. Int. Congr. on Appl. Syst. Res. and Cybern.*, Acapulco, Mexico, Dec.

1032 Hisdal, E. (1980). Generalized fuzzy set systems and particularization, *Fuzzy Sets Syst.*, **4** (3), Nov., 275–292.

1033 Hockney, D., Harper, W., and Freed, B. (1975). *Contemporary Research in Philosophical Logic and Linguistic Semantics*, P. Reidel, Dordrecht, Holland.

1034 Hoffman, L. J., and Clements, D. (1977). Fuzzy computer security metrics, A preliminary report, ERL Memo. M77-6, Electron. Res. Lab., Univ. of California, Berkeley.

1035 Hoffman, L. J., Michelman, E., and Clements, D. P. (1978). Securate-security evaluation and analysis using fuzzy metrics, in *Proc. Nat. Comput. Conf.*

1036 Hogan, M., and Oden, G. C. (1978). A fuzzy propositional model of linguistic modifiers, 1st Joint Psychometric Soc. and the Soc. for Math Psychol.

1037 Hogarth, R. M. (1975). Cognitive processes and the assessment of subjective probability distributions, *J. Amer. Stat. Assoc.*, **70**, 271–294.

1038 Hohle, U. (1977). Probabilistic uniformization of fuzzy topologies, *Fuzzy Sets Syst.*, **1**, 311–332.

1039 Hohle, U. (1979). Example a locally convex L-fuzzy topology on the set of all almost everywhere defined random variables, in *Proc. Int. Semin. on Fuzzy Set Theory*, Johannes Kepler Univ., Linz, Austria.

1040 Hohle, U. (1979). Minkowski functionals of L-fuzzy sets, 1st Int. Symp. on Policy Anal. and Inf. Syst., Duke Univ., Durham, NC, June.

1041 Hohle, U. (1979). Upper semicontinuous fuzzy sets and their applications, in *Proc. Int. Semin. on Fuzzy Set Theory*, Johannes Kepler Univ., Linz, Austria.

Key References in Fuzzy Pattern Recognition 261

1042 Hohle, U. (1980). A mathematical theory of uncertainty 1–Fuzzy experiments and their representation, in *Proc. Int. Congr. on Appl. Syst. Res. and Cybern.*, Acapulco, Mexico, Dec.

1043 Hohle, U. (1980). L-fuzzy real numbers, Table Ronde CNRS sur le Flou, Lyon, France, June.

1044 Hohle, U. (1980). Probabilistic metrization of fuzzy uniformities, *Bull. Stud. Exch. Fuzziness Appl.*, Printemps.

1045 Hohle, U. (1981). Representation theorems for L-fuzzy quantities, *Fuzzy Sets Syst.*, **5** (1), Jan., 83–108.

1046 Honda, N. (1971). Fuzzy sets, *J. IECE* (Japan), **54**, 1359–1363.

1047 Honda, N. (1975). Applications of fuzzy sets theory to automata and linguistics, *J. JAACE*, **19**, 249–254.

1048 Honda, N., and Aida, S. (1975). Environmental index by faces method, in *Summ. of Pap. on Gen. Fuzzy Problems*, The Working Group on Fuzzy Syst., Tokyo, Nov., pp. 19–22.

1049 Honda, N., and Aida, S. (1977). An approach to the production line by man-computer system, in *Summ. of Pap. on Gen. Fuzzy Problems*, The Working Group on Fuzzy Syst., Tokyo, pp. 156–166.

1050 Honda, N., and Aida, S. (1980). Applications of faces method to evaluation and expression of multidimensional data, in *Summ. of Pap. on Gen. Fuzzy Problems*, The Working Group on Fuzzy Syst., Tokyo, Dec.

1051 Honda, N., and Nasu, M. (1975). Recognition of fuzzy languages, in *Fuzzy Sets and Their Applications to Cognitive and Decision Processes*, (L. A. Zadeh, K. S. Fu, K. Tanaka, and M. Shimura, Eds.), Academic Press, New York, pp. 279–299.

1052 Honda, N., Nasu, M., and Hirose, S. (1977). F-recognition of fuzzy languages, in *Fuzzy Automata and Decision Processes*, (M. M. Gupta, G. N. Saridis, and B. R. Gaines, Eds.), North-Holland, Amsterdam, pp. 149–168.

1053 Horejs, J. (1965). Classifications and their relationship to a measure, Publ. de la Fac. des Sci. de L'Univ. J. E. Purkyne, No. 168, Brno, Czechoslavakia, pp. 475–493.

1054 Hormann, A. M. (1971). Machine-Aided value judgements using fuzzy set techniques, SP-3590, Syst. Dev. Corp., Santa Monica, CA.

1055 Horvath, M. J. (1978). An identification procedure in learning disability through fuzzy set modeling of a verbal theory, Ph.D. Thesis, The Univ. of Arizona, in *Diss. Abstr. Int.*, **39/02-A**, 813.

1056 Hoskney, D., Harper, W., and Freed, B. (1975). *Contemporary Research*, D. Reidel, Dordrecht, Holland.

1057 Hou, R. (1980). The classification of the fuzzy systems and its optimal control problem, Collect. Pap. on Fuzzy Math. (Abstr.), Huazhong Inst. of Tech., Huazhong, China, Apr.

1058 Hou, R. (1980). The mathematical description of fuzzy system and applications in control of environmental pollution, News on 1980 Ann. Rep. Meet. of Beijing Working Group on Fuzzy Sets, Beijing, China.

1059 Hu, C. (1980). A class of fuzzy topological spaces (I), Collect. Pap. on Fuzzy Math. (Abstr.), Huazhong Inst. of Tech., Huazhong, China, Apr.

1060 Hu, C. (1980). A class of fuzzy topological spaces (II), Collect. Pap. on Fuzzy Math. (Abstr.), Huazhong Inst. of Tech., Huazhong, China, Apr.

1061 Hu, C. (1980). An embedding theorem in fuzzy topology, Collect. Pap. on Fuzzy Math. (Abstr.), Huazhong Inst. of Tech., Huazhong, China, Apr.

1062 Hu, C. (1980). Fuzzy Tychonoff spaces, Collect. Pap. on Fuzzy Math. (Abstr.), Huazhong Inst. of Tech., Huazhong, China, Apr.

1063 Hu, C. (1980). On the fuzzy boundary of fuzzy sets, Collect. Pap. on Fuzzy Math. (Abstr.), Huazhong Inst. of Tech., Huazhong, China, Apr.

1064 Hu, C. (1981). Metrization of fuzzy topological spaces, Inner Mongolia Univ., A List of Lit. in China, The Univ. of Sci. and Tech. of China, Hofei, Anhwei, China.

1065 Hu, C. (1981). On fuzzy uniform space, Inner Mongolia Univ., A List of Lit. in China, The Univ. of Sci. and Tech. of China, Hofei, Anhwei, China.

1066 Huang, F., Ruan, Y., Liu, Z., and Chiu, F. (1980). Some method of computer aided diagnosis for acute abdominal pain, News on 1980 Ann. Rep. Meet. of Beijing Working Group on Fuzzy Sets, Beijing, China.

1067 Huang, J. (1980). Measure of fuzzy sets, *J. of Hunan Univ.* (3), A List of Lit. in China, The Univ. of Sci. and Tech. of China, Hofei, Anhwei, China.

1068 Hughes, G. E., and Creswell, M. J. (1986). *An Introduction to Modal Logic*, Methuen, London.

1069 Hughes, J. S., and Kandel, A. (1977). Applications of fuzzy algebra to hazard detection in combinational switching circuits, *Int. J. Comput. Inf. Sci.*, **6**, pp. 71–82.

1070 Hughes, P., and Brecht, G. (1976). *Vicious Circles and Infinity*, Jonathon Cape, London.

1071 Hung, N. T. (1975). Information fonctionelle et ensembles flous, Semin. on Questionaires, Univ. of Paris, Paris.

1072 Hutton, B. (1975). Normality in fuzzy topological spaces, *J. Math. Anal. Appl.*, **50**, 74–79.

1073 Hutton, B. (1977). Uniformities on fuzzy topological spaces, *J. Math. Anal. Appl.*, **58**, 559–571.

1074 Hutton, B., and Reilly, J. L. (1974). Separation axioms in fuzzy topological spaces, Univ. of Auckland, New Zealand, Mar.

1075 Iaifu, L. (1980). The application of fuzzy sets to recognizing whear parents, in *Selected Papers on Fuzzy Subsets*, Beijing Normal Univ., Beijing, China, Mar.

1076 Ichikawa, A. (1976). Structure of multidimensional criteria, in *Summ. of Pap. on Gen. Fuzzy Problems*, The Working Group on Fuzzy Syst., Tokyo, pp. 14–24.

1077 Ichikawa, A., Nakao, K., and Kobayashi, S. (1975). An analysis of social group behaviour by means of a threshold element network model, in *Summ. of Pap. on Gen. Fuzzy Problems*, The Working Group on Fuzzy Syst., Tokyo, Nov., pp. 23–28.

1078 Idesawa, M. (1975). Automatic input of line drawing and generation of solid figure, in *Summ. of Pap. on Gen. Fuzzy Problems*, The Working Group on Fuzzy Syst., Tokyo, Nov., pp. 29–33.

1079 Iiyama, Y., Yamauchi, K., and Yanagawa, K. (1977). Analytical study of the computer aided control system for the Shinkansen—An actual behavior, in *Proc. 1977 IEEE Conf. on Decis. and Control*, New Orleans.

1080 Imaoka, H., and Sugeno, M. (1979). A model of dialogue based on fuzzy sets concept, Int. Joint Conf. on Artif. Intell., Tokyo, Aug.

1081 Inagaki, Y. (1976). Mathematical foundation of fuzzy sets theory, 1976 Joint Conv. of Four Inst. of Elec. Eng. of Japan.

1082 Inagaki, Y., and Fukumura, T. (1975). On the description of fuzzy meaning of context-free language, in *Fuzzy Sets and their Applications to Cognitive and Decision Processes* (L. A. Zadeh, K. S. Fu, K. Tanaka, and M. Shimura, Eds.), Academic Press, New York, pp. 301–328.

Key References in Fuzzy Pattern Recognition

1083 Inagaki, Y., Tanaka, S., and Fukumura, T. (1976). Some consideration on problematic inference based on fuzzy logic, Tech. Rep. on Autom. and Lang. of IECE, AL76-3.

1084 Irtem, A. (1976). Fuzziness and human conciousness, ORSA/TIMS Conf., Miami, FL.

1085 Ishikawa, A. (1976). Feedforward control systems and fuzzy entropy, ORSA/TIMS Conf., Miami, FL.

1086 Ishikawa, A. (1977). Fuzzy function analysis, in *Proc. Symp. on Fuzzy Set Theory and Appl., IEEE Conf. on Decis. and Control*, New Orleans.

1087 Ishikawa, A. (1979). Design automation in creativity development, *ACM Sigma*, **9** (1), Mar.

1088 Ishikawa, A., and Mieno, H. (1975). Design of a video information system based upon the fuzzy information theory, *Bull., ORSA* **23**, Supp. 2, B375.

1089 Ishikawa, A., and Mieno, H. (1977). Fuzzy functions analysis, in *Proc. 1977 IEEE Conf. on Decision and Control*, New Orleans, pp. 1315–1317.

1090 Ishikawa, A., and Mieno, H. (1979). A video information system and the information theory, in *Proc. of the Eur. Comput. Congr.*, pp. 441–450.

1091 Ishikawa, A., and Mieno, H. (1979). The fuzzy entropy concept and its application, *Fuzzy Sets Syst.*, **2** (2), Apr.

1092 Isnat, C. (1978). Liapounoff stability of fuzzy sets, *Resume, Zentralblatt fur Math.*, **372**, 04002.

1093 Isomichi, Y. (1975). Segmentation-free recognition of time series patterns, Semin. on Stud. on Time Series Pattern Recognition Syst., Kyoto Univ., Kyoto, Japan.

1094 Ito, T., and Kizawa, M. (1976). On the semantic structure of natural language, *Trans. IECE* (Japan), **59-D**, 141–148.

1095 Itzinger, D. (1977). Measuring logical structures of small social systems, in *Proc. 1st. Int. Conf. of Math Modelling*, St. Louis, MO, pp. 2607–2616.

1096 Itzinger, O. (1974). Aspects of axiomatization of behaviour—Towards an application of Rasch's measurement model to fuzzy logic, in *COMSTAT 1974 (Proc. Symp. Comput. Stat., Univ. of Vienna)* (G. Bruckman, F. Fresche, and L. Schmatterer, Eds.), Physica-Verlag, Vienna, pp. 173–182.

1097 Jacobson, D. H. (1976). On fuzzy goals and maximizing decisions in stochastic optimal control, *J. Math. Anal. Appl.*, **55**, 434–440.

1098 Jacq, J., Defolle, M., Dussuyer, I., Maigrot, J. C., and Crocq, L. (1978). A contribution to the valuation theory of fuzzy sets as applied to the social sciences, Colloq. Int. Theorie et Appl. des Sous-Ensembles Flous. Journ. de Biomath. et d'Inf. Med., Marseille, France, Sept.

1099 Jacquet-Lagreze, E. (1975). La modelisation des preferences, pre-ordres, quasi-ordres et relations floues, Doc. Metra, Rapp. de Rech. No. 80, July.

1100 Jacquet-Lagreze, E. (1975). Modelling preferences among distributions using fuzzy relations, 5th Conf. on Subjective Probability, Utility and Decision-Making, Darmstadt, Germany, Sept.

1101 Jacquet-Lagreze, E. (1976). Explicative models in multi-criteria preference analysis, *Cah. Lansade*, (5), May.

1102 Jahn, K. U. (1971). Aufbau Einer 3-Wertigen Linearen Algebra und Affinen Geometrie auf Grundlage der Intervall-Arithmetik, Ph.D. Thesis, Univ. of Liepzig, Leipzig, Germany.

1103 Jahn, K. U. (1974). Eine Theorie der Gleichungesysteme, mit Intervall-Koeffizienten, *Z. Angew. Math. Mech.*, **54**, 405–412.

1104 Jahn, K. U. (1975). Eine auf der Intervall-Zahlen Fussende 3-Wertige Lineare Algebra, *Math. Nachr.*, **65**, 105-116.

1105 Jahn, K. U. (1975). Intervall-Wertige Mengen, *Math. Nachr.*, **68**, 115-132.

1106 Jahn, K. U. (1976). Anvendungen von Fuzzy Sets, Vortrage aus dem Problemseminar Automata—Und Algorithmen Theorie, Apr., Weissig, pp. 30-43.

1107 Jahn, K. U. (1977). Grundfragen Einer Mehrwertigen (Fuzzy) Analysis, in *Vortage Zu Grundlagen der Informatik*, Heft, Dresden, Germany, pp. 36-45.

1108 Jain, R. (1975). Outline of an approach for the analysis of fuzzy systems, Spec. Interest Disc. Sess. on Fuzzy Automata and Decis. Processes, 6th IFAC World Congr., Boston, MA, Aug.

1109 Jain, R. (1975). Pattern classification using property sets, Symp. on Circuits, Syst., and Comput., Univ. of Calcutta, Calcutta, India, Feb.

1110 Jain, R. (1976). Convolution of fuzzy variables, *JIETE*, **22**.

1111 Jain, R. (1976). Decision making in the presence of fuzzy variables, *IEEE Trans. Syst., Man, Cybern.*, **SMC-6**, 698-703.

1112 Jain, R. (1976). Decision making with fuzzy knowledge about the state of the system, Nat. Syst. Conf., Roorke, India, Feb.

1113 Jain, R. (1977). A procedure for multiple-aspect decision making using fuzzy sets, *Nat. J. Syst. Sci.*, **B**, 1-7.

1114 Jain, R. (1977). Analysis of fuzzy systems, in *Fuzzy Auatomata and Decision Processes*, (M. M. Gupta, G. N. Saridis, and B. R. Gaines, Eds.), North-Holland, Amsterdam, pp. 251-268.

1115 Jain, R. (1977). Decision-making in the presence of fuzziness and uncertainty, in *Proc. 1977 IEEE Conf. on Decis. and Control*, New Orleans, Dec., pp. 1318-1323.

1116 Jain, R. (1977). Tolerance analysis using fuzzy sets, *Int. J. Syst. Sci.*, **7** (12), 1393-1401.

1117 Jain, R. (1979). Application of fuzzy set theory for the analysis of complex scenes, Univ. of Texas, Austin, in *Proc 1978 IEEE Conf. on Decis. and Control*, includes the *17th Symp. on Adaptive Processes*, 78CH1392-OCS, San Diego, CA, Jan. 1979, pp. 1444-1449.

1118 Jain, R. (1980). Fuzzyism and real world problems, Fuzzy Sets Theory and Appl. to Policy Anal. and Inf. Syst. (P. P. Wang and S. K. Chang, Eds.), pp. 129-132, Plenum, New York.

1119 Jain, R., and Nagel, H. H. (1977). Analyzing a real world scene sequence using fuzziness, in *Proc. 1977 IEEE Conf. on Decis. and Control*, New Orleans, Dec., pp. 1367-1372.

1120 Jain, R., and Nagel, H. H. (1977). Analyzing a scene sequence using fuzziness in *Proc. 1977 IEEE Conf. on Decis. and Control*, Vol. 2, New Orleans, Dec., pp. 1367-1372.

1121 Jain, R., and Stallings, W. (1978). Comments on fuzzy set theory versus Bayesian statistics, *IEEE Trans. Syst., Man, Cybern.*, **SMC-8**, 332-333.

1122 Jajinski, R., Pelech, A., and Wojcicka, M. (1977). An analysis of weak-defined information from questionnaire-based research with the use of fuzzy sets languages, Komm. No. 219, Inst. Org. i Zarzadzania Politech. Wroclawskieg, Wroclaw, Poland.

1123 Jakubowski, R. (1977). Application of formal language and fuzzy automata in designing, in *Int. Conf. on Inf. Processing* (J. Madey, Ed.), IFIP-INFOPOL-76, North-Holland, Amsterdam.

1124 Jakubowski, R., and Kasprzak, A. (1972). Algorytm automatycznego projektowania procesow technologicnych obrokki skrawaniem, *Podstawy Sterowania*, **2** (Z.4).

1125 Jakubowski, R., and Kasprzak, A. (1973). Application of fuzzy programs to the design of machining technology, *Bull. Pol. Acad. Sci.*, **21** (21), 17–22.

1126 Jakubowski, R., and Szelc, A. (1977). Applications of formal languages and fuzzy automata in problem solving, *Podstawy Sterowania*, **7** (Z.1), 69–72.

1127 Jakubowski, R., and Szelc, A. (1977). Quasi-linguistic representation in description of microprograms, *Bull. Acad. Pol. Sci.*, **10**, Ser. des Sci. Tech., **XXV**, (12).

1128 Jakubowski, R., and Szelc, A. (1977). Quasi-linguistic representation of systems, *Bull. Acad. Pol. Sci.*, **9**, Ser. des Sci. Tech., **XXV** (12).

1129 Jarvis, R. A. (1975). Optimization strategies in adaptive control—A selective survey, *IEEE Trans. Syst., Man, Cybern.*, **SMC-5**, 83–94.

1130 Jaskowski, S. (1969). Propositional calculus for contradictory deductive systems, *Stud. Logica*, **24**, 143–159 (Trans. of 1948 Polish Pap.).

1131 Jen, P. (1980). Possibility theory on a probability space, Collect. Pap. on Fuzzy Math. (Abstr.), Huazhong Inst. of Tech., Huazhong, China, Apr.

1132 Jensen, J. H. (1976). Application for fuzzy logic control, No. 1, No. 7607, Elec. Power Eng. Dept., Tech. Univ. of Denmark, Lyngby, June.

1133 Jiang, J. (1979). Concerning separation axioms for induced fuzzy topological spaces, A List of Lit. in China, The Univ. of Sci. and Tech. of China, Hofei, Anhwei, China (in Chinese).

1134 Jiang, J. (1979). Separation axioms in fuzzy topological spaces and fuzzy compactness, *Sichuan Daxue Xuebao*, **3**, 1–10, A List of Lit. in China, The Univ. of Sci. and Tech. of China, Hofei, Anhwei, China (in Chinese).

1135 Jiang, J. (1980). Two-dimensional information currents and fuzzy computer, Collect. Pap. on Fuzzy Math. (Abstr.), Huazhong Inst. of Tech., Huazhong, China, Apr.

1136 Jin, C. (1980). Local compactness of fuzzy topological spaces, Collect. Pap. on Fuzzy Math. (Abstr.), Huazhong Inst. of Tech., Huazhong, China, Apr.

1137 Jinwen, Z. (1980). A kind of nonstandard models of the axiomatic set theory with urelements—The normal fuzzy set structure, *J. Huazhong Inst. Tech.*, **2** (1) (English ed.).

1138 Jinwen, Z. (1980). A unified treatment of fuzzy set theory and Boolean-valued set theory—Fuzzy set structures and normal fuzzy set structures, *J. Math. Anal. Appl.*, **76** (1), July.

1139 Jobe, W. H. (1962). Functional completeness and canonical forms in many-valued logics, *J. Symb. Logic*, **28**, 409–421.

1140 Johnson, R. W., and Shore, J. E. (1979). Solving fuzzy set problems using probability theory, Tech. Memo. NRL 7503-211, Naval Res. Lab., Washington, D.C.

1141 Jolles, E. (1975). Contribution a l'analyse de la decision floue, deux exemples d'applications, Mem. Fac. des Sci. Econ. et de Gestion, Univ. de Dijon, Dijon, France, Oct.

1142 Jolles, E. (1976). Processus de prise de decision par etudes successives dans un environment flou, resolution complete, *Rev. Econ. Polit.* (3), 416–436.

1143 Jolles, E. (1978). La theorie des sous-ensembles flous au service de la decision, deux exemples d'applications, Colloq. Int. Theorie et Appl. des Sous-Ensembles Flous, Journ. de Biomath. et d'Inf. Med., Marseille, France, Sept.

1144 Joly, H., Sanchez, E., Gouvernet, J., and Valty, J. (1980). Application of fuzzy set theory to the evaluation of the cardiac function, Medinfo, 80, Tokyo.

1145 Jones, A. (1974). Towards the right solution, *Int. J. Math. Educ. Sci. Tech.*, **5**, 337–357.

1146 Jones, A. (1976). L'adaptation des objectifs moyens et fonctions d'un systeme d'education, par la theorie des sous-ensembles flous, IDP 108, Centre Imago, Univ. Catholique de Louvain, Louvain, Belgium.

1147 Jones, W. T. (1976). A fuzzy set characterization of interaction in scientific research, *J. Amer. Soc. Inf. Sci.*, Sept.–Oct.

1148 Jones, W. T., and Ragade, R. K. (1978). Identification of stable journal interaction clusters, in *Proc. 22nd Meet. Soc. for Gen. Syst. Res.*, Washington D.C., pp. 431–442.

1149 Jordon, P. (1952). Algebraische Betrachtungen zur Theorie des Wirkungskvantum, *Math. Sem. Hamburg*, **18**, 99–119.

1150 Jordon, P. (1962). Halbgruppen von Idempotenten und Nichtkommutative Varbande, *J. Reine Angew, Math.*, **211**, 136–161.

1151 Jouault, J. P., and Luan, P. M. (1975). Application des concepts flous a la programmation en languages quasi-naturels, Inst. Inf. d'Entreprise, CNAM, Paris.

1152 Joyce, J. (1976). Fuzzy sets and the study of linguistics, *Pac. Coast Philo.*, **11**, 39–42.

1153 Jumarie, G. (1977). Some technical applications of relativistic information theory, Shannon information, fuzzy sets, linguistics, relativistic sets and communication, *Cybernetica*, **20** (2), 91–128.

1154 Jumarie, G. (1978). Relativistic fuzzy sets as a means to introduce human factors in pattern recognition systems, Dept. of Math., Univ. du Quebec, Montreal, in *Proc. of the 1978 Int. Conf. on Cybern. and Syst.*, pp. 930–935.

1155 Jumarie, G. (1979). New results in relativistic information and general systems. Observed probability, relativistic fuzzy sets, generative semantics, *Cybernetica*.

1156 Jumarie, G. (1980). On the definition of fuzziness and subjectivity in human communication, in *IXth Int. Congr. Cybern.*, Namur, Belgium, pp. 8–13.

1157 Kabbara, G. (1979). Optimisation floue, Dr. Math. Appl. Thesis, Univ. de Provence, Centre St-Charles, Marseille, France, June.

1158 Kacprzyk, J. (1976). Fuzzy set theoretic approach to the optimal assignment of work places, IFAC Symp. on Large Scale Syst. Theory and Appl.

1159 Kacprzyk, J. (1976). Linguistic variables and fuzzy conditional statements for the description of complex systems, in *Proc. III, Nat. Conf. Inf. Sci.*, Katowice, Poland.

1160 Kacprzyk, J. (1977). Decision-making in a fuzzy environment with fuzzy termination time, *Fuzzy Sets Syst.*, **1**, 169–179.

1161 Kacprzyk, J. (1977). Fuzzy integral as model of operator in complex automation systems, in *Proc. VII Nat. Conf. Autom. Control.*

1162 Kacprzyk, J. (1977). Control of a nonfuzzy system in a fuzzy environment with fuzzy termination time, *Syst. Sci.*, **3**.

1163 Kacprzyk, J. (1978). Branch-and-bound algorithms for the decision-making in a fuzzy environment, Syst. Res. Inst., Polish Acad. of Sci., Warzawa, Poland, *UL. Newelska*, **6**, 1–447.

1164 Kacprzyk, J. (1978). Fuzzy termination time on decision-making in a fuzzy environment, in *4th Int. Congr. Cybern. and Syst.*, Amsterdam, pp. 368–369.

1165 Kacprzyk, J. (1978). On some multistage decision-making problems in a fuzzy environment—Branch and bound approach, in *Proc. German (GDR)—Polish Symp. on Nonconventional Optimization Problems*, Mogilany, Poland.

1166 Kacprzyk, J. (1979). A branch-and-bound algorithm for the multistage control of a nonfuzzy system in a fuzzy environment, *Contr. Cybern.*, **8** (2), 139–147.

1167 Kacprzyk, J., and Shaszak, A. (1980). A fuzzy approach to the stability of integrated regional developments, in *Proc. Int. Congr. on Appl. Syst. Res. and Cybern.*, Acapulco, Mexico, Dec.

1168 Kacprzyk, J., and Staniewki, P. (1980). Long-term inventory policy-making through fuzzy decision making models. ORSA/TIMS Mil.

1169 Kacprzyk, J., and Staniewski, P. (1982). Control of a deterministic system in a fuzzy environment over infinite planning horizon, *Fuzzy Sets Syst.*

1170 Kacprzyk, J., and Staniewski, P. (1980). On a fuzzy inventory control problem with random demand, 4th Eur. Congr. on Oper. Res., July.

1171 Kacprzyk, J., and Straszak, A. (1980). Application of fuzzy decision-making models for determining optimal policies in "stable" integrated regional development, Fuzzy Sets Theory and Appl. to Policy Anal. and Inf. Syst.

1172 Kahne, S. (1975). A procedure for optimizing development decisions, *Automatica*, **11**, 261–269.

1173 Kaiser, D. (1979). Vers une modelisation du raionnement approximatif, Rap. de Rech. L.R.I., No. 47, Centre d'Orsay, Univ. de Paris, Paris.

1174 Kaji, H., and Nitta, Y. (1979). A fuzzy model of a document retrieval system and its implementation, *Trans. IECE '79/4*, **62-D** (4).

1175 Kalman, J. A. (1958). Lattices with involution, *Trans. Amer. Math. Soc.*, **87**, 485–491.

1176 Kalmanson, D., Stegall, F. (1973). Recherche cardio-vacuraire et theorie des ensembles flous, *Nouv. Presse Med.*, **41**, 2757–2760.

1177 Kalmanson, D., and Stegall, H. F. (1975). Cardiovascular investigations and fuzzy sets theory, *Amer. J. Cardiol.*, **35**, 30–34.

1178 Kalsaras, A. K., Liu, D. B. (1977). Fuzzy vector spaces and fuzzy topological vector spaces, *J. Math. Anal. Appl.*, **58**, 135–146.

1179 Kamburowski, J. (1979). Critical path analysis in a project with fuzzy estimates of activity duration time, Kommunikaty Seria Pre No. 39, Inst. Org. i Zarzadzania, Politech. Wroclawkiej, Wroclaw, Apr.

1180 Kameda, T., and Sadeh, E. (1977). Bounds on the number of fuzzy functions, *Inf. Control*, **35**, 139–145.

1181 Kampe, J., and Feriet, D. (1970). Mesure de l'information fourie par un evenement, *Colloq. Int. CNRS*, Paris, Vol. 186, pp. 191–221.

1182 Kampe, J., and Feriet, D. (1980). Independent fuzzy sets, in *Proc. Int. Congr. on Appl. Syst. Res. and Cybern.*, Acapulco, Mexico, Dec.

1183 Kampe, J., and Feriet, D. (1980). Ensembles flous et plausibilite au sens de shafer, Table Ronde CNRS sur le Flou, Lyon, France, June.

1184 Kandel, A. (1972). A new algorithm for minimizing incompletely specified fuzzy functions, CSR127, Comput. Sci. Dept., New Mexico Inst. of Mining and Tech., Socorro, Nov.

1185 Kandel, A. (1972). On coded grammars and fuzzy structures, Comput. Sci. Rep. 118, New Mexico Inst. of Mining and Tech., Socorro, Sept.

1186 Kandel, A. (1972). Toward simplification of fuzzy functions, Comput. Sci. Rep. 114, New Mexico Inst. of Mining and Tech., Socorro, June.

1187 Kandel, A. (1973). A new method for generating fuzzy prime implicants and an algorithm for the automatic minimization of inexact structures, Comput. Sci. Rep. 126, New Mexico Inst. Mining and Tech., Socorro, Oct.

1188 Kandel, A. (1973). Application of fuzzy logic to the detection of static hazards in combinational switching systems, Comput. Sci. Rep. 122, New Mexico Inst. of Mining and Tech., Socorro, Apr.

1189 Kandel, A. (1973). Comment on an algorithm that generates fuzzy prime implicants by Lee and Chang, *Inf. Control*, **22**, 279–282.

1190 Kandel, A. (1973). Comments on "Minimization of fuzzy functions," *IEEE Trans. Comput.*, **C-22**, 217.

1191 Kandel, A. (1973). Fuzzy chains—A new concept in decision-making under uncertainty, Comput. Sci. Rep. 123, New Mexico Inst. of Mining and Tech., Aug.

1192 Kandel, A. (1973). Fuzzy functions and their application to the analysis of switching hazards, in *Proc. 2nd Texas Conf. on Comput. Syst.*, Austin, Nov., pp. 42/1–6.

1193 Kandel, A. (1973). On minimization of fuzzy functions, *IEEE Trans. Comput.*, **C-22**, 826–832.

1194 Kandel, A. (1973). On the analysis of fuzzy logic, in *Proc. 6th Int. Conf. Syst. Sci.*, Honolulu, HI, Jan.

1195 Kandel, A. (1974). Application of fuzzy logic to the detection of static hazards in combinational switching systems, *Int. J. Comput. Inf. Sci.*, **3**, 129–139.

1196 Kandel, A. (1974). Codes over languages, *IEEE Trans. Syst., Man, Cybern.*, **SMC-4**, 135–138.

1197 Kandel, A. (1974). Fuzzy representation CNF minimization and their application to fuzzy transmission structures, *1974 Symp. on Multiple-Valued Logic*, IEEE 74CHO845-8C, pp. 361–379, Morgantown, West Virginia, May.

1198 Kandel, A. (1974). Generation of the set representing all fuzzy prime implicants, Comput. Sci. Rep., 136, New Mexico Inst. of Mining and Tech., Socorro, Oct.

1199 Kandel, A. (1974). On fuzzy maps—Some initial thoughts, Comput. Sci. Rep. 131, New Mexico Inst. of Mining and Tech., Socorro.

1200 Kandel, A. (1974). On the enumeration of fuzzy functions, 12th Holiday Symp. "Development in Combinatorics," New Mexico State Univ., Las Cruces, Dec.

1201 Kandel, A. (1974). On the minimization of incompletely specified fuzzy functions, *Inf. Control*, **26**, 141–153.

1202 Kandel, A. (1974). On the properties of fuzzy switching functions, *J. Cybern.*, **4**, 119–126.

1203 Kandel, A. (1974). On the theory of fuzzy matrices, Comput. Sci. Rep. 135, New Mexico Inst. of Mining and Tech., Socorro, Oct.

1204 Kandel, A. (1974). Simple disjunctive decompositions of fuzzy functions, Comput. Sci. Rep. 132, New Mexico Inst. of Mining and Tech., Socorro, July.

1205 Kandel, A. (1974). Synthesis of fuzzy logic with analog modules—Preliminary developments, *Comput. in Educ. Trans.* (ASEE Div.), **6**, 71–79.

1206 Kandel, A. (1975). Block decomposition of imprecise models, 9th Asilomar Conf. on Circuits, Syst. and Comput., Pacific Grove, CA, Nov.

1207 Kandel, A. (1975). Fuzzy hierarchical classifications of dynamic patterns, NATO ASI Pattern Recogn. and Classification, France, Sept.

1208 Kandel, A. (1975). Properties of fuzzy matrices and their applications to hierarchical structures, 9th Asilomar Conf. Circuits, Syst. and Comput., Pacific Grove, CA, Nov.

1209 Kandel, A. (1975). Fuzzy maps and their application in the simplification of fuzzy switching functions, in *6th Int. Symp. on Multiple-Valued Logic*, IEEE 76CH1111-4C, Logan, Utah, May.

1210 Kandel, A. (1976). Fuzzy systems and their applications to simulations, in *Proc. 9th Hawaii Int. Conf. on Syst. Sci.*, Honolulu, HI, Jan.
1211 Kandel, A. (1976). Inexact switching logic, *IEEE Trans. Syst., Man, Cybern.*, **SMC-6**, 215-219.
1212 Kandel, A. (1976). On the decomposition of fuzzy functions, *IEEE Trans. Comput.*, **C-25**, 1124-1130.
1213 Kandel, A. (1977). A note on the simplification of fuzzy functions, *Inf. Sci.*, **13**, 91-94.
1214 KAndel, A. (1977). Comments on Comments by Lee, *Inf. Control*, **35**, 109-113.
1215 Kandel, A. (1978). Fuzzy statistics and forecast evaluation, *IEEE Trans. Syst., Man, Cybern.*, **SMC-8** (5), 396-401.
1216 Kandel, A. (1978). Imprecise switching structures and their applications in models using approximate information, in *Proc. Int. Conf. on Cybern. and Soc.*, Tokyo, Nov., pp. 926-929.
1217 Kandel, A. (1978). On the compactification and enumeration of fuzzy switching functions, in *Proc. 8th Int. Symp. on Multiple-Valued Logic*, Chicago, May, pp. 87-90.
1218 Kandel, A. (1979). Evaluating procedures and fuzzy logic, The 9th Int. Symp. on Multi-Valued Logic, Beth, UK.
1219 Kandel, A. (1979). Fuzzy statistics and policy analysis, 1st Int. Symp. on Policy Anal. and Inf. Syst., Durham, NC, June.
1220 Kandel, A. (1979). On fuzzy statistics, in *Advances in Fuzzy Set Theory and Applications*, (M. M. Gupta, R. K. Ragade, and R. R. Yager, Eds.), North-Holland, Amsterdam.
1221 Kandel, A. (1979). On the theory of fuzzy switching mechanisms (FSM'S), in *Advances in Fuzzy Set Theory and Applications*, (M. M. Gupta, R. K. Ragade, and R. R. Yager, Eds.), North-Holland, Amsterdam.
1222 Kandel, A. (1979). On the theory of fuzzy switching mechanisms, in *Proc. of the 18th IEEE Conf. on Decis. and Control*, pp. 771-776, Ft. Lauderdale, FL, Dec.
1223 Kandel, A. (1980). Fuzzy dependencies for relational databases, ORSA-TIMS Mil.
1224 Kandel, A. (1980). Fuzzy dynamical systems and the nature of their solutions, Fuzzy Sets Theory and Appl. to Policy Anal. and Inf. Syst. (P. P. Wang and S. K. Chang, Eds.) pp. 93-122, Plenum, New York.
1225 Kandel, A. (1980). Improved bounds on the number of distinct fuzzy valued switching functions over N variables, Table Ronde du CNRS sur le Flou, France, June.
1226 Kandel, A. (1980). On the control and evaluation of uncertain processes, in *Proc. of JACC*, San Francisco, Aug.
1227 Kandel, A. (1980). On the modelling of uncertain systems, in *Proc. Int. Congr. on Appl. Syst. Res. and Cybern.*, Acapulco, Mexico, Dec.
1228 Kandel, A. (1980). On the theory of fuzzy logics, in *Proc. of JACC*, San Francisco, Aug.
1229 Kandel, A. (1980). On the theory of possibility of fuzzy strategy, Int. Conf. on Cybern. and Soc., Cambridge, MA, Oct.
1230 Kandel, A., and Byatt, W. J. (1978). Fuzzy differential equations, in *Proc. Int. Conf. on Cybern. and Soc.*, Tokyo, Japan, Nov., pp. 1213-1216.
1231 Kandel, A., and Byatt, W. J. (1978). Fuzzy sets, fuzzy algebra, and fuzzy statistics, *Proc. IEEE*, **66** (12), Dec., 1619-1639.
1232 Kandel, A., and Byatt, W. J. (1980). Fuzzy processes, *Fuzzy Sets Syst.*, **4** (2), Sept., 117-152.

1233 Kandel, A., and Clark, C. M. (1981). New results in the enumeration of minimized fuzzy-valued switching functions, in *Proc. 11th MVL Conf.*, Oklahoma City, OK, May.

1234 Kandel, A., and Davis, H. A. (1976). The first fuzzy decade (Bibliography on fuzzy sets and their applications), Comput. Sci. Rep. 140, New Mexico Inst. of Mining and Tech., Sorocco, Apr.

1235 Kandel, A., and Francioni, J. (1980). On the properties and applications of fuzzy-valued switching functions, *IEEE Trans. Comput.*, **C-29** (11), Nov.

1236 Kandel, A., and Hughes, J. S. (1977). Applications of fuzzy algebra to hazard detection in combinational switching circuits, *Int. J. Comput. Inf. Sci.*, **6** (1), Mar., 71–82.

1237 Kandel, A., and Lee, S. C. (1979). *Fuzzy Switching and Automata—Theory and Applications*, Crane, Russak and Co., Inc., New York, and Edward Arnold, London, 303 pp.

1238 Kandel, A., and Neff, T. P. (1977). Simplification of fuzzy switching functions, *Int. J. Comput. Inf. Sci.*, **6** (1), Mar. 55–70.

1239 Kandel, A., and Oberhauf, T. A. (1974). On fuzzy lattices, Comput. Sci. Rep. 128, New Mexico Inst. of Mining and Tech., Socorro.

1240 Kandel, A., and Rickman, S. M. (1976). Tabular minimization of fuzzy switching functions, *IEEE Trans. Syst., Man, Cybern.*, Nov., 766–769.

1241 Kandel, A., and Rickman, S. M. (1977). Column table approach for the minimization of fuzzy functions, *Inf. Sci.*, **12**, 118–128.

1242 Kandel, A., and Yager, R. R. (1979). A 1979 bibliography on fuzzy sets, their applications, and related topics, in *Advances in Fuzzy Set Theory and Applications* (M. M. Gupta, R. K. Ragade, and R. R. Yager, Eds.), North-Holland, Amsterdam.

1243 Kandel, A., and Yelowitz, L. (1974). Fuzzy chains, *IEEE Trans. Syst., Man, Cybern.*, **SMC-4**, 472–475.

1244 Kania, A. A. (1979). Sensitivity of formal fuzziness system, Int. Rep., Inst. of Autom. Control, Tech. Univ. of Kielce, Kielce, Poland.

1245 Kania, A. A., Kiszka, J. B., Gorzalczany, M. B., and Maj, J. R. (1980). On stability of formal fuzziness systems, *Inf. Sci. Int. J.*, **21**.

1246 Kania, A. A., and STachowicz, M. S. (1980). Robustness of fuzzy relation operator, in *Proc. Int. Congr. on Appl. Syst. Res. and Cybern.*, Acapulco, Mexico, Dec.

1247 Karsky, M., and Adamo, J. M. (1977). Application de la dynamique des systemes et de la logique floue a la modelisation d'un probleme de postes en raffinerie, Congr. Afcet Model and Maitrise Syst. (Versailles, France) Editions Hommes and Tech.

1248 Karttunen, L. (1972). Possible and must, in *Syntax and Semantics*, Vol. 1 (J. P. Kimball, Ed.), Seminar Press, New York, pp. 1–20.

1249 Katsaras, A. K. (1979). Fuzzy proximity spaces, *J. Math. Anal. Appl.*, **68** (1) Mar., 100–111.

1250 Katsaras, A. K., and Liu, D. B. (1977). Fuzzy vector spaces and fuzzy topological vector spaces, *J. Math. Anal. Appl.*, **58**, 135–146.

1251 Katz, J. J. (1962). *The Problem of Induction and its Solution*, Univ. of Chicago Press, Chicago.

1252 Kaufmann, A. (1973). *Introduction a la Theorie des Sous-Ensembles Flous, 1—Elements Theoretiques de Base*, Massom et Cie, Paris.

1253 Kaufmann, A. (1974). Ebauche d'une theorie de l'operateur humain. Proceeding learning management based on formal model of behaviour and aptitudes in computer assisted systems of instruction, UCODI Summer School, Centre Imago, Univ. Catholique de Louvain, Louvain, Belgium, pp. 11–14.

1254 Kaufmann, A. (1975). *Introduction a la Theorie des Sous-Ensembles Flous, 2—Applications a la Linguistique et a la Semantique*, Masson et Cie, Paris.

1255 Kaufmann, A. (1975). *Introduction a la Theorie des Sous-Ensembles Flous, 3—Applications a la Classification et la Reconnaisance des Formes, aux Automates et aux Systemes, aux Choix des Critares*, Masson et Cie, Paris.

1256 Kaufmann, A. (1975). Introduction to a fuzzy theory of the human operator, Spec. Interest Disc. Sess. on Fuzzy Automata and Decis. Processes, 6th IFAC World Congr., Boston, MA, Aug.

1257 Kaufmann, A. (1975). Les modeles dans un environment flou (systemes flous). Semin. "Contrib. des Syst. Flous a l'Autom.," Centre d'Autom. de Lille, Lille, June.

1258 Kaufmann, A. (1975). *Theory of Fuzzy Subsets*, Academic Press, New York.

1259 Kaufmann, A. (1976). *La Theorie des sous-Ensembles flous et ses Applications dans les Sciences Humains, Economie Appliques*, Tome 29, No. 3, Librairie Droz, Geneva, pp. 469–478.

1260 Kaufmann, A. (1976). Valuation et probabilisation des sous-ensembles flous, ORSA-TIMS Conf., Miami, FL, Nov.

1261 Kaufmann, A. (1977). *Mathematical Tools Used in the Informatic Education System Computer-Based Science Instruction*, (A. Jones and H. Weinstock, Eds), Nato Adv. Study Inst. Ser., Noordhoff-Leyden, pp. 29–52.

1262 Kaufmann, A. (1977). Progress in modeling of human reasoning of fuzzy logic, in *Fuzzy Automata and Decision Processes* (M. M. Gupta, G. N. Saridis, and B. R. Gaines, Eds.), North-Holland, Amsterdam, pp. 11–18.

1263 Kaufmann, A. (1978). Le calcul des admissibilities. Une idee nouvelle a partir de la theorie des sous-ensembles flous, Colloq. Int., Theorie et Appl. des Sous-Ensembles Flous, Journ. de Biomath. et d'Inf. Med., Marseille, France, Sept.

1264 Kaufmann, A. (1978). Le point sur la theorie des sous-ensembles flous, Colloq. Int., Theorie et Appl. des Sous-Ensembles Flous, Journ. de Biomath. et d'Inf. Med., Marseille, France, Sept.

1265 Kaufmann, A. (1979). Les mathematiques de l'imprecis, in *L'Approche du Systeme* (P. Valery, Ed.), Revue des Lettres Modernes, pp. 27–37.

1266 Kaufmann, A. (1979). *Modeles Mathematiques pour la Stimulation Inventive*, Albin-Michel, Paris.

1267 Kaufmann, A. (1979). Theorie d'un operateur humain dans un dialogue homme-machine, in *Les Conduites Simulees* (P. Mounoud and J. P. Bronckart, Eds.), Gallimard, Paris.

1268 Kaufmann, A. (1980). Bibliography on fuzzy sets and their applications, *Busefal*, No. 1–3, LSI Lab., Univ. Paul Sabatier, Toulouse, France.

1269 Kaufmann, A. (1980). La simulation des sous ensembles flous, Table Ronde CNRS sur le Flou, Lyon, France, June.

1270 Kaufmann, A., Cools, M., and Dubois, T. (1973). Stimulation inventive dans un dialogue homme-machine utilisant la methode des morphologies et la theorie des sous-ensembles flous, Imago Disc, Pap. 6, Univ. Catholique de Louvain, Louvain, Belgium.

1271 Kaufmann, A., Cools, M., and Dubois, T. (1975). *Exercise avec Solutions sur la Theorie des Sous-Ensembles Flous*, Masson et Cie, Paris.

1272 Kaufmann, A., and Sanchez, E. (1979). Complements sur les concepts flous, *Rech. Appl.*, **3**.

1273 Kay, M. (1975). Color categories as fuzzy sets, Working Pap. No. 44, Univ. of California, Berkeley.

1274 Ke, J. S. (1978). Database modelling techniques with applications to fuzzy query translation, Ph.D. Diss., Dept. Inf. Eng., Univ. of Illinois, Chicago, June, in *Diss. Abstr. Int.*, **39/06-B**, 2899.

1275 Kempf, J., Duckstein, L., and Casti, J. (1979). Polyedral dynamics and fuzzy sets as a multi-criteria decision making aid, Dept. Syst. and Ind. Eng., Univ. of Arizona, Tucson, ORSA-TIMS Nat. Meet., New Orleans.

1276 Kempton, W. M. (1977). Grading of category membership in the folk classification of ceramtics—Socially correlated cognitive variation and a test of the theory of fuzzy subsets, Ph.D. Thesis, Univ. of Texas, Austin, in *Diss. Abstr. Int.*, **38/12-A**, 7416.

1277 Kerre, E. E. (1980). Fuzzy Sierpinski space and its generalizations, *J. Math. Anal. Appl.*, **74** (1).

1278 Kerridge, D. F. (1961). Inaccuracy and inference, *J. Roy. Stat. Soc.* (Ser. B), 184–194.

1279 Khalili, S. (1979). Fuzzy measures and mappings. *J. Math. Anal. Appl.*, **68**, 92–99.

1280 Khalili, S. (1979). Independent fuzzy events, *J. Math. Anal. Appl.*, **67** (2), Feb., 412–421.

1281 Khatchadourian, H. (1965). Vagueness, meaning and absurdity, *Amer. Philos. Quart.*, **2**, 119–129.

1282 Kiang, T. R. (1979). Some system-theoretic properties of Zadeh-fuzzy sets, Ph.D. Thesis, New Mexico State Univ. in *Diss. Abstr. Int.*, **40/04-B** 1858.

1283 Kickert, W. J. M. (1974). Application of fuzzy set theory to warm water control, Thesis, Delft Univ. of Tech., Delft, Holland, (in Dutch).

1284 Kickert, W. J. M. (1974). Een toepassing van fuzzy talen op patroon herkennen, Lab. voor regelttechniek Afdeling Electrotexhniek, TH Delft, Delft, Holland, May.

1285 Kickert, W. J. M. (1974). Toepassing van fuzzy sets op warm water regeling, Afatudee-Rverslag, Lab. voor Regeltechniek, Sfdeling Electrotechniek, TH Delft, Delft, Holland, Dec.

1286 Kickert, W. J. M. (1975). Analysis of fuzzy logic controller, Fuzzy Logic Working Group Rep. F/WKI/75, Queen Mary College, Univ. of London, London, June.

1287 Kickert, W. J. M. (1975). Further analysis and application of fuzzy logic, Fuzzy Logic Working Group Rep. F/WK2/75, Queen Mary College, Univ. of London, London, Aug.

1288 Kickert, W. J. M. (1975). Off-line analysis of the fuzzy rules, Fuzzy Logic Working Group Rep., Queen Mary College, Univ. of London, London, July.

1289 Kickert, W. J. M. (1975). Review of applications of fuzzy sets, Int. Rep., Dept. of Elec. Eng., Queen Mary College, Univ. of London, London, Jan.

1290 Kickert, W. J. M. (1976). An example of linguistic modeling, a 2nd attempt at simulating Mulder's theory, Rep. No. 30, Univ. of Tech., Eindhoven, Holland.

1291 Kickert, W. J. M. (1976). Fuzzy theories on decision making—A critical survey, Rep. No. 29, Univ. of Tech. Eindhoven, Holland, Sept.

1292 Kickert, W. J. M. (1978). *Fuzzy Theories on Decision-Making Frontiers in Systems Research*, Vol. 3, Martinus, Nijhoff, Leiden, Holland.

1293 Kickert, W. J. M. (1978). Towards an analysis of linguistic modelling, Dept. of Ind. Eng., Univ. of Tech., Eindhoven, Holland, 4th Int. Congr. on Cybern. and Syst., Amsterdam.

1294 Kickert, W. J. M. (1979). An example of linguistic modelling, the case of Mulder's

Key References in Fuzzy Pattern Recognition

theory of power, in *Advances in Fuzzy Set Theory and Applications*, (M. M. Gupta, R. K. Ragade, and R. R. Yager, Eds.), North-Holland, Amsterdam.

1295 Kickert, W. J. M., and Koppelaar, H. (1976). Applications of fuzzy set theory to syntactic pattern recognition of handwritten capitals, *IEEE Trans. Syst., Man, Cybern.*, **SMC-6**, 148–151.

1296 Kickert, W. J. M., and Mamdani, E. H. (1978). Analysis of a fuzzy logic controller, *Fuzzy Sets Systems*, **1** (1), 29–44.

1297 Kickert, W. J. M., and Van Nauta Lemke, H. R. (1976). Application of a fuzzy controller in a warm water plant, *Automatica*, **12**, 301–308.

1298 Killing, R. (1973). Fuzzy planner, Tech. Rep. No. 168, Comput. Sci. Dept., Univ. of Wisconsin, Madison, Feb.

1299 Kim, H. H., Mizumoto, M., Toyoda, J., and Tanaka, K. (1974). Automated editing of fuzzy line drawings for picture description, *Trans. IECE*, (Japan), **57-A**, 216–223.

1300 Kim, H. H., Mizumoto, M., Toyoda, J., and Tanaka, K. (1974). Lattice Grammars, *Syst., Comput., Controls*, **5**, 1–9, (Orig. *Trans. IECE* **57-D**, 253–260).

1301 Kim, H. H., Mizumoto, M., Toyoda, J., and Tanaka, K. (1975). L-fuzzy grammars, *Inf. Sci.*, **8**, 123–140.

1302 Kim, J. B. (1977). Note on the semigroup of fuzzy matrices, *Acta Math. Hung.*

1303 Kim, J. B. (1978). On circulant fuzzy matrices, *Math. Jap.*

1304 Kim, J. B. (1978). On the semigroup of the circulant fuzzy matrices, *J. Linear Alg. Appl.*

1305 Kim, K. H., and Roush, F. W. (1980). Generalized Fuzzy Matrices, *Fuzzy Sets Syst.*, vol. 4 (3), Nov., 293–316.

1306 Kimball, J. P (Ed.) *Syntax and Semantics*, (1972). Vol. 1, Seminar Press, New York.

1307 Kimball, J. P. (Ed.) (1973). *Syntax and Semantics*, Vol. 2, Seminar Press, New York.

1308 Kimball, J. P. (Ed.) (1975). *Syntax and Semantics*, Vol. 4, Academic Press, New York.

1309 King, P. J., and Mamdani, E. H. (1977). The application of fuzzy control systems to industrial processes, in *Fuzzy Automata and Decision Processes* (M. M. Gupta, G. N. Saridis, and B. R. Gaines, Eds.), North-Holland, Amsterdam, pp. 321–330.

1310 Kise, V. A., and Osis, J. J. (1969). Search methods for establishing of maximal separability of fuzzy sets, in *Kibernetika i Diagnostika*, Vol. 3 (D. S. Kritinkov, J. J. Osis, and L. A. Rastrigin, Eds.), Zinatne, Riga, U.S.S.R., pp. 79–88 (in Russian).

1311 Kiszka, J., Stachowicz, M., Kania, A., and Gorzalczany, M. (1980). Analysis and synthesis of fuzzy control systems, The 10th Meet. of the EWG on Fuzzy Sets, June.

1312 Kiszka, J. B. (1979). Comparison of fuzzy logic controller with optimal DDC controller, in *24 Int. Wiss., Kiloqium Ilmenqu—Alalyse, Synth., Optimierung und Steverung von Tech. und Nichttech. Syst.*, Oct., pp. 22–26.

1313 Kiszka, J. B. (1979). Fuzzy control of an oxygen converter process, Internal Rep., Inst. of Autom. Control, Tech. Univ. of Kiele.

1314 Kiszka, J. B. (1980). Fuzzy model and fuzzy logic controllers of oxygen converter process, Internal Rep., Inst. of Autom. Control, Tech. Univ. of Kiele, Kiele.

1315 Kiszka, J. B. (1980). On stability of formal fuzziness systems, *Inf. Sci.*, **22**, 51–68.

1316 Kiszka, J. B., and Stachowicz, M. S. (1980). Some problems of studying adequacy of fuzzy models, Int. Congr. on Appl. Syst. Res. and Cybern., Dec.

1317 Kitagawa, T. (1973). Biorobots for simulation studies of learning and intelligent controls, US-Jap. Semin. on Learning Control and Intell. Control, Gainesville, FL.

1318 Kitagawa, T. (1973). Three coordinate systems for information science approaches, *Inf. Sci.*, **15**, 159–169.

1319 Kitagawa, T. (1975). Fuzziness in informative logics, in *Fuzzy Sets and their Applications to Cognitive and Decision Processes*, (L. A. Zadeh, K. S. Fu, K. Tanaka, and M. Shimura, Eds.) Academic Press, New York, pp. 97–124.

1320 Kitahashi, T. (1975). A survey of studies on applications of many-valued logic in Japan, in *Proc. 1975 Int. Symp. Multiple-Valued Logic*, Logan, Utah, IEEE 78CH0959-7C, pp. 462–467.

1321 Kitajima, S., and Asai, K. (1970). Learning controls by fuzzy automata, *J. JAACE*, **14**, 551–559.

1322 Kitajima, S., and Asai, K. (1972). Learning model of fuzzy automaton with state-dependent output (3), Ann. Joint Conf. Rec. of JAACE.

1323 Kitajima, S., and Asai, K. (1974). A method of learning control varying search domain by fuzzy automata, in *Learning Systems and Intelligent Robots* (K. S. Fu and J. T. Tou, Eds.) Plenum Press, New York, pp. 249–262.

1324 Kitajimi, T. (1973). Biorobots for simulation studies of learning and intelligent control, US-Jap. Semin. on Learning Control and Intelligent Control, Gainesville, FL.

1325 Klabbers, J. H. G. (1975). General system theory and social systems—A methology for the social sciences, *Ned. Tijd. Psychol.*, **30**, 493–514.

1326 Klaua, D. (1965). Uber Einen Ansatz zur Mehrwertigen Mengenlehre, *Monatsb. Deutsch. Akad. Wiss.* (Berlin), **7**, 859–867.

1327 Klaua, D. (1966). Grundbegriffe Einer Mehrwertigen Mengenlehre, *Monatsb. Deutsch. Akad. Wiss.* (Berlin), **8**, 782–802.

1328 Klaua, D. (1966). Uber Einen Zweiten Ansatz zur Mehrwertigen Mengenlehre, *Monatsb. Deutsch. Akad. Wiss.* (Berlin), **8**, 161–177.

1329 Klaua, D. (1967). Ein Ansatz zur Mehrwertigen Menlehre, *Math. Nachr.*, **33**, 273–296.

1330 Klaua, D. (1967). Einbettung der Klassischen Mengenlehre in die *Mehrwertige*, *Monatsb. Deutsch. Akad. Wiss.* (Berlin), **9**, 258–272.

1331 Klaua, D. (1968). Partiell Aefinlerte Mengen, *Monatsb. Deutsch. Akad. Wiss.* (Berlin), **10**, 571–578.

1332 Klaua, D. (1969). Partielle Mengen mit Mehrwertigen Grundbeziehunger, *Monatsb. Deutsch. Akad. Wiss.* (Berlin), **11**, 585–599.

1333 Klaua, D. (1970). Stetige Gleichmachtigkeiten Kontinuierlich-Wertiger Mengen, *Monatsb. Deutsch. Akad. Wiss.* (Berlin), **12**, 749–758.

1334 Klaua, D. (1972). Zum Kardinalzahlbegriff in der Mehrwertigen Mengenlehre, in *Theory of Sets and Topology*, Deutscher Verlag der Wissenschaften, Berlin, pp. 313–325.

1335 Klaua, D. (1973). Zur Arithmetik Mehrwertigen Zahlen, *Math. Nachr.*, **57**, 275–306.

1336 Klement, E. P. (1978). Fuzzy sigma-algebras, Colloq. Int., Theorie et Appl. des Sous-Ensembles Flous, Journ. de Biomath. et d'Inf. Med., Marseille, France, Sept.

1337 Klement, E. P. (1978). Some relations between integration on fuzzy sets and fuzzy measures, *EMCSR–PCSR*, **8**, to be published.

1338 Klement, E. P. (1979). Characterization of finite fuzzy measures using Markoff-kernals, Memo. No. UCB/ERL M79/40, Electron. Res. Lab., Univ. of California, Berkeley.

1339 Klement, E. P. (1979). Characterization of fuzzy measures by classical measures, 1st Int. Symp. on Policy Anal. and Inf. Syst., Durham, NC, June.

1340 Klement, E. P. (1979). Extension of probability measures to fuzzy measures and their characterization, in *Proc. of the Int. Semin. on Fuzzy Set Theory*, Johannes Kepler Univ., Linz, Austria, Sept.

1341 Klement, E. P. (1979). *Proc. of the Int. Semin. on Fuzzy Set Theory*, Johannes Kepler Univ., Linz, Austria, Sept.

1342 Klement, E. P. (1980). Fuzzy sigma-algebras and fuzzy measurable functions. *Fuzzy Sets Syst.*

1343 Klement, E. P. (1980). A mathematical theory of uncertainty 2—Possibility measures with respect to norms and fuzzy experiments, in *Proc. Int. Congr. on Appl. Syst. Res. and Cybern.*, Acapulco, Mexico, Dec.

1344 Klement, E. P. (1980). Axiomatic approach to fuzzy measures, Table Ronde CNRS sur le flou, Lyon, France, June.

1345 Klement, E. P. (1980). Characterization of fuzzy measures constructed by means of triangular norms and construction of fuzzy sigma-algebras using triangular norms, Rep. 180, 179, Institutsber., Johannes Kepler Univ., Linz, Austria.

1346 Klement, E. P. (1980). Fuzzy sigma-algebras and fuzzy measures with respect to T-norms, Table Ronde du CNRS sur le Flou, Lyon, France, June.

1347 Klement, E. P., Lowen, R., and Schwyhla, W. (1981). Fuzzy probability measures, *Fuzzy Sets Syst.*, **5** (1), Jan., 21–30.

1348 Klement, E. P., and Schwyhla, W. (1980). Correspondence between fuzzy measures and classical measures, Rep. 150, Institutsber., Johannes Kepler Univ., Linz, Austria.

1349 Klement, P. (1980). Fuzzy algebras and fuzzy measures with respect to T-norms, Table Ronde du CNRS sur le Flou, Lyon, France, June.

1350 Kling, R. (1973). Fuzzy planner, Tech. Rep. 168, Comput. Sci. Dept., Univ. of Wisconsin, Madison.

1351 Kling, R. (1974). Fuzzy-planner-reasoning with inexact concepts in a procedural problem-solving language, *J. Cybern.*, **4**, 105–122.

1352 Klir, G. J. (1971). On universal logic primitives, *IEEE Trans. Comput.*, **C-20** (4), Apr., 457–467.

1353 Klir, G. J. (1975). On the representation of activity arrays, *Int. J. Gen. Syst.*, **2**, 149–168.

1354 Klir, G. J. (1975). Processing of fuzzy activities of neural systems, in *Progress in Cybernetics and Systems Research*, Vol. 1 (R. Trappel and F. R. Pichler, Eds.), pp. 21–24.

1356 Klir, G. J. (1976). Identification of generative structures in empirical data, *Int. J. Gen. Syst.*, **3**, 89–104.

1356 Klir, G. J., and Uttenhove, H. J. J. (1976). Computerized methodology for structure modelling, in *Annals of Systems Research*, Vol. 4, (H. E. Stenfert, Ed.), Kroese, Leiden, Holland.

1357 Klir, G. J., and Uttenhove, H. J. J. (1976). Procedure of generating hypothical structures in the structure identification problem, 3rd Eur. Meet. Cybern. Syst. Res., Vienna.

1358 Kloeden, P. E. (1982). Fuzzy dynamic systems, *Fuzzy Sets Syst.*, to appear.

1359 Kloeden, P. E. (1980). Compact supported endographs and fuzzy sets, *Fuzzy Sets Syst.*, **4** (2), Sept., 193–202.

1360 Kluska-Nawarecka, S., Mysona-Byrska, E., and Nawarecki, E. (1975). Algorithmes de commande pour certaines classes de problemes operationnels construits avec utilisation de la simulation numerique, Doc. Inst. de Fonterie, Craco-Vie, et Inst. d'Inf. de l'Acad. des mines et Metall., Jan.

1361 Kneale, W., and Kneale, M. (1962). *The Development of Logic*, Clarendon Press, Oxford, UK.

1362 Knopfmacher, K. (1975). On measures of fuzziness, *J. Math. Anal. Appl.*, **49**, 529–534.

1363 Knutson, T. J., and Holdredge, W. E. (1975). Orientation behavior, leadership, and consensus—A possible functional relationship, *Speech Mono.*, 107–114.

1364 Kobayashi, S. (1976). On interactive solution for multiple criteria problems—An approach by pairwise comparison, in *Summ. of Pap. on Gen. Fuzzy Problems*, The Working Group on Fuzzy Systems, Tokyo, pp. 42–47.

1365 Kochen, M. (1975). Applications of fuzzy sets in psychology, in *Fuzzy Sets and their Applications to Cognitive and Decision Processes*, (L. A. Zadeh, K. S. Fu, K. Tanaka, and M. Shimura, Eds.), Academic Press, New York, pp. 395–408.

1366 Kochen, M. (1977). Imprecision in coping and attending processes, in *Proc. IEEE Symp. on Fuzzy Sets*, New Orleans, pp. 1324–1329.

1367 Kochen, M. (1977). On fundamentals, theory, and applications of fuzzy sets, in *Proc. Symp. on Fuzzy Set Theory and Appl., IEEE Conf. on Decis. and Control*, New Orleans.

1368 Kochen, M. (1979). Enhancement of coping through blurring, *Fuzzy Sets Syst.*, **2** (1), Jan., 37–52.

1369 Kochen, M. (1980). Appropriate precision in concept genesis, in *Proc. Int. Congr. on Appl. Syst. Res. and Cybern.*, Acapulco, Mexico, Dec.

1370 Kochen, M., and Badre, A. N. (1974). On the precision of adjectives which denote fuzzy sets, *J. Cybern.*, **4**, 49–59.

1371 Kochen, M., Dreyfuss-Raimi, G. (1974). On the psycholinguistic reality of fuzzy sets—Effect of context and set, Univ. of Michigan Mental Health Res. Inst., Ann Arbor, June.

1372 Koczy, L. T. (1974). A fuzzy halmazok nehany elmeleti kerdese (Some theoretical questions of fuzzy sets), Rep., Dept. of Process Control, Tech. Univ., Budapest, Jan.

1373 Koczy, L. T. (1975). R-fuzzy algebra as a generalized formulation of the intuitive logic, Dept. of Process Control, Tech. Univ., Budapest.

1374 Koczy, L. T. (1976). Some questions of sigma-algebras of fuzzy objects of type N, 3rd Eur. Meet. Cybern. Syst. Res., Vienna.

1375 Koczy, L. T. (1976). Fuzzy algebrak es muszaki alkalmazasaik neheny kerdese (Fuzzy algebra and some questions of their technical applications), Ph.D. Thesis, Dept. of Process Control, Tech. Univ., Budapest.

1376 Koczy, L. T. (1977). On some basic theoretical problems of fuzzy mathematics, *Atica Cybern.*, **3**, 225–237.

1377 Koczy, L. T. (1978). Interactive sigma-algebras and fuzzy objects of type N, *J. Cybern.*, **8**.

1378 Koczy, L. T. (1978), Vector valued fuzzy sets, *Busefal*.

1379 Koczy, L. T. (1979). Some questions of B-algebras of fuzzy objects of type N, *IEEE Trans. Syst., Man, Cybern.*, **SMC-9** (9), Sept.

1380 Koczy, L. T., and Hajnal, M. (1975). A new fuzzy calculus and its application as a pattern recognition technique, in *Proc. 3rd Int. Congr. of Cybern. and Syst.*, Bucharest, Aug.

1381 Koczy, L. T., and Hajnal, M. (1976). A karometric classification algorithm based on R-fuzzy set calculus, in *Proc. 2nd Nat. Meet. on Biophys. and Biotech.*, Espoo, Finland, Feb.

1382 Koczy, L. T., and Hajnal, M. (1977). Cluster analysis in karyometry applying a new fuzzy algebra, in *Biomedical Computing* (W. J. Perkins, Ed.), Pitman Medical Publ. Co. Ltd., Tunbridge Wells, UK.

1383 Kogan, A. B., and Tchorayan, O. G. (1979). Elements d'algorithmes flous pour representer l'activite de l'intellect. *Physiol. Homme* (URSS), **5** (3).

1384 Kohout, L. J. (1974). The Pinkava many-valued complete logic systems and their applications in the design of many-valued switching circuits, in *Proc. 1974 Int. Symp. on Multiple-Valued Logic*, IEEE 74CH0845-8C, May, pp. 261–284, Morgantown, West Virginia.

1385 Kohout, L. J. (1975). Generalized topologies and their relevance to general systems, *Int. J. Gen. Syst.*, **2**, 25–34.

1386 Kohout, L. J. (1976). Application of multi-valued logics to the study of human movement control and of movement disorders, in *Proc. 6th Int. Symp. Multiple-Valued Logic*, IEEE 76CH1111-4C, pp. 224–231.

1387 Kohout, L. J. (1976). Automata and topology, in *Discrete Systems and Fuzzy Reasoning* (E. H. Mamdani, and B. R. Gaines, Eds.), EES-MMS-DSFR-76, Queen Mary College, Univ. of London, London, (Workshop Proc.).

1388 Kohout, L. J. (1976). Representation of functional hierarchies of movement in the brain, *Int. J. Man-Mach. Stud.*, **8**, 699–709.

1389 Kohout, K. J. (1980). Possibilistic structures as a conceptual tool for analysis of operations research problems, 4th Eur. Congr. on Oper. Res, July.

1390 Kohout, L. J., and Pinkava, V. (1976). The functional completeness of pi-algebras and its relevance to biological modelling and to technological applications of many-valued logics, in Mamdani, *Discrete Systems and Fuzzy Reasoning*, (E. H. Mamdani and B. R. Gaines, Eds.), EES-MMS-DSFR-76, Queen Mary College, Univ. of London, London (Workshop Proc.).

1391 Kokawa, M. (1977). Fuzzy theoretic and concept-formational approaches to inference and information experiments in human decision processes, in *Proc. Symp. on Fuzzy Set Theory and Appl., IEEE Conf. on Decis. and Control*, New Orleans.

1392 Kokawa, M., Nakamura, K., and Oda, M. (1972). A formulation of human decision-making process, Res. Rep. 19, Autom. Control Lab., Nagoya Univ., Nagoya, Japan, pp. 3–10.

1393 Kokawa, M., Nakamura, K., and Oda, M. (1973). Fuzzy expression of human experience-to memory process, Res. Rep. 20, Autom. Control Lab., Nagoya Univ., Nagoya Japan, June pp. 27–33.

1394 Kokawa, M., Nakamura, K., and Oda, M. (1974). Fuzzy process of memory behaviour, *Trans. Soc. Instrum. Control Eng.*, **10**, 385–386.

1395 Kokawa, M., Nakamura, K., and Oda, M. (1974). Fuzzy theoretical and concept formational approaches to memory and inference experiments, *Trans. IECE* (Japan), **57-D**, 487–493.

1396 Kokawa, M., Nakamura, K., and Oda, M. (1974). Fuzzy-theoretical approaches to forgetting processes and inference, Res. Rep. 21, Autom. Control Lab., Nagoya Univ., Nagoya, Japan, pp. 1–10.

1397 Kokawa, M., Nakamura, K., and Oda, M. (1975). Effect of a jump of logic in a decision process, *Trans. IECE* (Japan), **58-D** (5).

1398 Kokawa, M., Nakamura, K., and Oda, M. (1975). Experimental approach to fuzzy simulation of memorizing, forgetting and inference process, in *Fuzzy Sets and their*

Applications to Cognitive and Decision Processes (L. A. Zadeh, K. S. Fu, K. Tanaka, and M. Shimura, Eds.), Academic Press, New York, pp. 409–428.

1399 Kokawa, M., Nakamura, K., and Oda, M. (1975). Hint effect and jump of logic in a decision process, *Trans. IECE* (Japan), **58-D**.

1400 Kokawa, M., Oda, M., and Nakamura, K. (1975). Fuzzy theoretical one-dimensionalizing method of multidimensional quantity, *Trans. Soc. Instrum. Control Eng.*, **11** (5), 8–14.

1401 Kokawa, M., Nakamura, K., and Oda, M. (1976). Fuzzy description of decision-making process and experimental approach to fuzzy simulation of memorizing, forgetting and inference process, 1976 Joint Conv. of Four Inst. of Elec. Eng.

1402 Kokawa, M., Oda, M., and Nakamura, K. (1977). Fuzzy-theoretical dimensionality reduction method of multi-dimensional quantity, in *Fuzzy Automata and Decision Processes* (M. M. Gupta, G. N. Saridis, and B. R. Gaines, Eds.), North-Holland, Amsterdam, pp. 235–250.

1403 Kokawa, M., Nakamura, K., and Oda, M. (1979). Fuzzy theoretic and concept formational approaches to hint effect experiments in human decision processes, *Fuzzy Sets Syst.*, **2** (1), Jan., 25–36.

1404 Kokawa, M., Nakamura, K., and Oda, M. (1979). Fuzziness and catastrope in estimation and decision processes, in *Proc. 18th IEEE Conf. on Decis. and Control*, Dec.

1405 Kokawa, M., Oda, M., and Nakamura, K. (1979). A study of handling method of fuzzy data, *Trans. IECE* (Japan), **62-A** (1), 97–102.

1406 Kolibiar, M. (1972). Distributive sublattices of a lattice, *Proc. Amer. Math. Soc.*, **34**, 359–364.

1407 Kolomov, S. V., Makeev, S. P., Serov, G. P., and Shakhnov, I. F. (1979). Optimal control of a finite automation with fuzzy constraints and a fuzzy target, *Cybernetics* (URSS), **15** (6), 805–810.

1408 Kong, Y. (1980). The composite forecast based on multiple factors by using fuzzy principle, News on 1980 Annu. Rep. Meet. of Beijing Working Group on Fuzzy Sets, Beijing, China.

1409 Kinrad, E., and Bollamn, P. (1976). Fuzzy document retrieval, 3rd Eur. Meet. Cybern. Syst. Res., Vienna.

1410 Korner, S. (1957). Reference, vagueness and necessity, *Philos. Rev.*, **66**, July.

1411 Korner, S. (1959). *Conceptual Thinking*, New York.

1412 Korner, S. (1966). *Experience and Theory*, Routledge and Kegan Paul, London.

1413 Korner, S. (1970). *Categorial Frameworks*, Basil Blackwell, Oxford, UK.

1414 Korner, S. (1971). *Fundamental Questions of Philosophy*, Penguin Books.

1415 Korner, S. (1976). *Experience and Conduct*, Cambridge Univ. Press, Cambridge, UK.

1416 Korner, S. (1976). *Philosophy of Logic*, Basil Blackwell, Oxford, UK.

1417 Kotas, J. (1963). Axioms for Birkhoff-V. Neumann quantum logic, *Bull. Acad. Pol. Sci.*, Ser. Math., Astr. Phys., **11**, 629–632.

1418 Kotoh, K., and Hiramatsu, K. (1973). A representation of pattern classes using the fuzzy sets, *Syst., Comput., Controls*, 1–8, (Originally in *Trans. IECE* (Japan), **56-D**, 275–282).

1419 Koutsky, K. (1947). Sur les lattices topologies, C. R. (Paris), **225**, 659–661.

1420 Koutsky, K. (1952). Theorie des lattices topolgiques, Publ. de la Fac. des Sci. de l'Univ. Masaryk, No. 337, Brno, Czechoslovakia, pp. 133–171.

1421 Kovacs, F. (1977). Defining membership functions of fuzzy sets, Thesis, Iona College, New Rochelle, NY.

1422 Kraft, M. L. (1979). *Vage Konzepte in der Okonomie*, Verlag Schonburg, Paderborn, 162 Seiten.

1423 Kramosil, I. (1975). A probabilistic approach to automaton-environment systems, *Kybernetika* (Prague), **11**, 173–206.

1424 Kramosil, I., and Michalek, J. (1975). Fuzzy metrices and statistical metric spaces, *Kybernetika* (Prague), **11**, 336–344.

1425 Krantz, D. H., Luce, R. D., Suppes, P., and Tversky, A. (1971). *Foundations of Measurement*, Academic Press, New York.

1426 Krivine, J. L. (1974). Langages a valeurs reelles et applications, *Fundam. Math.*, **81**, 213–253.

1427 Kubinski, T. (1958). Nazwy nieostre (Vague terms), *Stud. Logica*, **7**, 115–179.

1428 Kubinski, T. (1959). Systemy pozorie spreczne, Zeszyty Naukowe Uniwersytetu Wroclawskiego, Seria B, Matematyka, Fizyka, Astronomia (1959), pp. 53–61.

1429 Kubinski, T. (1960). An attempt to bring logic nearer to colloquial language, *Stud. Logica*, **10**, 61–75.

1430 Kulikov, V. F., and Gurjev, E. K. (1977). Compromise decision making in the two-level control system, in *Proc. 1977 IEEE Conf. on Decis. and Control*, New Orleans.

1431 Kumar, A. (1977). A real time system for pattern recognition of human sleep stages by fuzzy systems analysis, *Pattern Recogn.*, **9**, 43–46.

1432 Kummer, B., and Straube, B. (1977). Uber den Entwruf Eines Unscharfen Reglers, in *Unscharfe Modellbilung Steuerung*, Lectures Notes, TH Karl-Marx-Stadt, June, 34–38.

1433 Kunii, T. L. (1976). Dataplan—An interface generator for database semantics, *Inf. Sci.*, **10**, 279–298.

1434 Kuo-Jun, W. (1980). A more reasonable fuzzy compactness, Rep. Prepared by the Group of Topology in Sichuan Univ., Sichuan, Nov. (in Chinese).

1435 Kuroki, N. (1979). Fuzzy bi-ideals in semigroups, *Comm. Math. Univ. St. Pauli*, **28**, 17–21.

1436 Kuroki, N. (1981). On fuzzy ideals and fuzzy bi-ideals in semigroups, *FSS*, 203–215.

1437 Kurosu, K., Murayama, Y., Kobayashi, M., and Inasaka, F. (1976). Display method using tree pattern, in *Summ. of Pap. on Gen. Fuzzy Problems*, The Working Group on Fuzzy Syst., Toyko, pp. 35–41.

1438 Kuz'min, V. B. (1979). A "reference approach to obtaining fuzzy preference relations and problem of choice, Doc. Inst. of Syst. Sci., Moscow.

1439 Kuz'min, V. B. (1979). A parametric approach to description of linguistic values of variables and hedges, Doc. Inst. of Syst. Sci., Moscow.

1440 Kuz'min, V. B. (1980). A parametric approach to the description of linguistic values of variables and hedges, in *Proc. Int. Congr. on Appl. Syst. Res. and Cybern.*, Acapulco, Mexico, Dec.

1441 Kuz'min, V. B., and Ovchinn Ikov, S. V. (1979). Group decisions in arbitrary spaces of fuzzy binary relations, Doc. Inst. of Syst. Sci., Moscow.

1442 Kuz'min, V. B., and Ovchinnikov, S. V. (1979). Design of group decisions in spaces of partial order fuzzy relations, Doc. Inst. of Syst. Sci., Moscow.

1443 Kwakernaak, H. (1978). Fuzzy random variables, Part 1, Definitions and theorems, Memo. No. 193, TH Twente, pp. 1–33.

1444 Kwakernaak, H. (1978). Fuzzy random variables, Part 2, Algorithms and examples for the discrete case, Memo. No, 212, TH Twente, pp. 1–25.

1445 Kwakernaak, H. (1979). An algorithm for rating multiple-aspect alternatives using fuzzy sets, *Automatica*, **15**, 615–616.

1446 Kwakernaak, H. (1979). Fuzzy random variables, algorithms and examples for the discrete case, *Inf. Sci.*, **17**, 253–278.

1447 Kyberg, H. E. (1970). *Probability and Inductive Logic*, Macmillan, London.

1448 Labov, W. (1973). The boundaries of words and their meanings, in *New Ways of Analysing Variations in English*, (Bailey and Shuy, Eds.), Georgetown Univ. Press, Washington, D.C.

1449 Lafosse-Marin, J. (1977). Ultrametric distance function on grammatical deformed configurations by the way of fuzzy relations, Doc. Dept. Math., Univ. d'Alger, Feb.

1450 Lafosse-Marin, J. (1978). Moore closure of fuzzy graphs, Colloq. Int. Theorie des Sous-Ensembles Flous et Appl., Journ. de Biomath. et d'Inf. Med., Marseille, France, Sept.

1451 Lafosse-Marin, J. (1980). Topologies on fuzzy graphs, Table Ronde du CNRS sur le Flou, Lyon, France, June.

1452 Lafosse-Marin, J. M. (1976). Les sous-ensembles flous dans la reconnaissance et la classification des structures, Doct. Thesis, Univ. de Paris, Paris, Feb.

1453 Lafosse-Marin, J. M. (1977). Sous-ensembles flous en reconnaissance et classification, Semin. IRIA, Rocquencourt, Recon. des Structures, Feb.

1454 Lake, J. (1974). Fuzzy sets and bald men, Dept. of Math., Polytech. of the South Bank, London.

1455 Lake, J. (1974). Sets, fuzzy sets, multi-sets and functions, Dept. of Math., Polytech. of the South Bank, London.

1456 Lakoff, G. (1971). Linguistic and natural logic, in *Semantics of Natural Languages*, (Davidson and Harman, Eds.), D. Reidel, Dordrecht, Holland.

1457 Lakoff, G. (1972). A study in meaning criteria and the logic of fuzzy concepts, in *Proc. 8th Reg. Meet. Chicago Linguist. Soc.*, Chicago.

1458 Lakoff, G. (1973). Fuzzy grammar and the performance/competence terminology game, in *Proc. Meet. of Chicago Linguist. Soc.*, Chicago, pp. 271–291.

1459 Lakoff, G. (1973). Hedges—A study in meaning criteria and the logic of fuzzy concepts, *J. Philos. Logic*, **2**, 458–508.

1460 Lakoff, G. (1973). Notes on what it would take to understand how one adverb works, *Monist*, **57**, 328–343.

1461 Lakoff, G. (1973). Pragmatics in natural logic, in *Formal Semantics of Natural Language* (E. L. Keenan, Ed.), Cambridge Univ. Press, Cambridge, UK, pp. 253–286.

1462 Lakov, D. R., and Naplatanoff, N. (1977). Decision-making in vague conditions, *Kybernetes*, **6**, 91–93.

1463 Lakov, D. V. (1978). Ergodic, control systems, Ph.D. diss., Bulgarian Acad. of Sci., Inst. Eng. Cybern., Sofia.

1464 Lakshmivarahan, S., and Rajasethupathy, K. S. (1974). Considerations for fuzzifying formal languages and synthesis of fuzzy grammars, Indian Inst. of Tech., Madras.

1465 Lakshmivarahan, S., and Rajasethupathy, K. S. (1978). Considerations for fuzzifying formal languages and synthesis for fuzzy games, *J. Cybern.*, **8**, 83–100.

1466 Lao, S., and Chen, H. (1980). Application of fuzzy cluster to weather forecasting, Collect. Pap. on Fuzzy Math. (Abstr.), Huazhong Inst. of Tech., Huazhong, China, Apr.

1467 Larsen, J. (1976). A multi-step formulation of variable valued logic hypotheses, in *Proc. 6th Int. Symp. on Multiple-Valued Logic*, IEEE 76CH1111-4C, pp. 157–163.
1468 Larsen, L. E., Ruspini, E. H., McNew, J. J., Walter, D. O., and Adey, W. R. (1972). A test of sleep staging systems in the unrestrained chimpanzee, *Brain Res.*, **40**, 319–343.
1469 Larsen, P. M. (1980). Industrial applications of fuzzy logic control, *Int. Man-Mach. Stud.*, **12** (1).
1470 Lasker, G. E. (1981). Possibility analysis, CORS-TIMS-ORSA Joint Nat. Meet., Toronto, Canada, May.
1471 Lassibille, G., and Parron, C. (1975). Analyse multi-critere dans un contexte imprecis, Doc. Inst. Ath. Econ., Fac. Sci. Econ. de Gestion, Univ. de Dijon, Dijon, France, July.
1472 Law, H. Y. H., Wong, J., Kodani, M. (1977). Laboratory source selection under a fuzzy environment, ORSA/TIMS Meet., Atlanta, GA.
1473 Lawvere, F. W. (Ed.) (1972). *Toposes, Algebraic Geometry and Logic*, Springer-Verlag, Berlin.
1474 Lawvere, F. W., Maurer, C., and Wraith, G. C. (Eds.) (1975). *Model Theory and Topoi*, Lecture Notes in Mathematics, Springer-Verlag, Berlin, 445 pp.
1475 Lazak, D. (1977). Fuzzy sets and artificial intelligence, 6th Eur. Working Group on Fuzzy Sets, Abo, Finland.
1476 Leal, A., and Pearl, J. (1976). A computer system for conventional elicitation of problem structures, UCLA-ENG-7665, School of Eng. and Appl. Sci., Univ. of California, Los Angeles, June.
1477 Leal, A., and Pearl, J. (1977). An interactive program for conversational elicitation of decision structures, *IEEE Trans. Syst., Man, Cybern.*, **7** (5).
1478 Leberling, H. (1980). Determining compromise solutions in a linear vector maximum problem, using the fuzzy min operator and nonlinear membership functions, 4th Eur. Congr. on Oper. Res, July.
1479 Lee, C., and Ternano, T. (1980). Algorithm of creating the story, in *Summ. of Pap. on Gen. Fuzzy Problems*, The Working Group on Fuzzy Syst., Tokyo, Dec.
1480 Lee, E. T. (1972). Fuzzy languages and their relation to automata, Ph.D. Thesis, Dept. of Elect. Eng. and Comput. Sci., Univ. of California, Berkeley.
1481 Lee, E. T. (1972). Proximity measures for the classification of geometric figures, *J. Cybern.*, **2**, 43–59.
1482 Lee, E. T. (1974). An application of fuzzy sets to the classification of geometric figures and chromosome images, in US-Jap. Semin. on Fuzzy Sets and their Applications, Berkeley, CA.
1483 Lee, E. T. (1975). Shape-oriented chromosome classification, *IEEE Trans. Syst., Man, Cybern.*, **SMC-5**, 629–632.
1484 Lee, E. T. (1977). Application of fuzzy languages to pattern recognition, *Kybernetes*, **6**, 167–173.
1485 Lee, E. T. (1978). Applications of fuzzy sets to pattern description classification, recognition, storage and retrieval, *Proc. 1978 JACC*, **3**, 61–68.
1486 Lee, E. T. (1978). Application of fuzzy languages and fuzzy tree automata, pattern description, classification, recognition, storage and pattern retrieval, Colloq. Int. Theorie et Applications des Sous-Ensembles Flous, Journ. de Biomath. et d'Inf. Med., Marseille, France, Sept.
1487 Lee, E. T. (1980). Applications of fuzzy set theory to image sciences. *J. Cybern.*, **10**.
1488 Lee, E. T., and Zadeh, L. A. (1969). Notes on fuzzy languages, *Inf. Sci.*, **1**, 421–434.

1489 Lee, E. T., and Zadeh, L. A. (1970). Fuzzy languages and their acceptance by automata, in *4th Princeton Conf. on Inf. Sci. and Syst.*, Princeton, NJ, p. 399.

1490 Lee, J. (1980). Fault analysis of analog electronic systems—Algorithms based on fuzzy sets, Ph.D. Thesis, Univ. of Pennsylvania, Diss. Abstr. Int., **41/10-B**, 3865.

1491 Lee, R. C. T. (1972). Fuzzy logic and the resolution principle, *J. Assoc. Comput. Mach.*, **19**, 109–119.

1492 Lee, R. C. T., and Chang, C. L. (1971). Some properties of fuzzy logic, *Inf. Control*, **19**, 417–431.

1493 Lee, S. C. (1969). Analysis and synthesis of sequential fuzzy logic circuits, in *Proc. 7th Annu. Allecton Conf. on Circuit and Syst. Theory*, pp. 692–701.

1494 Lee, S. C., and Lee, E. T. (1970). Fuzzy neurons and automata, in *Proc. of 4th Princeton Conf. on Inf. Sci. and Syst.*, Princeton, NJ, pp. 381–385.

1495 Lee, S. C., and Lee, E. T. (1974). Fuzzy sets and neural networks, *J. Cybern.*, **4**, 83–103.

1496 Lee, S. C., and Lee, E. T. (1975). Fuzzy neural networks, *Math. Biosci.*, **23**, 151–177.

1497 Leenders, J. H. (1974). Vage Veramelingen—Een kritische benandering, *Kwartaalschr. Wetenschappelijk Onderwijs Limburg* (Belgium), 4, 441–455.

1498 Leenders, J. H. (1977). Some remarks on an article by Raymond T. Yeh and S. Y. Bang dealing with fuzzy relations, *Simon Stevin, Wis-Nat., Tijdschr.*, **51**, 93–100.

1499 Lefaivre, R. A. (1974). FUZZY—A programming language for fuzzy problem solving, Rep. PB-231813/7, Wisconsin Univ., Madison.

1500 Lefaivre, R. A. (1974). Fuzzy problem solving, Tech. Rep. 37, Madison Acad. Comput. Center, Univ. of Wisconsin, Madison, Aug.

1501 Lefaivre, R. A. (1974). The representation of fuzzy knowledge, *J. Cybern.*, **4**, 57–66.

1502 Lefaivre, R. A. (1976). Procedural representation in fuzzy problem solving systems, in *Proc. NCC*.

1503 Lemke, H. R., and Kickert, W. J. M. (1976). The application of fuzzy sets theory to control a warm water process, Delft Univ. of Tech., Delft, Holland.

1504 Lemmon, E. J. (1966). Algebraic semantics for modal logics I, *J. Symb. Logic*, **31**, 46–65.

1505 Lemmon, E. J. (1966). Algebraic semantics for modal logics II, *J. Symb. Logic*, **31**, 191–218.

1506 Lemmon, E. J., Meredith, C. A., Meredith, D., Prior, A. N., and Thomas, I. (1969). Calculi of pure strict implication, in *Philosophical Logic*, (J. W. Davis, D. J. Hockney, and W. K. Freed, Eds.), D. Reidel, Dordrecht, Holland, pp. 215–250.

1507 Lendaris, G. G., and Martinez, A. J. (1976). Bibliography on fuzzy sets and their applications, Rep. of Syst. Sci. Program, SS-11, Portland State Univ., Portland, OR.

1508 Lerman, I. C. (1979). Criosement de classification floues, Rap. No. 108, Fac. des Sci. Univ. de Rennes, Rennes, France.

1509 Lerman, I. C., Hardouin, M., and Chantrel, T. (1979). Analyse de la situation relative entre deux classification floues. 2nd Journ. Int. Anal. des Donnees et Inf., Versailles, France, Oct.

1510 Leroy, C. L., McRice, V., and Heaulme, M. (1978). Les diagnostics psychiatriques en tant qu'ensembles flous, Colloq. Int. Theorie et Appl. des Sous-Ensembles Flous, Journ. de Biomath. et d'Inf. Med., Marseille, France, Sept.

1511 Leung, Y. (1979). A fuzzy set procedure for project selection of an urban policy with hierarchical objectives, 1st Int. Symp. on Policy Anal. and Inf. Syst., June.

1512 Leung, Y. (1980). A fuzzy set procedure for project selection with hierarchical objectives, Fuzzy Sets Theory and Appl. to Policy Anal. and Inf. Syst., Durham, NC.
1513 Leung, Y. (1980). Empirical analysis of fuzzy decision making process, in *Proc. Int. Congr. on Appl. Syst. Res. and Cybern.*, Acapulco, Mexico, Dec.
1514 Levchenkov, V. (1980). Systems with subjective variables and fuzzy sets, The 10th Meet. of the EWG on Fuzzy Sets, June.
1515 Levi, I. (1967). *Gambling with Truth*, MIT Press, Cambridge, MA.
1516 Lewis, D. K. (1969). *Convention—A Philosophical Study*, Harvard Univ. Press, Cambridge, MA.
1517 Lewis, D. K. (1973). *Counterfactuals*, Basil Blackwell, Oxford, UK.
1518 Li, B. (1980). Fuzzy logic systems in the computers, News on 1980 Ann. Rep. Meet. of Beijing Working Group on Fuzzy Sets, Beijing, China.
1519 Li, B. (1980). Matrix theory of fuzzy subsets, Collect. Pap. on Fuzzy Math. (Abstr.), Huazhong Inst. of Tech., Huazhong, China, Apr.
1520 Li, B. (1980). Solving operator equation on the conditional fuzzy subset, Collect. Pap. on Fuzzy Math. (Abstr.), Huazhong Inst. of Tech., Huazhong, China, Apr.
1521 Li, B. (1980). The fuzzy representations of the modality system $S(5)$, News on 1980 Annu. Rep. Meeting of Beijing Working Group on Fuzzy Sets, Beijing, China.
1522 Li, B., and Liu, Z. (1980). Application of fuzzy set theory to identification of system models, *Inf. Control*, **9**, 3, A List of Lit. in China, The Univ. of Sci. and Tech. of China, Hofei, Anhwei, China.
1523 Li, B., and Liu, Z. (1980). The application of fuzzy set theory to the design of a class controllers, *Acta Autom. Sin.*, **6**, A List of Lit. in China, The Univ. of Sci. and Tech. of China, Hofei, Anhwei, China.
1524 Li, B., and Wong, P. (1980). The rules for resolving the fuzzy relation equations, News on 1980 Annu. Rep. Meeting of Beijing Working Group on Fuzzy Sets, Beijing, China.
1525 Li, B., and Wong, P. (1980). The simple method of resolving a family of special fuzzy equations, News on 1980 Annu. Rep. Meet. of Beijing Working Group on Fuzzy Sets, Beijing, China.
1526 Li, C., and Yu, H. (1980). Some problems concerning the resolution of fuzzy equation, *J. Huazhong Inst. Tech.*, **2**, A List of Lit. in China, The Univ. of Sci. and Tech. of China, Hofei, Anhwei, China.
1527 Li, J. (1980). Establishment of the subordinate function for basic geomorphological types, News on 1980 Annu. Rep. Meet. of Beijing Working Group on Fuzzy Sets, Beijing, China.
1528 Liao, Z. (1980). Mutually inverse properties of fuzzy matrix and its inverse, News on 1980 Annu. Rep. Meet. of Beijing Working Group on Fuzzy Sets, Beijing, China.
1529 Liao, Z. (1980). Sufficient and necessary conditions of existence of solutions of fuzzy equations, News on 1980 Annu. Rep. Meet. of Beijing Working Group on Fuzzy Sets, Beijing, China.
1530 Lientz, B. P. (1972). On time dependent fuzzy sets, *Inf. Sci.*, **4**, 367–376.
1531 Lientz, B. P. (1977). On the analysis of complex softly designed systems, Rep. Grad. School of Man., Univ. of California, Los Angeles.
1532 Lipp, H. P. (1977). Anwendung von Unscharfen Steueroperationen Beim Fuhren Technologischer Anlagen, in *Unscharfe Modellbilung and Steuerung*, TH Karl-Marx-Stadt, Germany, June, pp. 130–143.

1533 Liu, E. (1980). Applications of fuzzy mathematics to the classification of rivers, News on 1980 Annu. Rep. Meet. of Beijing Working Group on Fuzzy Sets, Beijing, China.

1534 Liu, G., and Li, B. (1980). Some properties of fuzzy space and fuzzy logic, *J. Huazhong Inst. Tech.*, 2, A List of Lit. in China, The Univ. of Sci. and Tech. of China, Hofei, Anhwei, China.

1535 Liu, L. (1980). The applications of fuzzy sets to the classification of larvas of genus Dermacentor, News on 1980 Annu. Rep. Meet. of Beijing Working Group on Fuzzy Sets, Beijing, China.

1536 Liu, Y. (1980). A generation of fuzzy category and constructions of inverse in the fuzzy category, Collect. Pap. on Fuzzy Math. (Abstr.), Huazhong Inst. of Tech., Huazhong, China, Apr.

1537 Liu, Y. (1980). A note on compactness in fuzzy unit interval, Collect. Pap. on Fuzzy Math. (Abstr.), Huazhong Inst. of Tech., Huazhong, China, Apr.

1538 Liu, Y. (1980). Compactness and Tychonoff theorem in fuzzy topological spaces, Collect. Pap. on Fuzzy Math. (Abstr.), Huazhong Inst. of Tech., Huazhong, China, Apr.

1539 Liu, Y. (1980). Intersection operation on union-preserving mappings in completely distributive lattices, Collect. Pap. on Fuzzy Math. (Abstr.), Huazhong Inst. of Tech., Huazhong, China, Apr.

1540 Loginov, V. I. (1966). Probability treatment of Zadeh membership function and their use in pattern recognition, *Eng. Cybern.*, 68–69.

1541 Lombaerde, J. (1974). Mesures d'entropie en theorie des sous-ensembles flous, Imago Disc. Pap. IDP-12, Centre Interfac. Imago, Univ. Catholique de Louvain, Louvain, Belgium, Jan.

1542 Lombaerde, J. (1976). Une mesure d'information comme outil de composition pour l'evaluation formative dans une inter-action eleve-machine, Communication Reunion G.P.I., Centre Imago, Univ. Catholique de Louvain, Louvain, Belgium.

1543 Long, S., He, K., Zhao, X., and Zhou, Y. (1980). Determinization of man's fuzzy concepts in man-machine systems, News on 1980 Ann. Rep. Meet. of Beijing Working Group on Fuzzy Sets, Beijing, China.

1544 Longo, G. (1975). Fuzzy set, graphs and source coding, in *New Directions in Signal Processing in Communications and Control*, (J. K. Swirzynski, Ed.), Noordhoff, Leyden, Holland, pp. 27–33.

1545 Loo, S. G. (1977) Measures of fuzziness, *Cybernetica*, 3, 201–207.

1546 Loo, S. G. (1978). Fuzzy relations in social and behavioral sciences, *J. Cybern.*, 8, 1–16.

1547 Lopez, R. (1980). Learning the meaning of imprecise concepts by adaptation of their possibility distributions, *Proc. Int. Congr. on Appl. Syst. Res. and Cybern.*, Acapulco, Mexico, Dec.

1548 Los, J., and Ryll-Nardzewski, C. (1951). On the application of Tychnoff's theorem in mathematical proofs, *Fundam. Math.*, 38, 233–237.

1549 Lou, C. (1980). Fuzzy relation equation on infinite sets, Rep. Beijing Normal Univ., 11, Beijing, China, A List of Lit. in China, The Univ. of Sci. and Tech. of China, Hofei, Anhwei, China.

1550 Lou, C. (1980). Fuzzy relation equations on finite sets, Rep. Beijing Normal Univ., 11, Beijing, China, A List of Lit. in China, The Univ. of Sci. and Tech. of China, Hofei, Anhwei, China.

1551 Lou, C. (1980). Reachable solution set of a fuzzy relation equation, Rep. Beijing Normal Univ. 11, Beijing, China, A List of lit. in China, The Univ. of Sci. and Tech. of China, Hofei, Anhwei, China.

Key References in Fuzzy Pattern Recognition

1552 Lour, C. (1980). The image and inverse of the fuzzy subgroup for a homomorphism of group G onto G, Rep. Beijing Normal Univ. 11, Beijing, China, A List of Lit. in China, The Univ. of Sci. and Tech. of China, Hofei, Anhwei, China.

1553 Lou, S. (1980). Fuzzy mapping, Rep. Shanghai Railway Inst. Shanghai, A List of Lit. in China, The Univ. of Sci. and Tech. of China, Hofei, Anhwei, China.

1554 Lou, S. (1980). The method of fuzzy graph in the classification of chemicals, Rep. Shanghai Railway Inst., Shanghai, A List of Lit. in China, The Univ. of Sci. and Tech. of China, Hofei, Anhwei, China.

1555 Lou, S. P., and Pan, S. H. (1980). Fuzzy structure, *J. Math. Anal. Appl.*, **76** (2), 631–642.

1556 Lowe, E. A., and Tinker, A. M. (1977). Regulating jumping fuzzy sets, Int. Conf. on Appl. Gen. Syst. Res., Binghamton, NY.

1557 Lowen, R. (1974). A theory of fuzzy topologies, Ph.D. Thesis, Free Univ. of Brussels, Brussels, Belgium.

1558 Lowen, R. (1974). Topologies flous, *C.R. Acad. Sci.* (Paris), **278A**, 925–928.

1559 Lowen, R. (1975). Convergence flous, *C.R. Acad. Sci.* (Paris), **280**, 1181–1183.

1560 Lowen, R. (1976). A comparison of different compactness notions in fuzzy topology, Free Univ. of Brussels, Brussels, Belgium.

1561 Lowen, R. (1976). Fuzzy topological spaces and fuzzy compactness, *J. Math. Anal. Appl.*

1562 Lowen, R. (1976). Initial and final fuzziness topologies and the fuzzy Tychnoff theorem, *J. Math. Anal. Appl.*

1563 Lowen, R. (1978). A comparison of different compactness notions in fuzzy topological spaces 1, Notices of the AMS, Oct. 1976, *J. Math. Anal. Appl.* **64**, 446–454,

1564 Lowen, R. (1977). Fuzzy topology, in *Proc. IEEE Symp. On Fuzzy Set Theory and Appl.*, *IEEE Conf. on Decis. and Control*, New Orleans.

1565 Lowen, R. (1977). Initial and fuzzy topologies and the fuzzy Tychonoff Theorem, *J. Math. Anal. Appl.*, **58**, 11–21.

1566 Lowen, R. (1977). Lattice convergence in fuzzy topological spaces, Notices of the AMS, Nov. 1977.

1567 Lowen, R. (1977). On fuzzy complements and convergence in fuzzy topological spaces, in *Proc. Symp. on Fuzzy Set Theory and Appl., IEEE Conf. on Decis. and Control*, New Orleans, pp. 1338–1372.

1568 Lowen, R. (1978). Convergence in fuzzy topological spaces, in *Proc. 4th Prague Conf. on General Topology*, Prague.

1569 Lowen, R. (1979). Compact Hausdorff fuzzy topological spaces are topological, in *Proc. Int. Semin. on Fuzzy Set Theory*, Johannes Kepler Univ. Linz, Austria.

1570 Lowen, R. (1980). Convex fuzzy sets, *Fuzzy Sets Syst.*, **3**, (3), May, 291–310.

1571 Lowen, R. (1980). Fuzzy topology—How and why, in *Proc. Int. Congr. on Appl. Syst. Res. and Cybern.*, Acapulco, Mexico, Dec.

1572 Lu, L. (1980). The application of fuzzy sets to recognizing wheat parents, Rep. Beijing Normal Univ., 3, Beijing, China, A List of Lit. in China, The Univ. of Sci. and Tech. of China, Hofei, Anhwei, China.

1573 Lucas, D., Ekman, K., and Weete, F. (1979). The application of fuzzy pointers in multisensor and multitarget integration, in *Proc. 17th IEEE Conf. on Decis. and Control*, pp. 1217–1219.

1574 Ludescher, H., and Roventa, E. (1976). Surles topologies floues definies a l'aide des voisinages, *C.R. Acad. Sci.* (Paris) 283, 575–577.

1575 Lukaciewicz, J. (1920). Logike trojwartesciowog, *Ruch. Filos.*, **169**, Varsovie.
1576 Lukaciewicz, J., and Tarski, A. (1930). Untersuchungen uber den Aussagenkalkul, *C. R. Soc. Lett., Varsovie*, **XXIII**, 30–50.
1577 Luo, C. (1980). Fuzzy relation equations on finite sets, News on 1980 Ann. Rep. Meet. of Beijing Working Group on Fuzzy Sets, Beijing, China.
1578 Luo, C. (1980). Fuzzy relation equations on infinite sets, News on 1980 Ann. Rep. Meet. of Beijing Working Group on Fuzzy Sets, Beijing, China.
1579 Luo, C. (1980). On the sup-property of fuzzy subgroups, News on 1980 Ann. Rep. Meet. of Beijing Working Group on Fuzzy Sets, Beijing, China.
1580 Luschei, E. C. (1962). *The Logical Systems of Lesniewski*, North-Holland, Amsterdam.
1581 Lusk, E. J. (1981). Evaluating performance statistics used to monitor performance—A fuzzy approach, *Fuzzy Sets Syst.*, 149–158.
1582 Lysvag, B. (1975). Verbs of Hedging, in *Syntax and Semantics*, Vol. 4, (J. P. Kimball, Ed.), Academic Press, New York, pp. 125–154.
1583 Ma, M. (1980). Human Recognition of the feature of Chinese characters—An elementary research on recognition of different words with similar structure, Psychol. Inst., A List of Lit. in China, The Univ. of Sci. and Tech. of China, Hofei, Anhwei, China.
1584 Ma, M. (1980). The measurement on the psychological fuzziness—(I) The characteristic curve of the psychological conceptualization, Collect. Pap. on Fuzzy Math. (Abstr.), Huazhong Inst. of Tech., Huazhong, China, Apr.
1585 Ma, M. (1980). The measurement on the psychological fuzziness—(II) The theory of fuzzy entropy and the inference process, Collect. Pap. on Fuzzy Math. (Abstr.), Huazhong Inst. of Tech., Huazhong, China, Apr.
1586 Ma, M. (1980). The measurement on the psychological fuzziness—(III) A tentative exploration of functioning fuzzy semantics, Collect. Pap. on Fuzzy Math. (Abstr.), Huazhong Inst. of Tech., Huazhong, China, Apr.
1587 Ma, M., and Cao, Z. (1980). Decision making for preference multidimension scaling and its mathematical model, news on 1980 Annu. Rep. Meet. of Beijing Working Group on Fuzzy Sets, Beijing, China.
1588 Ma, M., and Cao, Z. (1980). The fuzziness statistical and the probability statistical approaches for a psychological measurement, News on 1980 Ann. Rep. Meet. of Beijing Working Group on Fuzzy Sets, Beijing, China.
1589 Ma, M., and Wang, P. (1980). Empirical tests for feature selection based on a psychological rule of ambiguous shaped Chinese word recognition, Collect. Pap. on Fuzzy Math. (Abstr.), Huazhong Inst. of Tech., Huazhong, China, Apr.
1590 Maarschalk, C. G. D. (1975). Exact and fuzzy concepts superimposed on the GST (A meta theory), in *Proc. 3rd Int. Congr. of Cybern. and Syst.*, Bucharest, Aug.
1591 Maarschalk, C. G. D. (1976). Metology in systems thinking and systems language—An approach to formalized and conceptual systems, exact and fuzzy concepts and Systol (system oriented language), 3rd Eur. Meet. Cybern. Syst. Res., Vienna.
1592 Machina, K. F. (1972). Vague predicates, *Amer. Philos. Quart.*, **9**, 225–233.
1593 Machina, K. F. (1976). Truth, belief and vagueness, *J. Philos. Logic*, **5**, 47–77.
1594 Mackie, J. L. (1973). *Truth, Probability and Paradox*, Clarendon Press, Oxford, UK.
1595 MacLane, S. (1971). *Categories for the Working Mathematician*, Springer-Verlag, Berlin.
1596 MacVicar-Whelan, P. J. (1974). Fuzzy sets, the concept of height, and the hedge very, Tech. Memo. 1, Phys. Dept., Grand Valley State Colleges, Allendale, MI.

1597 MacVicar-Whelan, P. J. (1975). Un modele de signification de termes quantifiant les dimensions—Application a la taille humaine, LAAS-SMA4 75.I.49, Lab. d'Autom. et d'Anal. des Syst., Toulouse, France. Dec.

1598 MacVicar-Whelan, P. J. (1976). Fuzzy sets for man-machine interaction, *Int. J. Man-Mach. Stud.*, **8**.

1599 MacVicar-Whelan, P. J. (1977). Fuzzy and Multivalued logic, in *Proc. 7th Int. Symp. Multiple-Valued Logic*, IEEE 77CH1222-9C, pp. 98–102.

1600 MacVicar-Whelan, P. J. (1978). Fuzzy filtering, Lab. d'Autom. et d'Anal. des Syst. du CNRS Toulouse, France.

1601 MacVicar-Whelan, P. J. (1978). Fuzzy logic—An alternative approach, FST-3-78, Univ. of California, Los Angeles, pp. 1–21.

1602 MacVicar-Whelan, P. J. (1980). Fuzzy classification of human height, *Int. Symp. Multiple-Valued Logic*.

1603 Majumder, D. (1978). A fuzzy set theoretic approach for recognition of patterns generated from bio-social systems, *EMCSR—PCSR*, **11**, to be published.

1604 Majumder, D. D., and Chadhuri, B. B. (1979). Fuzzy sets and possibility theory in reliability studies of man-machine system.

1605 Majumder, D. D., Datta, A. K., and Pal, S. K. (1976). Computer recognition of vowel speech sound using 3-dimensional weighted discriminant function, IEEE Comput. Group Repository.

1606 Majumder, D. D., and Pal, S. K. (1976). The concept of fuzzy sets and its application in pattern recognition problems, in *Proc. CSI, 76 Conv.*, Hyderabad, India, No. SD 02, Jan.

1607 Majumder, D. D., and Pal, S. K. (1977). On Fuzzification, fuzzy language, and multicategory fuzzy classifiers, in *Proc. 7th Int. Conf. on Cybern. and Soc.*, Washington D.C.

1608 Majumder, D. D., and Pal, S. K. (1977). On some applications of fuzzy algorithm in man-machine communication research, *J. Inst. Telecom. Electron. Eng.*, **23**, 117–120.

1609 Majumder, D. D., and Pal, S. K. (1978). Fuzzy recognition systems for patterns of biological origin, Electron. and Comm. Sci. Unit, Indian Stat. Inst., Calcutta, India.

1610 Majumder, D. D., and Pal, S. K. (1979). A self adaptive algorithm and fuzzy filter for pattern recognition, 1st Symp. on Policy Anal. and Inf. Syst. Durham, NC, June.

1611 Majumder, D. D., and Pal, S. K. (1979). A self adaptive fuzzy recognition system for speech sounds, 1st Int. Symp. on Policy Anal. and Inf. Syst., Durham, NC, June.

1612 Majumder, D. D., Pal, S. K., and Choudhuri, B. B. (1977). Fuzzy set in handwritten character recognition, in *Proc. Interdiscpl. Symp. on Digital Comm. and Pattern Recogn.*, Indian Stat. Inst., Calcutta, India.

1613 Makarovitsch, A. (1975). Valuation visuelle d'une matrice, Doc. Interne, Cie CII-Honeywell Bull., Paris, Sept.

1614 Makarovitsch, A. (1976). How to build fuzzy visual symbols, *Comput. Graphics Art*, Feb.

1615 Makarovitsch, A. (1977). Visual fuzziness, *Comput. Graphics Art*, Nov.

1616 Malvache, N. (1975). Analyse et identification des systemes visuel et manual en vision frontale et peripherique chez l'homme, Ph.D. Thesis, Univ. de Lille, Lille, France, Apr.

1617 Malvache, N. (1977). Utilization of fuzzy sets for system modelling and control, in *Proc. Symp. on Fuzzy Set Theory and Appl., IEEE Conf. on Decis. and Control*, New Orleans.

1618 Malvache, N., Milbred, G., Angue, J. C., and Fernandez, G. (1975). Contribution d'un interface flou dans une liaison homme-machine, 3rd Congr. Nat. d'Inf. et d'Automat., Madrid, Spain, Oct.

1619 Malvache, N., Milbred, G., and Vidal, P. (1973). Perception visuelle—Champ de vision laterale, modele de la fonction du regard, Rapp. de Synthese, Contrat DRME No. 71-251, Paris.

1620 Malvache, N., and Vidal, P. (1974). Application des systemes flous a la modelisation des phenomenes de prise de decision et d'apprehension des informations visuelles chex l'homme, ATP-CNRS 1K05, Paris.

1621 Malvache, N., and Willayes, D. (1974). Representation et minimisation de fonctions flous, Doc. Centre Univ. de Valenciennes, Valenciennes, France.

1622 Malvache, N., and Willayes, D. (1979). Contribution of the fuzzy sets theory to man machine systems, in *Advances in Fuzzy Sets Theory and Applications* (M. M. Gupta, R. K. Ragade, and R. R. Yager, Eds.), North-Holland, Amsterdam.

1623 Mamdani, A., and Kaufmann, A. (1975). Recherche inventive sur les problemes du 4th age en utilisant des notions de la theorie des sous-ensembles flous, Ler Symp. Nat. sur le 4th Age, Nice, France, Apr.

1624 Mamdani, E. H. (1974). Applications of fuzzy algorithms for control of simple dynamic plant, *Proc. IEEE*, **121**, 1585–1588.

1625 Mamdani, E. H. (1975). FLCS, a control system for fuzzy logic, Queen Mary College, Univ. of London, London.

1626 Mamdani, E. H. (1976). Advances in the linguistic synthesis of fuzzy controllers, *Int. J. Man-Mach. Stud.*, **8**, 669–678.

1627 Mamdani, E. H. (1976). Application of fuzzy logic to approximate reasoning using linguistic synthesis, in *Proc. 6th Int. Symp. on Multiple-Valued Logic*, IEEE 76CH1111-4C, May, 196–202.

1628 Mamdani, E. H. (1977). Applications of fuzzy set theory to control systems—A survey, in *Fuzzy Automata and Decision Processes* (M. M. Gupta, G. N. Saridis, and B. R. Gaines, Eds.), North-Holland, Amsterdam, pp. 77–88.

1629 Mamdani, E. H. (1980). Fuzzy logic controllers with industrial applications, in *Proc. JACC*, San Fransisco, Aug.

1630 Mamdani, E. H., and Assilian, S. (1975). An experiment in linguistic synthesis with a fuzzy logic controller, *Int. J. Man-Mach. Stud.*, **7**, 1–13.

1631 Mamdani, E. H., and Baaklini, N. (1975). Prescriptive method for deriving control policy in a fuzzy-logic controller, *Electron. Lett.*, **11**, 625–626.

1632 Mamdani, E. H., and Gaines, B. R. (Eds.), (1976). Discrete systems and fuzzy reasoning, EES-MMS-DSFR-76, Queen Mary College, Univ. of London, London (Workskop Proc.).

1633 Mamdani, E. H., and King, P. J. (1975). The application of fuzzy control systems to industrial processes, Round Table at 6th IFAC World Congr., Boston, MA.

1634 Mamdani, E. H., and Procyk, T. J. (1976). Application of fuzzy logic to controller design based on linguistic protocol, 3rd Eur. Meet. Cybern. Syst. Res., Vienna.

1635 Mamdani, E. H., and Sembi, B. S. (1979). Process control using fuzzy logic, 1st Int. Symp. on Policy Anal. and Inf. Syst., Durham, NC, June.

1636 Mandic, N. J., and Mamdani, E. H. (1980). A fuzzy linguistic calculator, 4th Eur. Congr. on Operations Res., Cambridge, UK, July.

1637 Manek, W., and Traczyk, T. (1969). Generalized Lukasiewicz algebras, *Bull. de l'Acad. Pol. Sci.*, Ser. Math., Astr. and Phys., **17**, 789–792.

1638 Manes, E. G. (1977). Topas methods for quantal systems and fuzzy systems, Int. Conf. on Appl. Gen. Syst. Res., Binghamton, NY.
1639 Manes, E. G. (1980). A class of fuzzy theories, Coins Tech. Rep. 80-15, Univ. of Massachusetts, Amherst.
1640 Manteras, L. D. (1980). Learning the meaning of imprecise concepts by adaptation of their possibility distribution, Table Ronde du CNRS sur le Flou, Lyon, France, June.
1641 Marcus, R. B. (1953). Strict implication, deductability and the deduction theorem, *J. Symb. Logic*, **18**, 234–236.
1642 Mares, M. (1977). How to handle quantities, *Kybernetica*, **13** (1), 22–40.
1643 Mares, M. (1977). On fuzzy quantities with real and integer values, *Kybernetica*, **13** (1) 41–56.
1644 Margolis, I. B., and Kandel, A. (1980). Absolutely dispensable prime implicants of fuzzy switching functions, in *Proc. 11th, multiple-valued logic Conf.*, Oklahoma City, OK, May.
1645 Marinos, P. N. (1966). Fuzzy logic, Tech. Memo. 66-3344-1, Bell Telephone Labs., Holmdel, NJ, Aug.
1646 Marinos, P. N. (1969). Application of fuzzy logic to analog and hybrid systems, in *Proc. 8th Ann. IEEE Region III Conv.*, Huntsville, AL, pp. 108–113.
1647 Marinos, P. N. (1969). Fuzzy logic and its application to switching systems, *IEEE Trans. Comput.*, **C-18**, 343–348.
1648 Markowitz, J. A. (1977). A look at fuzzy categories, Ph.D. Thesis, Northwestern Univ., Evanston, IL, 38-09-A *Diss. Abstr. Int.*, **38/09-A**, 5436.
1649 Marks, P. (1975). FLCS—A control system for fuzzy logic, Fuzzy Logic Working Group, Rep. 3, Queen Mary College, Univ. of London, London, Nov.
1650 Marks, P. (1976). A fuzzy logic control software, Intern. Rep., Queen Mary College, Univ. of London, London.
1651 Marku, I., and Nurminen, L. (1976). Studies on systemring in fuzziness in the analysis of information systems, Thesis, Inst. for Appl. Math., Univ. Turku, Turku, Finland.
1652 Maronna, R. (1964). A characterization of the Morgan lattices, *Port. Math.*, **23**.
1653 Martin, H. W. (1979). A Stone-Cech ultra-fuzzy compactification.
1654 Martin, J. F. (1980). Une approache des codages flous. Quelques proprietes. Publ. Lab. Stat. and Prob., No. 04, Univ. Paul Sabatier, Toulouse, France.
1655 Martin, J. K., and Turksen, I. B. (1975). Formative evaluation of information need analysis, Dept. Ind. Eng., Univ. of Toronto, Toronto, Canada.
1656 Martin, J. K., and Turksen, I. B. (1978). A fuzzy stimulus response theory for information system design, Working Paper No. 78-004, Dept. of Ind. Eng., Univ. of Toronto, Toronto, Canada.
1657 Martin, J. N. (1975). A syntactic characteristic of Kleene's strong connectives with two designated values, *Z. Math. Logik Grundlagen Math.*, **21**, 181–184.
1658 Martin, R. L. (Ed.) (1970). *The Paradox of the Liar*, Yale Univ. Press, New Haven, CT.
1659 Martin, T. (1966). Fuzzy algorithmische schemata, *Vortrage Prolemseminar Automat.—Und Algorithmentheorie*, Apr., Weissig, 44–51.
1660 Mas, M. T., Lem, H., Vigneron, M. J., and Lamotte, M. (1977). Classification de formes au moyen d'une relation de dissimilitude floue. Application a la classification des formes, Doc. Lab. Elec. et Autom., Univ. de Nancy, Nancy, France, Apr.
1661 Massonie, J. (1976). L'utilisation des sous-ensembles flous en geographie, *Cah. Geogr. Univ. Besancon*.

1662 Massonie, J., and Wieber, J. C. (1976). Application des sous-ensembles flous a l'etude d'un espace factoriel—Exemple des paysages, *Cah. Univ. Besancon.*

1663 Masuda, S., Sugeno, M., and Ternano, T. (1980). Control based on a new algorithm—Experimental study, in *Summ. of Pap. on Gen. Fuzzy Problems*, The Working Group on Fuzzy Syst., Tokyo, Dec.

1664 Masuda, S., Sugeno, M., and Ternano, T. (1980). Fuzzy control based on a new algorithm. Experimental study, Table Ronde du CNRS sur le Flou, Lyon, France, June.

1665 Materna, P. (1972). Intensional semantics of vague constants. An application of Tichy's concept of semantics, *Theory Decis.*, **2**, 267–273.

1666 Mathai, A. M., and Rathie, P. N. (1975). *Basic Concepts in Information Theory and Statistics—Axiomatic Foundation and Applications*, Wiley Eastern Ltd., New Delhi, India.

1667 Mathesius, V. (1911). 0 potencialnosti jevu jazykovych (On the potientiality of the phenomena of language) *Vestnik Kral., Ceske Spolecnosti Nauk, Trida Filo-Soficko-Historicka*, Prague, English translation in *Prague School Reader in Linguistics* (J. Vachet, Ed.), Indiana Univ. Press, Bloomington, 1964.

1668 *Matrose, E., and Mukundan, R. (1971). Planning of a health care system modeled as a two-person game, *Int. J. Syst. Sci.*, **3** (4), 375–383.

1669 Maurer, W. D. (1974). Input-output correctness and fuzzy correctness, George Washington Univ., Washington, D.C.

1670 Mauro, V., Bona, B., and Inaudi, D. (1976). A fuzzy approach to residential location theory, 3rd Eur. Meet. Cybern. Syst. Res., Vienna.

1671 Maydole, R. E. (1972). Many-valued logic as a basis for set theory, Ph.D. Thesis, Boston Univ., Boston MA.

1672 Maydole, R. E. (1975). Paradoxes and many-valued set theory, *J. Philos. Logic*, **4**, 269–291.

1673 McCall, S. S. (Ed.) (1967). *Polish Logic 1920–1939*, Clarendon Press, Oxford, UK.

1674 McCawley (1975). Fuzzy logic and restricted quantifiers, Univ. of Chicago, Chicago.

1675 McCloskey, M. E., and Glucksberg, S. (1978). Natural categories well defined or fuzzy sets? *Mem. Cognition*, **6**, 462–472.

1676 McGovern, D. (1979). Fuzzy logic and non-distributive truth variations, 1st Int. Symp. on Policy Anal. and Inf. Syst., Durham, NC, June.

1677 McKay, A. F., and Merrill, D. D. (Eds.) (1976). *Issues in the Philosophy of Language*, Yale Univ. Press, New Haven, CT.

1678 McKinsey, J. C. C. (1941). A solution of the decision problem for the Lewis Systems $S2$ and $S4$, with an application to topology, *J. Symb. Logic*, **6**, 117–134.

1679 McKinsey, J. C. C. (1945). On the syntactical construction of systems of modal logic, *J. Symb. Logic*, **10**, 83–94.

1680 McKinsey, J. C. C., and Tarski, A. (1944). The algebra of topology—Ann. Math., **45**, 141–191.

1681 McKinsey, J. C. C., and Tarski, A. (1948). Some theorems about the sentimental calculi of Lewis and Heyting, *J. Symb. Logic*, **13**, 1–15.

1682 McNaughten, R. (1951). A theorem about infinite-valued sentential logic, *J. Symb. Logic*, **16**, 1–13.

1683 Mehlburg, H. (1958). *The Reach of Science*, Univ. of Toronto Press, Toronto, Canada.

1684 Menger, K. (1951). Ensembles flous et fonctions aleatoires, C.R. *Acad. Sci.* (Paris), **232** (22), May 28, 2001-2003.
1685 Menges, G. (1970). On subjective probability and related problems, *Theory Decis.*, **1**, 40-60.
1686 Menges, G. (Ed.) (1974). *Information, Inference and Decision*, D. Reidel, Dordrecht, Holland.
1687 Menges, G., and Kofler, E. (1976). Linear partial information as fuzziness, in *Systems Theory in the Social Sciences*, (H. Bossel, S. Klaczko, and N. Muller, Eds.), Birkhauser Verlag, Basel, Switzerland, pp. 307-322.
1688 Menges, G., and Skala, H. J. (1974). On the problem of vagueness in the social sciences, in *Information, Inference and Decision*, (G. Menges, Ed.), D. Reidel, Dordrecht, Holland, pp. 51-61.
1689 Meredith, C. A. (1958). The dependence of an axiom of Lukasiewicz, *Trans. Amer. Math. Soc.*, **87**, 54.
1690 Meseguer, J., and Sols, I. (1974). Automata in semimodule categories, in *Proc. 1st Int. Symp. Category Theory Appl. to Comput. and Control*, pp. 196-202.
1691 Meseguer, J., and Sols, I. (1975). Fuzzy semantics in higher order logic and universal algebra, Univ. of Zaragoza, Zaragoza, Spain.
1692 Meseguer, J., and Sols, I. (1975). Topology in complete lattices and continuous fuzzy relations, Univ. of Zaragoza, Zaragoza, Spain.
1693 Mesnard, L. (1978). La dominance regionale et son imprecision, Colloq. Int. Theorie et Appl. des Sous-Ensembles Flous, Journ. de Biomath. et d'Inf. Med., Marseille, France, Sept.
1694 Mesnard, L. (1978). La dominance regionale et son imprecision, Traitement dans le type general de structure, Doc. Inst. Math. Econ., Univ. de Dijon, Dijon, France.
1695 Michalek, J. (1975). Fuzzy topologies, *Kybernetika* (Prague), **11**, 345-354.
1696 Michalos, A. C. (1971). *The Popper-Carnap Controversy*, M. Nijhoff, The Hague.
1697 Michalski, R. S. (1975). Variable-valued logic and its applications to pattern recognition and machine learning, in *Multiple-Valued Logic and Computer Science*, North-Holland, Amsterdam.
1698 Michelman, E. H., and Hoffman, L. J. (1977). Securate—A security evaluation and analysis system using fuzzy metrics, Memo. No. UCB/ERL M77/36, Electron. Res. Lab., Univ. of California, Berkeley, July.
1699 Miemo, H., and Iskikawa, A. (1978). Some thought on fuzziness in creativity development, Colloq. Int. Theorie et Appl. des Sous-Ensembles Flous, Journ. de Biomath. et d'Inf. Med., Marseille, France, Sept.
1700 Miller, D. (1974). Popper's qualitative theory of verisimilitude, *Brit. J. Philos. Sci.*, **25**, 166-188.
1701 Mira, C. (1979). Frontiere floue separant les domaines d'attraction de deux attracteurs, *C.R. Acad. Sci.* (Paris), Ser. A., **288**, 591-594.
1702 Miroshnikov, V. V. (1979). Planning of technological systems on the basis of application of fuzzy sets and fuzzy algorithms, *Eng. Cybern.* (*Sov. J. Comput. Syst. Sci.*), May-June, 124-135; English translation (Abstr.), 95.
1703 Mishima, H., Katoh, N., Hattori, T., Nomura, Y., and Nakamura, M. (1980). A fuzzy decision analysis for the management of pancreatic cancer, in *Medinfo 80, Proc. 3rd World Conf. on Med. Inf.*, Tokyo, Sept. pp. 830-834.

1704 Miura, S. (1972). Probabilistic models of modal logics, *Bull. Nagoya Inst. Tech.*, **24** 67–72.

1705 Miyano, H. (1980). A representation method in a humanistic environment, in *Proc. Int. Congr. on Appl. Syst. Res. and Cybern.*, Acapulco, Mexico, Dec.

1706 Mizumoto, M. (1971). Fuzzy automata and fuzzy grammars, Ph.D. Thesis, Fac. of Eng. Sci., Osaka Univ., Osaka, Japan.

1707 Mizumoto, M. (1971). Fuzzy sets theory, 11TH Prof. Group Meet. on Control Theory of SICE, Japan.

1708 Mizumoto, M. (1980). Fuzzy reasoning with "if. then. else.," in *Proc. Int. Congr. on Appl. Syst. Res. and Cybern.*, Acapulco, Mexico, Dec.

1709 Mizumoto, M., Fukami, S., and Tanaka, K. (1978). Fuzzy reasoning methods by Zadeh and Mamdani and improved methods, 3rd Fuzzy Symp. on Fuzzy Reasoning, Queen Mary College, Univ. of London, London.

1710 Mizumoto, M., Fukami, S., and Tanaka, K. (1979). Fuzzy conditional inferences and fuzzy inferences with fuzzy quantifiers, in *Int. Joint Conf. on Artif. Intell.*, Tokyo, Aug., 589–591.

1711 Mizumoto, M., Fukami, S., and Tanaka, K. (1979). Several methods for fuzzy conditional inferences, in *Proc. 18th IEEE Conf. on Decis. and Control*, Fort Lauderdale, FL, Dec.

1712 Mizumoto, M., Fukami, S., and Tanaka, K. (1979). Some methods of fuzzy reasoning, in *Advances in Fuzzy Set Theory and Applications* (M. M. Gupta, R. K. Ragade, and R. R. Yager, Eds.), North-Holland, Amsterdam.

1713 Mizumoto, M., and Tanaka, K. (1975). Algebraic structures of fuzzy-fuzzy sets, *Trans. IECE* (Japan), **58-D**, 421–428.

1714 Mizumoto, M., and Tanaka, K. (1976). Algebraic properties of fuzzy numbers, 1976 Int. Conf. on Cybern. and Soc., Washington, D.C.

1715 Mizumoto, M., and Tanaka, K. (1976). Bounded-sum and bounded-difference for fuzzy sets, *Trans. IECE* (Japan), **59-D**, 905–912.

1716 Mizumoto, M., and Tanaka, K. (1976). Four arithmetic operations of fuzzy numbers, *Trans. IECE* (Japan), **59-D**, 703–710.

1717 Mizumoto, M., and Tanaka, K. (1976). Fuzzy data structure and fuzzy artificial intelligence language, 1976 Joint Conv. of Elec. Eng.

1718 Mizumoto, M., and Tanaka, K. (1976). Fuzzy-fuzzy automata, *Kybernetes*, **5**, 107–112.

1719 Mizumoto, M., and Tanaka, K. (1976). Some properties of fuzzy sets of type 2, *Inform. Control*, **31**, 312/340.

1720 Mizumoto, M., and Tanaka, K. (1976). Various kinds of automata with weights, *J. Comput. Syst. Sci.*

1721 Mizumoto, M., and Tanaka, K. (1978). Algebraic product and algebraic sum of fuzzy sets of type 2, 4th Eur. Meet. on Cybern. and Syst. Res., Linz, Austria.

1722 Mizumoto, M., and Tanaka, K. (1978). Fuzzy sets of type 2 under algebraic product and algebraic sum, in *Proc. Int. Colloq. on Fuzzy Sets*, Marseille, France.

1723 Mizumoto, M., and Tanaka, K. (1978). Fuzzy sets under various operations, Dept. of Inf. and Comput. Sci., Fac. of Eng. Sci., Osaka Univ., Toyonaka, Osaka, Japan.

1724 Mizumoto, M., and Tanaka, K. (1979). Some properties of fuzzy numbers, in *Advances in Fuzzy Set Theory and Applications* (M. M. Gupta, R. K. Ragade, and R. R. Yager, Eds.), North-Holland, Amsterdam.

1725 Mizumoto, M., and Tanaka, K. (1979). Algebraic product and algebraic sum of fuzzy grades, *Trans. IECE* (Japan).

1726 Mizumoto, M., Toyoda, J., and Tanaka, K. (1969). Some considerations on fuzzy automata, *J. Comput. Syst. Sci.*, **3**, 409–422.

1727 Mizumoto, M., Toyoda, J., and Tanaka, K. (1970). Fuzzy languages, *Syst., Comput., Controls*, **1**, 36; originally *Trans. IECE*, **53-C**, 333–340.

1728 Mizumoto, M., Toyoda, J., and Tanaka, K. (1971). Fuzzy algebra, Res. Rep. on *Many-Valued Logic and its Appl.*, Math. and Sci. Res. Rec., Kyoto Univ., Kyoto, Japan, Mar.

1729 Mizumoto, M., Toyoda, J., and Tanaka, K. (1972). General information of formal grammars, *Inf. Sci.*, **4**, 87–100.

1730 Mizumoto, M., Toyoda, J., and Tanaka, K. (1972). L-fuzzy logic, in research on many-valued logic and its applications, Kyoto Univ., Kyoto, Japan.

1731 Mizumoto, M., Toyoda, J., and Tanaka, K. (1972). Normal grammars with weights, *Trans. IECE* (Japan), **55D**, 292–293.

1732 Mizumoto, M., Toyoda, J., and Tanaka, K. (1973). Examples of formal grammars with weight, *Inf. Processing. Lett.*, **2**, 74–78.

1733 Mizumoto, M., Toyoda, J., and Tanaka, K. (1973). N-fold fuzzy grammars, *Inf. Sci.*, **5**, 25–43.

1734 Mizumoto, M., Toyoda, J., and Tanaka, K. (1975). B-fuzzy grammars, *Comput. Math.*, **4**, 343–368.

1735 Mizumoto, M., Umano, M., and Tanaka, K. (1977). Implementation of a fuzzy-set theoretic data structure system, 3rd Int. Conf. on Very Large Data Bases, Tokyo.

1736 Moisil, G. C. (1935). Recherches sur l'algebre de la logique, *Ann. Sci. Univ. Jassy* (Rumania), **22**, 1–77.

1737 Moisil, G. C. (1940). Recherches sur les logiques non-chrysipiennes, *Ann. Sci. Univ. Jassy* (Rumania), **26**, 431–436.

1738 Moisil, G. C. (1941). Notes sur les logiques non-chrysipiennes, *Ann. Sci. Univ. Jassy* (Rumania), **27**.

1739 Moisil, G. C. (1942). Logique modale, *Disquisitiones Math. Phys.*, **III** (1), 3–98.

1740 Moisil, G. C. (1964). Sur les logiques de Lukaciewicz a un nombre fini de variables, *Rev. Roum. Math. Pures Appl.*, **IX**, 905.

1741 Moisil, G. C. (1965). *Icercari Vechi Si Noi De Logica Neclasica*, Scientific Editions, Bucharest.

1742 Moisil, G. C. (1968). Lukasiewiczian algebras, Centre de Calcul, Univ. of Bucharest, Oct.

1743 Moisil, G. C. (1971). Role of computers in the evolution of science, in *Proc. of Int. Conf. on Sci. and Soc.*, Belgrade, Yugoslavia, pp.134–136.

1744 Moisil, G. C. (1972). La logique des concepts nuances, in *Essais sur les Logiques Non Chrysippiennes*, Editions of Acad. de la Republique Socialiste de Roumanie, Bucharest, pp. 157–163.

1745 Moisil, G. C. (1972). Sur les algebras de Lukasiewicz O-Valentes in *Essais sur les Logiques Non Chrysipiennes*, Editions of Acad. de la Republique Socialiste de Roumanie, Bucharest, pp. 311–324.

1746 Moisil, G. C. (1972). Essais sur les logiques non cripeppieus Pul, Roum. Acad. of *Sci.*, Bucharest.

1747 Moisil, G. C. (1973) Ensembles flous et logiques a plusieurs valeurs, CRM-286, Centre de Rech. Math., Univ. de Montreal, Montreal, Canada, May.

1748 Moisil, G. C. (1975). Lectii despre logica ratinamentuliul nuantat, Scientific and Encyclopaedic Editions, Bucharest (in Rumanian).

1749 Moisil, G. C. (1975). *Lectures on Fuzzy Logic*, Scientific and Encyclopaedic Editions, Bucharest (in Rumanian).

1750 Moisil, G. C. (1975). Sur une logique Intuitioniste, *An. Mat. Pura ed Appl.* (Bologna, Italy).

1751 Moisil, G. C. (1976). Sur l'emploi des mathematiques dans les sciences de l'homme, Acad. Naz. dei Lincei, Contrib. des Centro Linceo Interdisciplinare di Sci. Mat. e loro Appl., Rome.

1752 Molzen, N. (1975). Fuzzy logic control, Ph.D. Thesis, Tech. Univ. of Denmark (in Danish).

1753 Monteiro, A. A., and Ribeiro, H. (1972). L'operation de fermeture et ses invarients dans les systemes partiellement ordonnes, *Port. Math.*, **3**, 171–184.

1754 Montes, C. G., Camacho, E. F., and Aracil, J. (1976). A fuzzy algorithm for nonlinear system identification, 3rd Eur. Meet. Cybern. Syst. Res., Vienna.

1755 Montes, C. G., Camacho, E. F., and Aracil, J. (1976). Estimcion de parametros en systemas nonlineares, IX Reun. Nac. de Invest. Operat., Escuela Sup. Ing. Industr., Barcelona, Spain, Sept.

1756 Moon, R., Jordanov, S., Turksen, I. B., and Perez (1978). Human-like reasoning capacity in a medical diagnosis system., The Application of fuzzy sets to computer diagnosis, Working Pap. No. 78-001, Dept. of Ind. Eng., Univ. of Toronto, Toronto, Canada.

1757 Morgan, C., and Pelletier, F. (1977) Some notes concerning fuzzy logics, *Linguist. Philos.*, **1** (1).

1758 Morgan, C. G. (1975). Similarity as a theory of graded equality for a class of many-valued predicate calculi, in *Proc. 1975 Int. Symp. Multiple-Valued Logic*, Logan, Utah, IEEE-75CH0959-7C, pp. 436–449.

1759 Morgan, C. G. (1976). Many-valued propositional intuitionism, in *Proc. 6th Int. Symp. Multiple-Valued Logic*, IEEE 76CH1111-4C, pp. 150–156.

1760 Morgan, C. G. (1976). Methods for automated theorem proving in non-classical logics, *IEEE Trans. Comput.*, **C-25**, 852–862.

1761 Morita, Y., and Iida, H. (1975). Measurement, information and human subjectivity described by an order relationship, in *Summ. of Pap. on Gen. Fuzzy Problems*, The Working Group on Fuzzy Syst., Tokyo, Nov., pp. 34–39.

1762 Morita, Y., and Oka, Y. (1976). On a loop in fuzzy evaluation and measurement, in *Summ. of Pap. on Gen. Fuzzy Problems*, The Working Group on Fuzzy Syst., Tokyo, pp. 87–100.

1763 Morita, Y., Oka, Y., and Ogata, Y. (1978). Intransitive ordering and fuzziness, Rep. No. 4, The Working Group on Fuzzy Syst., Tokyo, Dec, pp. 36–39.

1764 Morozov, A. (1975). Some problems of decision theory, *Ekon. Math. Metody*, **11**, 252–262 (in Russian).

1765 Morton, A. (1975). Complex individuals and multigrade relations, *Nous*, **9**, 309–318.

1766 Morviller, M. S., and Lepage, D. (1974). Applications des concepts flous—Description dynamique d'un ensemble de donnees et son utilisation en langage naturel, Mem. Inst. d'Inf. d'Entrep., CNAM, Paris, June.

1767 Moscarola, J., and Roy, B. (1976). Procedure automatique d'examen des dossiers fondee sur un classement trichotomique en presence de critere multiples, Doc. Lamsade, Univ. de Paris, Paris, Jan.

1768 Mostowski, A. (1957). On a generalization of quantifiers, *Fundam. Math.*, **44**, 12–36.

1769 Mostowski, A. (1961). Axiomatizability of many valued predicate calculi, *Fundam. Math.*, **50**, 165–190.

1770 Mostowski, A. (1966). *Thirty Years of Foundational Studies*, Basil Blackwell, Oxford, UK.

1771 Mukaidono, M. (1972). On some properties of fuzzy logic, Tech. Rep. on Autom. of IECE.

1772 Mukaidono, M. (1972). On the B-ternary logical function—A ternary logic with consideration of ambiguity, Trans. *IECE* (Japan), **55-D**, 355–362.

1773 Mukaidono, M. (1974). Prime implicants of fuzzy logic functions and their minimization, *Pap. of Tech. Group on Electron. Comput.* (IECE Japan), **EC-73**, Jan., 58.

1774 Mukaidono, M. (1975). An algebraic structure of fuzzy logical functions and its minimal and irredundant form, *Trans. IECE* (Japan), **58-D**, 748–755.

1775 Mukaidono, M. (1975). An application of fuzzy logical functions to pattern classification, Tech. Rep. on Pattern Recogn. and Learning of IECE, PRL75-67.

1776 Mukaidono, M. (1975). Some properties of fuzzy logics, *Trans. IECE* (Japan), **58-D**, 150–157.

1777 Mukaidono, M. (1976). Some properties on the resolvent in fuzzy logic, Tech. Rep. on Pattern Recogn. and Learning of IECE, PRL76-3.

1778 Mukaidono, M. (1977). On some properties of a quantization in fuzzy logic, in *Proc. of 7th Symp. on Multi-Valued Logic*, May.

1779 Mukaidono, M. (1979). A necessary and sufficient condition for fuzzy logic functions, The 9th Int. Symp. on Multi-Valued Logic, Beth, UK.

1780 Mukaidono, M. (1980). Fuzzy inference of resolution style, in *Proc. Int. Congr. on Appl. Syst. Res. and Cybern.*, Acapulco, Mexico, Dec.

1781 Mulhall, D. J. (1979). Scaling techniques and the membership function of fuzzy sets, submitted to *Int. J. Man-Mach. Stud.*

1782 Murakami, Y. (1966). Formal structures of majority decision, *Econometrica*, **34**, (3), July.

1783 Muzynski, W., and Jacak, W. (1976). Conception of describing the behaviour of the eventistic system by means of the formalism of fuzzy sets and relations, 3rd Eur. Meet. Cybern. Syst. Res., Vienna.

1784 Nadiu, G. S. (1971). Sur la logique de heytung, in *Logique, Automatique, Informatique*, Acad. Rep. Soc. de Rumania, Bucharest, pp. 42–70.

1785 Nagai, S. (1973). On a semantics for non-classical logics, *Proc. Jap. Acad.*, **49**, 337–340.

1786 Nahmias, S. (1974). Discrete fuzzy random variables, Univ. of Pittsburgh, Pittsburgh, PA.

1787 Nahmias, S. (1977). Fuzzy variables, *Fuzzy Sets Syst.*, **1**, 97–110.

1788 Nahmias, S. (1978). Fuzzy variables in a random environment, Tech. Rep. No. 39, School of Eng., Univ. of Pittsburgh, Pittsburgh, PA.

1789 Nahmias, S. (1980). Compte rendu sur "Fuzzy automata and decision processes," *Interfaces*, 117–118.

1790 Nakagawa, Y., and Rosenfeld, A. (1978). A note on the use of logical minimum and maximum operations in digital processing, *IEEE Trans. Syst., Man, and Cybern.*, **SMC-8**, 632-635.

1791 Nakamura, K. (1980). Preference relations between fuzzy outcomes, in *Summ. of Pap. on Gen. Fuzzy Problems*, The Working Group on Fuzzy Syst., Tokyo, Dec.

1792 Nakamura, K., and Yoshioka, M. (1975). A simulation model of pedestrian flow and its investigation, in *Summ. of Pap. on Gen. Fuzzy Problems*, The Working Group on Fuzzy Syst., Tokyo.

1793 Nakamura, K., and Yosioka, M. (1976). Some macroscopic models of collective human behaviour, in *Summ. of Pap. on Gen. Fuzzy Problems*, The Working Group on Fuzzy Syst., Toyko.

1794 Nakamura, M. (1941). Closure in general lattices, *Proc. Imper. Acad.* (Tokyo), **17**, 5-6.

1795 Nakata, H., Mizumoto, M., Toyoda, J., and Tanaka, K. (1972). Some characteristics of N-fold fuzzy of grammars, *Trans. IECE* (Japan), **55-D**, 287-288.

1796 Naptalanoff, N., and Lakov, D. (1977). Decision making in vague conditions, *Kybernites*, 91-93.

1797 Naranyani, A. S. (1980). Sub-definite set—A formal model of uncompletely specified aggregate, in *Proc. Int. Congr. on Appl. Syst. Res. and Cybern.*, Acapulco, Mexico, Dec.

1798 Naranyani, A. S. (1980). Survey of fuzzy sets in the USSR, in *Proc. Int. Congr. on Appl. Syst. Res. and Cybern.*, Acapulco, Mexico, Dec.

1799 Nasu, M., and Honda, N. (1968). Fuzzy events realized by finite probabilistic automata, *Inf. Control*, **12**, 284-303.

1800 Nasu, M., and Honda, N. (1969). Mapping induced by PGSM—Mapping and some recursively unsolvable problems of finite probabilistic automata, *Inf. Control*, **15**, 250-273.

1801 Nazaroff, G. J. (1973). Fuzzy topological polysystems, *J. Math. Anal. Appl.*, **41**, 478-485.

1802 Neff, T. P., and Kandel, A. (1977). Simplification of fuzzy switching functions, *Int. J. Comput. Inf. Sci.*, **6**, 55-70.

1803 Negoita, C. V. (1969). Informational retrieval systems, Ph.D. Thesis, Polytech. Inst. of Bucharest, Bucharest (in Rumanian).

1804 Negoita, C. V. (1970). On the strategies in automatic information systems, 6th Int. Congr. Cybern. Syst., Namur, Belgium.

1805 Negoita, C. V. (1971). *Information Storage and Retrieval*, Editura Academiei, Bucharest (in Rumanian).

1806 Negoita, C. V. (1973). Linear and nonlinear information retrieval, *Stud. Cercet. Doc.*, 21-57.

1807 Negoita, C. V. (1973). On the application of the fuzzy sets separation theorem for automatic classification in information retrieval systems, *Inf. Sci.*, **5**, 279-286.

1808 Negoita, C. V. (1973). On the decision process in information retrieval, *Stud. Cercet. Doc.*, 369-381.

1809 Negoita, C. V. (1973). On the notion of relevance in information retrieval, *Kybernetes*, **2**, 161-165.

1810 Negoita, C. V. (1976). Fuzziness in management, ORSA/TIMS Meet., Miami, FL, Nov.

1811 Negoita, C. V. (1976). Fuzzy Models for social processes, in *Systems Theory in the Social Sciences* (H. Bossel, S. Klaczko, and N. Muller, Eds.), Birkhauser Verlag, Basel, Switzerland, pp. 283-291.

Key References in Fuzzy Pattern Recognition

1812 Negoita, C. V. (1976). Fuzzy systems and management science, 3rd Eur. Meet. Cybern. Syst. Res., Vienna.

1813 Negoita, C. V. (1976). Overlapping tendencies in operations research system theory and cybernetics, in *Proc. Int. Symp.* (Univ. of Fribourg, Fribourg, Switzerland), Oct., Birkhauser-Verlag, Basel, Switzerland.

1814 Negoita, C. V. (1977). Fuzzy systems and soft sciences, The Meet. on Math. Syst. and Informatics in Contemp. Res., Unesco Sci. Coop. Bur. in Europe, Bucharest, Sept.

1815 Negoita, C. V. (1977). On dynamics and fuzziness in management systems, in *Modern Trends in Cybernetics and Systems* (J. Rose, and C. Belciu, Ed.), Springer Verlag, Berlin.

1816 Negoita, C. V. (1977). On the internal model principle, in *Proc. Symp. on Fuzzy Set Theory and Appl., IEEE Conf. on Decis. and Control*, New Orleans.

1817 Negoita, C. V. (1977). Review of fuzzy sets and their application to cognition and decision processes, *IEEE Trans. Syst., Man, Cybern.*, **SMC-7** (2).

1818 Negoita, C. V. (1978). Fuzzy mathematics—A new paradigm, in *Proc. 22nd Annu. Meet. of the Soc. of Gen. Syst. Res.*, Washington, D. C., Feb.

1819 Negoita, C. V. (1978). On fuzzy systems, 4th Int. Congr. of Cybern. and Syst., Amsterdam.

1820 Negoita, C. V. (1978). On the stability of fuzzy systems, in *Proc. 1978 Conf. on Cybern. and Soc.*, Tokyo, Japan, pp. 936–937.

1821 Negoita, C. V. (1978). *Fuzzy Systems*, Abacus Press, London.

1822 Negoita, C. V. (1979). Fuzzy sets, system theory and management, in E. Billeter, M. Cuenod, and S. Klazko, Eds.

1823 Negoita, C. V. (1979). *Management Applications of System Theory*, Birkhauser Verlag, Basel, Switzerland, and Stuttgart, Germany

1824 Negoita, C. V. (1980). Fuzzy sets in Topoi, Table Ronde du CNRS sur le Flou, Lyon, France, June.

1825 Negoita, C. V. (1980). Fuzzy systems as models for time-variant systems, in *Proc. Int. Congr. on Appl. Syst. Res. and Cybern.*, Acapulco, Mexico, Dec.

1826 Negoita, C. V. (1980). The current interest in fuzzy optimization, *Bull. Stud. Exch. Fuzziness Appl.*, Printemps.

1827 Negoita, C. V., and Flondor, P. (1976). On fuzziness in information retrieval, *Int. J. Man-Mach. Stud.*, **8**, 711–716.

1828 Negoita, C. V., and Flondor, P. (1979). Le concept fuzzy dans processus de recherche des informations, *Probl. Inf. Doc.*, **13** (3), 136–141.

1829 Negoita, C. V., Flondor, P., and Sularia, M. (1977). On fuzzy environments in optimization problems, *Econ. Comput. Econ. Cybern. Stud. Res.*, 13–24.

1830 Negoita, C. V., Keleman, M., and Stefanescu, A. C. (1980). The internalization of the internal model principle (IMP), *Bull. Sous Ensembles Flous Appl.*, Automne.

1831 Negoita, C. V., Minoui, S., and Stan, E. (1976). On considering impression in dynamic linear programming, *Econ. Comput. Econ. Cybern.*, **3**.

1832 Negoita, C. V., and Ralescu, D. A. (1974). Fuzzy systems and artificial intelligence, *Kybernetes*, **3**, 173–178.

1833 Negoita, C. V., and Ralescu, D. A. (1974). Inexactness in dynamic systems, *Econ. Comput. Econ. Cybern. Stud. Res.*, **4**, 69–81.

1834 Negoita, C. V., and Ralescu, D. A. (1974). *Multini Vagi Applicabile Lor*, Editura Technica, Bucharest, Rumania.

1835 Negoita, C. V., and Ralescu, D. A. (1975). *Applications of Fuzzy Sets to Systems Analysis*, John Wiley, New York.

1836 Negoita, C. V., and Ralescu, D. A. (1975). Relations on monoids and minimal realization theory for dynamic systems; Applications for fuzzy systems, in *Proc. 3rd Int. Congr. of Cybern. and Syst.*, Bucharest, Rumania, Aug.

1837 Negoita, C. V., and Ralescu, D. A. (1975). Representation theorems for fuzzy concepts, *Kybernetes*, **4**, 169–174.

1838 Negoita, C. V., and Ralescu, D. A. (1976). Comment on a comment on an algorithm that generates fuzzy prime implicants by Lee and Chang, *Inf. Contr.*, **30**, 199–201.

1839 Negoita, C. V., and Ralescu, D. A. (1977). On fuzzy optimization, *Kybernetics*, **6**, 193–195.

1840 Negoita, C. V., and Ralescu, D. A. (1977). Some results in fuzzy systems theory, in *Modern Trends in Cybernetics and Systems*, (J. Rose and C. Bilciu, Eds.) Springer-Verlag, Berlin.

1841 Negoita, C. V., and Stefanescu, A. C. (1975). On the state equation of fuzzy systems, *Kybernetes*, **4**, 231–214.

1842 Negoita, C. V., and Sularia, M. (1976). Fuzzy linear programming and tolerances in planning, *Econ. Comput. Econ. Cybern. Stud. Res.*, **1**, 3–15.

1843 Negoita, C. V., and Sularia, M. (1978). A selection method of non-dominated points in multi-criteria decision problems, *Econ. Comp. Econ. Cybern. Stud. Res.* (1), 19–23.

1844 Negoita, C. V., and Suluriu, M. (1976). On fuzzy mathematical programming and tolerances in planning, *Econ. Comput. Econ. Cybern. Stud. Res.*, Bucharest, Romania.

1845 Neitzel, L. A., and Hoffman, L. J. (1979). Fuzzy cost/benefit analysis, 1st Int. Symp. on Policy Anal. and Inf. Syst., Durham, NC, June.

1846 Nelson, J. H. (1978). A man-machine interactive technique for solution of large-scale, computationally intractable, and/or "fuzzy goal" routing, scheduling, and dispatching problems, Ph.D. Thesis, Vanderbilt Univ., Nashville, TN, *Diss. Abstr. Int.*, **39/06-B**, 2921.

1847 Netto, A. B. (1970). Fuzzy classes, *Not. Amer. Math. Soc.*, **68T-H28**, 945.

1848 Neuhaus, N. J., and Spevack, M. (1975). Shakespeare dictionary—Some preliminaries for a semanic description, *Comput. Humanities*, **9**, 263–270.

1849 Neustupny, J. V. (1966). On the analysis of linguistic vagueness, *Trav. Linguist. Prague*, **2**, 39–51.

1850 Newpeck, F. (1977). Fuzzy sets, in *Proc. 1977 AIDS Conf.*, Chicago.

1851 Newton, L. K. (1978). Fuzzy set theory in a decision analysis formulation of the Acc. Materiality Dec., Joint Nat. ORSA/TIMS Meet., Los Angeles.

1852 Nguyen, C.-H. (1973). Generalized post algebras and their application to some infinitary many-valued logics, *Diss. Math.* (Warsaw), **107**.

1853 Nguyen, H. T. (1974). Sur les mesures d'information du Type Inf., in *Lectures Notes in Mathematics*, No. 398, Springer Verlag, Berlin, pp. 62–75.

1854 Nguyen, H. T. (1975). Information fonctionnelle et ensembles flous, Semin. sur les Questionnaires, Univ. de Paris, Paris.

1855 Nguyen, H. T. (1976). A note on the extension principle for fuzzy sets, Electr. Res. Lab. Memo. M-611, Univ. of California, Berkeley.

1856 Nguyen, H. T. (1977). On fuzziness and linguistic probabilities, *J. Math. Anal. Appl.*, **61** (3), 658–671.

1857 Nguyen, H. T. (1977). On random sets and belief functions, Memo. ERL M77/14, Electron. Res. Lab., Univ. of California, Berkeley.

1858 Nguyen, H. T. (1977). Some mathematical tools for linguistic probabilities, in *Proc. IEEE Symp. on Fuzzy Sets*, New Orleans, pp. 1345–1350.

1859 Nguyen, H. T. (1978) Conditioning in possibility theory, Univ. of Massachusetts, in *Proc. 1978 IEEE Conf. on Decis. and Control, Includes the 17th Symp. on Adaptive Processes*, 78CH1392-OCS, Jan., 1979, San Diego, CA, pp. 1450–1453.

1860 Nguyen, H. T. (1978). On conditional possibility distributions, *Fuzzy Sets Syst.*, **1**, 299–309.

1861 Nguyen, H. T. (1978). On the non-interaction and conditioning in possibility theory, Dept. of Math. and Stat., Univ. of Massachusetts, in *Proc. 1978 Int. Conf. on Cybern. and Soc.*, Vol. 2, pp. 1210–1212.

1862 Nguyen, H. T. (1979). Toward a calculus of the mathematical notion of possibility, in *Proc. 18th IEEE Conf. on Decis. and Control*, Fort Lauderdale, FL, Dec.

1863 Nguyen, H. T. (1980). Amherst contribution to the analysis of evidence, Table Ronde du CNRS, Lyon, France, June.

1864 Nguyen, H. T. (1980). Computation with linguistic probabilities, in *Proc. Int. Congr. on Appl. Syst. Res. and Cybern.*, Acapulco, Mexico, Dec.

1865 Nguyen, H. T., and Ohsuga, S. (1977). Study of linguistic probabilities, in *Proc. Symp. on Fuzzy Set Theory and Appl., IEEE Conf. on Decis. and Control*, New Orleans.

1866 Nicolau, E., and Popovici, A. (1972). *Introduction in the Cybernetics of Continuous Systems*, Editura Tehnica, Bucharest (in Rumanian).

1867 Nie, Y. (1980). Pansystem logic conservations of fuzzy set, *J. Huazhong Inst. Tech.* **2**, A List of Lit. in China, The Univ. of Sci. and Tech. of China, Hofei, Anhwei, China.

1868 Nieminen, J. (1977). On the algebraic structure of fuzzy sets of type 2, *Kybernetika*, **13**, 261–273.

1869 Nieminen, J. (1978). Fuzzy mappings and algebraic structures, *Fuzzy Sets Syst.*, **1**, 231–235.

1870 Ning, D., and Wu, Y. (1980). Drawing graph automatically with space interaction, News on 1980 Annu. Rep. Meet. of Beijing Working Group on Fuzzy Sets, Beijing, China.

1871 Nishibe, T. (1977). A method of placing vertices of a graph on plane in auto-graph-drawing, in *Summ. of Pap. on Gen. Fuzzy Problems*, The Working Group on Fuzzy Syst., Tokyo, pp. 118–125.

1872 Nitta, Y., Kaji, H., Mori, K., and Matsuoka, H. (1977). A practical approach to pictorial data retrieval, in *Proc. of 1977 Workshop on Pattern Database Syst.*, Dec.

1873 Noguchi, K., Umano, M., Mizumoto, M., and Tanaka, K. (1976). Implementation of fuzzy artificial intelligence language flou, Tech. Rep. on Autom. and Lang.

1874 Noguchi, K., Umano, M., Mizumoto, M., and Tanaka, K. (1978). Improvement of a fuzzy artificial language flou, *Tech. Rep. of IECE of Japan*, Vol. 77 (on Autom. and Lang.), pp. 133–142.

1875 Noguchi, Y. (1972). A pattern clustering method on the basis of association schemes, *Bull. Electrotech. Lab.*, **36**, 753–767.

1876 Nojiri, H. (1979). A model of fuzzy team decision, *Fuzzy Sets Syst.*, **2** (3), July.

1877 Nojiri, H. (1980). On the fuzzy team decision in a changing environment, *Fuzzy Sets Syst.*, **3** (2), Mar.

1878 Nojiri, H. (1980). A formulation of fuzzy-fuzzy team decision problems, in *Summ. of Pap. on Gen. Fuzzy Problems*, The Working Group on Fuzzy Syst., Tokyo, Dec.

1879 Nojiri, H. (1980). On the fuzzy team decision in a changing environment, *Fuzzy Sets Syst.*, **3**, (2).

1880 Nola, A. D., and Ventre, A. G. S. (1979). On some sequences of fuzzy sets, RAIRO *Inf./Comput. Sci.*, **12**(2), 199–204.

1881 Norwich, A. M. (1981). Stochastic fuzziness, CORS-TIMS-ORSA Joint Nat. Meet., May, Toronto, Canada.

1882 Norwich, A. M., and Turksen, I. B. (1978). The membership function of fuzzy set theory—Representation, uniqueness and construction, Working Pap. No. 78-011, Dept. of Ind. Eng., Univ. of Toronto, Toronto, Canada.

1883 Norwich, A. M., and Turksen, I. B. (1980). Meaningfulness in fuzzy set theory, Working Pap. No. 80-024, Dept. of Ind. Eng., Univ. of Toronto, Toronto, Canada.

1884 Norwich, A. M., and Turksen, I. B. (1980). Measurement and scaling of membership functions, in *Proc. Int. Congr. on Appl. Syst. Res. and Cybern.*, Acapulco, Mexico, Dec.

1885 Norwich, A. M., and Turksen, I. B. (1980). The fundamental measurement of fuzziness, Working Pap. No. 80-022, Dept. of Ind. Eng., Univ. of Toronto, Toronto, Canada.

1886 Norwich, A. M., and Turksen, I. B. (1981). A model for the measurement of membership functions and the consequences of its empirical implementation, Working Paper No. 81-003, Dept. of Ind. Eng., Univ. of Toronto, Toronto, Canada.

1887 Novak, J. (1968). On probability defined on certain classes of non-Boolean algebra, *Nachr. Osterreichischen Math. Ges.*, **23**, 89–90.

1888 Novak, V. (1980). An attempt at Godel-Bernays-like axiomatization of fuzzy sets, *Fuzzy Sets Syst.*, **3** (3).

1889 Nowakowska, M. (1976). Formal theory of actions and its application to social sciences, 3rd Eur. Meet. Cybern. Syst. Res., Vienna.

1890 Nowakowska, M. (1976). Methodological problems of measurement of fuzzy concepts in social sciences, *Behav. Sci.*

1891 Nowakowska, M. (1976). Towards a formal theory of dialogues, *Semiotics*, **18**.

1892 Nowakowska, M. (1977). Fuzzy concepts in the social sciences, *Behav. Sci.*, **22**, 107–115.

1893 Nowakowska, M. (1979). New ideas in decision-theory, *Int. J. Man-Mach. Stud.*, **11**, 213–234.

1894 Nowakowska, M. (1978). Fuzzy reasoning and dialogues, Marzalkowska 140-100, 00-061, Inst. of Philos. and Sociol., Polish Acad. of Sci., Warsaw.

1895 Nowakowska, M. (1979). Fuzzy concepts their structure and problems of measurement, *Advances in Fuzzy Set Theory and Applications* (M. M. Gupta, R. K. Ragade, and R. R. Yager, Eds.), North-Holland, Amsterdam.

1896 Nowakowska, M. (1979). Possibility distributions in the linguistic theory of actions, Working Paper, Tech. Univ., Warsaw.

1897 Nowakowska, M. (1980). Semiotic systems—A formal approach, in *Proc. Int. Congr. on Appl. Syst. Res. and Cybern.*, Acapulco, Mexico, Dec.

1898 Nurmi, H. (1976). On fuzzy games, 3rd Eur. Meet. Cybern. Syst. Res., Vienna.

1899 Nurmi, H. (1977). Probability and fuzziness—Some methodological considerations, 6th Res. Conf. on Subj. Probab., Utility, and Decis. Making, Warsaw, Sept.

1900 Nurmi, H. (1978). Modelling impreciseness in human systems, 4th Eur. Meet. on Cybern. and Syst. Res., Linz, Austria, Mar.

1901 Nurmi, H. (1978). Modelling political vagueness, impreciseness and ambiguity, ECPR Joint Sess. Workshops, Grenoble, France, Apr.

1902 Nurmi, H. (1980). Approaches to collective decision-making with fuzzy preference relations, 4th Eur. Congr. on Oper. Res, July.

1903 Nurmi, H. (1980). Modelling uncertainty in political decision making, in *Models of Political Economy* (P. Whitely, Ed.), Sage, London.

1904 Nurmi, H. (1981). A fuzzy solution to a majority voting game, *Fuzzy Sets Syst.*, **5** (2), 187–198.

1905 Nurminen, M. I. (1976). About the fuzziness in the analysis of information systems, 3rd Eur. Meet. Cybern. Syst. Res., Vienna.

1906 Nurminen, M. I. (1976). Studies in systemeering on fuzziness in the analysis of information systems, Diss., Inst. for Appl. Math., Univ. of Turku, Finland.

1907 Nurminen, M. I., and Paasio, A. (1976). Some remarks on the fuzzy approach to mutligoal devision making, *Finnish J. Bus. Econ.* (Special Edition) **3**.

1908 Nyang, S., and Tesiere, J. C. (1974). Applications des relations floues aux problemes multicriteria, le cas de la classification des semin. de rech. a Paris IX-Dauphine, Mem. Univ., Paris-Dauphine, Paris, Sept.

1909 Oda, M. (1977). Structure analysis and applications of fuzzy-and-uncertain reasoning methods, in *Proc. Symp. on Fuzzy Set Theory and Appl., IEEE Conf. on Decis. and Control*, New Orleans.

1910 Oda, M., and Shimomura, T. (1978). A study of fuzzy distortion process of concept communication-and-formation, Res. Rep. of Autom. Control Lab., Fac. of Eng., Nagoya Univ., Nagoya, Japan, Vol. 25, June.

1911 Oda, M., Shimomura, H., Chimura, H., and Womack, B. F. (1977). Measurement, evaluation, and control of communication-and-formation process of morality concept, in *Proc. 1977 IEEE Conf. on Decis. and Control*, New Orleans.

1912 Oda, M., Shimomura, T., and Womack, B. F. (1980). Concept structure and its distortion in the communication and formation process of morality concepts, Fuzzy Sets Theory and Appl. to Policy Anal. and Inf. Syst., Durham, NC.

1913 Oden, G. (1978). Applications of fuzzy set theory and fuzzy logic to psycholinguist problems, in *Proc. 22nd Meet. Soc. for Gen. Syst. Res.*, Washington, D.C., pp. 431–439.

1914 Oden, G. (1978). On the use of semantic constraints in guiding syntactic analysis, Whipp Rep. No. 3, Univ. of Wisconsin, Madison.

1915 Oden, G., and Hogan, M. E. (1977). A fuzzy propositional model of negation on semantic continuation, Midwestern Psychol. Assoc.

1916 Oden, G., and Massaro, D. W. (1977). Integer of place and voicing information in identifying synthetic stop—Con. Syll. Wisconsin Human Inf. Proc. Program. Rep., Whip No. 1.

1917 Oden, G. C. (1977). Fuzziness in semantic memory—Choosing exemplars of subjective categories, *Mem. Cognition*, **5**, 198–204.

1918 Oden, G. C. (1977). Integration of fuzzy logical information, *J. Exp. Psychol.*, **3**, 565–575.

1919 Oden, G. C. (1979). Fuzzy propositional approach to psycholinguistic problems an application of fuzzy set theory in cognitive science, *Advances in fuzzy set theory and applications* (M. M. Gupta, R. K. Ragade, and R. R. Yager, Eds.), North-Holland, Amsterdam.

1920 Oden, G. C. (1980). A fuzzy logical model of letter identification, *J. Exp. Psychol. Human Percept. Performance*, **5**.

1921 Oden, G. C. (1980). A fuzzy propositional model of concept structure and use—A case study in object identification, in *Proc. Int. Congr. On Appl. Syst. Res. and Cybern.*, Acapulco, Mexico, Dec.

1922 Oden, G. C., and Erson, N. H. (1974). Integration of semantic constraints, *J. Verbal Learn. Behav.*, **13**, 138–148.

1923 Oguntade, O. O., and Gero, J. S. (1978). Context, Meaning and preference—A conceptual basis for the use of fuzzy sets in the design and evaluation of multi-attribute multi-objective systems in architecture, Colloq. Int., Theorie et Appl. des Sous-Ensembles Flous, Journ. de Biomath. et d'Inf. Med., Marseille, France, Sept.

1924 Oguntade, O. O., and Gero, J. S. (1981). Evaluation of architectural design profiles using fuzzy sets, *Fuzzy Sets Syst.*, **5** (3), 221–234.

1925 Ohnishi, M., and Matsumoto, K. (1957). Gentzen method in calculi, *Osaka Math. J.*, **9**, 113–130.

1926 Ohsato, A., Hiki, T., Itoh, S., Sekiguchi, T., and Ohta, T. (1980). Solar-hydrogen energy system-model applied to isolated island. The 3rd World Hydrogen Energy Conf., Tokyo, June.

1927 Ohsato, A., Sekiguchi, T., Nakai, T., and Yamazaki, K. (1980). Information processing for nursing by fuzzy relation, The 3rd World Conf. on Med. Informatics, Tokyo, Sept.–Oct.

1928 Ohsuga, S. (1977). Semantic information processing in man-machine systems, in *Proc. 1977 IEEE Conf. on Decis. and Control*, New Orleans.

1929 Ohsumi, N. (1979). Evaluation procedure of agglomerative hierarchical clustering methods by fuzzy relations, Anal. des Donnees et Inf., Journ. Int. 2, Versailles, France.

1930 Okada, N., and Tamachi, T. (1974). Automated editing of fuzzy line drawings for picture description, *Trans. IECE* (Japan), **57-A**, 216–223.

1931 Okada, N., and Tamachi, T. (1974). Theory of fuzzy integrals and its applications, Thesis, Tokyo Inst. of Tech., Tokyo.

1932 Okamoto, M. B. (1979). A measure of closeness of weak implication to strict implication, Int. Joint Conf. on Artif. Intell., Tokyo, Aug.

1933 Okuda, S. (1977). Structural analysis of scientific information usage, in *Summ. of Pap. on Gen. Fuzzy Problems*, The Working Group on Fuzzy Syst., Tokyo, pp. 11–18.

1934 Okuda, T., Tanaka, H., and Asai, K. (1974). Decision-making and information in fuzzy events, *Bull. Univ. Osaka Prefect.*, **23A**.

1935 Okuda, T., Tanaka, H., and Asai, K. (1976). Decision problems and quantity of information in fuzzy events, *Trans. SICE*, **12**, 63–68.

1936 Okuda, T., Tanaka, H., and Asai, K. (1978). A formulation of fuzzy decision problems with fuzzy information using probability measures of fuzzy events *Inf. Control*, **38**, 135–147.

1937 Ollero, A., Camacho, E. F., and Aracil, J. (1979). Optimal policies for dynamic socio-economic models, IFAC/IFORS Symp., Toulouse, France, Mar.

1938 Ollero, A., and Freire, E. (1978). The structure of fuzzy relations between finite sets, Colloq. Int., Theorie et Appl. des Sous-Ensembles Flous, Journ. de Biomath. et d'Inf. Med., Marseille, France, Sept.

1939 Ollero, A., and Freire, E. (1981). The structure of relations in personnel management, *FSS*, Vol. 5, No. 2, 115–126.

1940 Ollier, C. (1981). Fuzzy sets, Les Nouvelles Literaires No. 2768, Jan.

1941 Onaga, K., and Mayeda, W. (1976). Boolean flow theory and its applications to cluster analysis, fuzzy logics and particle transmission, Allerton Conf. at Univ. of Illinois, Urbana.

1942 Onicescu, O. (1971). Principles de logique et de philosophie mathematique, Rum. Acad. of Sci., Bucharest.

1943 Oniga, T. (1975). Developpements de la logique trivalente, Rev. Cybern., Namur, Belgium.

1944 Oppenchaim, S. (1980). Classification et distance energetique entre sous ensembles flous, Theme 3rd Cycle, Math. et Inf. Univ. de Compiegne, Compiegne, France.

1945 Oppenchaim, S., and Dubuisson, B. (1978). A definition of a metric on the class of fuzzy sets based on an energy measure. Notion of an energy balance, *EMCSR—PCSR*, **8**, to be published.

1946 Oppenchaim, S., and Dubuisson, B. (1980). Distances entre sous ensembles flous et classification hierarchique relative ascendante, Table Ronde du CNRS sur le Flou, Lyon, France, June.

1947 Ore, O. (1942). Theory of equivalence relations, *Duke Math. J.*, **9**, 573–627.

1948 Orlovsky, S. A. (1977). Decision making with a fuzzy preference relation, *Fuzzy Sets Syst.*, **1**, 155–167.

1949 Orlovsky, S. A. (1977). On programming with fuzzy constraint sets, *Kybernetes*, **6**, 197–201.

1950 Orlovsky, S. A. (1978). Decision making with a fuzzy preference relation in a fuzzy set of alterations, in *Proc. 4th Int. Congr. on Cybern. and Syst.*, Amsterdam, pp. 375–376.

1951 Orlovsky, S. A. (1980). On formalization of a general fuzzy mathematical problem, *Fuzzy Sets Syst.*, **3** (3), May, 311–322.

1952 Orlovsky, S. A., and Shapiro, D. I. (1979). The 1st Sov. Semin. on Control in a Fuzzy Environment, *Fuzzy Sets Syst.*, **2** (4), Oct.

1953 Orlowska, E. (1967). Mechanical proof procedure for the *N*-valued propositional calculus, *Bull. Acad. Pol. Sci.*, Ser. Sci. Math., Astr., Phys., **15**, 537–541.

1954 Orlowska, E. (1973). Theorem-proving systems, *Diss. Math.* (Warsaw) **103**.

1955 Orlowsky, S. A. (1980). On formalization of a general fuzzy mathematical problem, *Fuzzy Sets Syst.*, **3** (3).

1956 Osgood, C. E., Suci, G. J., and Tannenbaum, P. H. (1964). *The Measurement of Meaning*, Univ. of Illinois Press, Urbana.

1957 Osis, J. J. (1968). Fault detection in complex systems using theory of fuzzy sets, in *Kibernetika i Diagnostika*, Vol. 2, (D. S. Kristinkov, J. J. Osis, and L. A. Ravtrigin, Eds.), Zinatne, Riga, U.S.S.R., pp. 13–18 (in Russian).

1958 Ostergaard, J. J. (1977). Fuzzy logic control of a heat exchanger process, in *Fuzzy Automata and Decision Processes*, (M. M. Gupta, G. N. Saridis, and B. R. Gaines, Eds.), North-Holland, Amsterdam, pp. 285–320.

1959 Ostergaard, J. J. (1980). Fuzzy logic controllers, in *Proc. of JACC*, San Fransisco, Aug.

1960 Otsuki, S. (1970). A model for learning and recognizing machine, *Inf. Process.*, **11**, 664–671.

1961 Pal, S. K., Datta, A. K., and Majumder, D. D. (1978). Adaptive learning algorithms in classification of fuzzy patterns, An application to vowels in CNC context, *Int. J. Syst. Sci.*, **9**, 887–897.

1962 Pal, S. K., and Majumder, D. D. (1977). Fuzzy sets and decision-making approaches in vowel and speaker recognition, *IEEE Trans. Syst. Man, Cybern.*, **SMC-7**, 625–629.

1963 Pal, S. K., and Majumder, D. D. (1978). Correction to "On automatic plosive identification using fuzziness in property sets," *IEEE Trans. Syst., Man, Cybern.*, **SMC-8** (12), 907.

1964 Pal, S. K., and Majumder, D. D. (1978). On automatic plosive identification using fuzziness in property sets, *IEEE Trans. Syst., Man, Cybern.*, **SMC-8**, 302-307.

1965 Pal, S. K., and Majumder, D. D. (1978). Effect of fuzzification and the plosive cognition system, *Int. J. Syst. Sci.*, **9**, 873-886.

1966 Pal, S. K., and Majumder, D. D. (1980). A self-adaptive fuzzy recognition system for speech sounds, Fuzzy Sets Theory and Appl. to Policy Anal. and Inf. Syst., Durham, NC.

1967 Pal, S. K., Majumder, D. D., and Chaudhury, B. B. (1980). Fuzzy sets in handwritten character recognition, in *Proc. All India Interdisciplinary Symp. on Digital Tech. and Pattern Recog.*, ISI, Calcutta, India, Feb., pp. 15-17.

1968 Panda, S. R. (1971). Inverse problem for linear systems containing uncertain parameters, ASME, Pap. 71-WA/AUT-14 for Meet. Nov. 28-Dec. 2, p. 12.

1969 Pappis, C. P., and Mamdani, E. H. (1976). A fuzzy logic controller for a traffic junction, Res. Rep., Dept. of Elec. Eng., Queen Mary College, Univ. of London, London.

1970 Pappis, C. P., and Sugeno, M. (1976). Fuzzy relational equations and the inverse problem, Internal Rep., Queen Mary College, Univ. of London, London.

1971 Parret, H. (1974). *Discussing Language*, Mouton, The Hague (see dialogue with G. Lakoff).

1972 Parrish, E. A. (1977). Electromagnetic interference source identification through fuzzy clustering, in *Proc. 1977 IEEE Conf. on Decis. and Control*, New Orleans.

1973 Parsons, C. (1974). The liar paradox, *J. Philos. Logic*, **3**, 381-412.

1974 Pask, G. (1975). *Conversation, Cognition and Learning*, Elsevier, Amsterdam.

1975 Pask, G. (1975). *The Cybernetics of Human Learning and Performance*, Hutchison, London.

1976 Pask, G. (Ed.) (1976). Current scientific approaches to decision making in complex systems, Syst. Res. Ltd., Richmond, UK, Apr.

1977 Pavlidis, T. (1975). Fuzzy representations as means of overcoming the over-commitment of segmentation, in *Proc. Conf. on Comput. Graphics, Pattern Recog. and Data Struct.*, Los Angeles.

1978 Pavlidis, T. (1977). Application of fuzzy sets in curve fitting, in *Proc. 1977 IEEE Conf. on Decis. and Control*, New Orleans.

1979 Paz, A. (1967). Fuzzy star functions, probabilistic automata and their approximation by nonprobabilistic automata, *J. Comput. Syst. Sci.*, **1**, 371-389.

1980 Pearl, J. (1974). Problem presentation research, UCLA-Eng-7404, School of Eng. and Appl. Sci., Univ. of California, Los Angeles.

1981 Pearl, J. (1975). An economic basis for certain methods of evaluating probabilistic forecasts, UCLA-Eng-Rep-7561, School of Eng. and Appl. Sci., Univ. of California, Los Angeles, July.

1982 Pearl, J. (1975). On the complexity of computing probabilistic assertions, UCLA-Eng-7562, School of Eng. and Appl. Sci., Univ. of California, Los Angeles, July.

1983 Pearl, J. (1975). On the complexity of inexact computations, UCLA-Eng-Pap.-0775, School of Eng. and Appl. Sci., Univ. of California, Los Angeles, July.

1984 Pearl, J. (1975). On the storage economy of inferential question-answering systems, *IEEE Trans. Syst., Man, Cybern.*, **SMC-5**, 595-602.

1985 Pearl, J. (1975). State complexity of imprecise casual models, UCLA-Eng-Rep-7560, School of Engineering and Applied Sci., Univ. of California, Los Angeles, Dec.

1986 Pearl, J. (1976). A framework for processing value judgements UCLA-Rep-7622, School of Eng. and Appl. Sci., Univ. of California, Los Angeles, Mar.

1987 Pearl, J. (1976). A note on the management of probability assessors, UCLA-Eng-Rep-7664, School of Eng. and Appl. Sci., Univ. of California, Los Angeles, Feb.

1988 Pelech, A. (1975). A dynamic model of the project. The case of the fuzzy event, Res. Rep. No. 84, Inst. of Organ. Manage., Tech. Univ. Wroclaw, Wroclaw, Poland.

1989 Pelech, A., and Chanas, S. (1975). A concept of the project modelling by means of Zadeh's fuzzy sets languages, Res. Rep. No. 120, Inst. of Organ. Manage., Tech. Univ. Wroclaw, Wroclaw, Poland.

1990 Peng, Z. (1980). Apply fuzzy mathematics to classification of rocks, Wuhan Water Conserv. and Power Inst., A List of Lit. in China, The Univ. of Sci. and Tech. of China, Hofei, Anhwei, China.

1991 Peng, Z. (1980). Fuzzy mathematics applied to the classification of the rock, Collected Pap. on Fuzzy Math. (Abstr.), Huazhong Inst. of Tech., Huazhong, China, Apr.

1992 Peng, Z. (1980). Neartude of fuzzy subsets and fuzzy distance space, *J. Wuhan Water Conserv. Power Inst.*, **4**, A List of Lit. in China, The Univ. of Sci. and Tech. of China, Hofei, Anhwei, China.

1993 Peng, Z. (1981). Fuzzy cluster, Wuhan Water Conserv. and Power Inst., A List of Lit. in China, The Univ. of Sci. and Tech. of China, Hofei, Anhwei, China.

1994 Perry, K. E., and Waddell, J. J. (1972). *The Rotary Cement Kiln*, The Chemical Publishing Co., New York.

1995 Peschel, M. (1975). Some remarks to "Fuzzy Systems" as a complement to the topic paper from L. A. Zadeh, Berlin, Feb.

1996 Peschel, M. (1977). Beitrag zum Modularen Aufbau der Methodologie fur Modellbilder—Klassifizieren und Steueren, in *Unscharfe Modellbildung und Steuerung*, Lecture Notes, TH Karl-Marx-Stadt, June, pp. 3–15.

1997 Peschel, M. (1977). Kombination der Unscharfen Beschreibungen mit der Poly-Otimierung bei Modernen Modellbildungstechniken, in *Unscharfe Modellbildung und Steuerung*, Lecture Notes, TH Karl-Marx-Stadt, June, pp. 15–21.

1998 Peschel, M. (1978). Model reference fuzzy identification method, *PCSR*, **6**.

1999 Peschel, M., and Ester, J. (1979). Die Methode der Richtungsdiagramme als Ausdruck Unscharfer Lokaler Ziele bei der Polyoptimierung, Material zur Vorlesung "Kennwertermittlung und Modellbildung," Teil II—Unscharfe Modellbildung und Steuerung, TH Karl-Marx-Stadt, Oct.

2000 Petrescu, I. (1971). Algebres de Morgen Injectives, in *Logique, Automatique, Informatique* (G. C. Moisil, Ed.), Bucharest, pp. 171–176.

2001 Petrov, E. G., and Zatov, V. G. (1977). A self theoretical method for estimation of the efficiency of complex systems, *Sov. Autom. Controls*, **9**, 57–65.

2002 Pfeilsticker, A. (1980). Fuzzy sets used in macroeconomic modelling, The 10th Meet. of the Eur. Working Group on Fuzzy Sets, June.

2003 Philips, R. J., Beaumont, M. J., and Richardson, D. (1979). Aesop—An architectural relational database, *Comput.-Aided-Des.*, **11** (4), 217–226.

2004 Pieters, L. (1977). La theorie des sous-ensembles flous et son application en docimologie linguistique, vers une evaluation plus objective des examensde traduction, Doc. Centre Imago, Inst. des Lang. Vivantes, Univ. Catholique de Louvain, Louvain, Belgium.

2005 Pinkava, V. (1965). On the nature of some logical paradoxes, *Kybernetika* (Prague), **1**, 111-121 (in Czech., Eng. summ.).

2006 Pinkava, V. (1975). Some further properties of the pi-logics, in *Proc. 1975 Int. Symp. on Multiple-Valued Logic*, Logan, Utah, IEEE 75CH0959-7C, pp. 20-26.

2007 Pinkava, V. (1976). "Fuzzification" of binary and finite multivalued logical calculi, *Int. J. Man-Mach. Stud.*, **8**, 71-730.

2008 Pinkava, V. (1976). On the nature of some logical paradoxes, *Int. J. Man-Mach. Stud.*

2009 Pinkava, V. (1980). On potential tautoligies in K-valued calculi and their assessment by means of normal forms, Fuzzy Sets Theory and Appl. to Policy Anal. and Inf. Syst., Durham, NC.

2010 Pinkava, V., and Kohout, L. J. (1976). Enumerably infinite-valued functionally complete pi-logic algebras, in *Discrete Systems and Fuzzy Reasoning* (E. H. Mamdani and B. R. Gaines, Eds.), EES-MMS-DSFR-76, Queen Mary College, Univ. of London, London (Workshop Proc.).

2011 Pipino, L. L. (1975). The application of fuzzy sets to system diagnosis and the design of a conceptual diagnostic procedure, Ph.D. Thesis, Univ. of Massachusetts, in *Diss. Abstr. Int.*, **36/09-A**, 6196.

2012 Poch, F. A. (1979). Algunas aplicaciones de los conjuntes borrosos a la estadistica, Inst. Nac. de Estad., Madrid.

2013 Poch, F. A. (1980). Glossary of fuzzy sets with respect to statistics, Minist. de Econ., Inst. Nac. de Estad., Madrid.

2014 Poch, F. A. (1980). Uncertainity and fuzziness in primary data and estimation procedures, in *Proc. Int. Congr. on Appl. Syst. Res. and Cybern.*, Acapulco, Mexico, Dec.

2015 Pollatschek, M. A. (1977). Hierarchical systems and fuzzy-set theory, in *Proc. 1977 IEEE Conf. on Decis. and Control*, New Orleans.

2016 Ponasse, D. (1975). Theorie des sous-ensembles flous, Conf. Univ. Claude Bernard, Lyon, France, May.

2017 Ponasse, D. (1977). Generalities sur les structures floues, Semin. Math. Floues, Dept. of Math, Univ. de Lyon, Lyon, France, Aug., pp. 1-27.

2018 Ponasse, D. (1978). Algebre floue et algebre de Lukasiewicz, *Rev. Roum. Math. Pures Appl.* (23).

2019 Ponasse, D. (1978). Representations des algebres de Lukasiewicz par des algebres floues, Semin. "Math. Floue," Lyon, France.

2020 Ponasse, D. (1979). Seminaire "Math. Floue," Annee 78-79, Univ. Claude, Bernard, Lyon, France.

2021 Ponasse, D. (1980). Catagorie des ensembles flous, to be published.

2022 Ponsard, C. (1975). Contribution a une theorie des espaces economiques imprecis, Doc. de Trav. IME, Univ. of Dijon, Dijon, France.

2023 Ponsard, C. (1975). L'impression et son traitment en analyse economique, Doc. de Trav. IME, Univ. of Dijon, Dijon, France.

2024 Ponsard, C. (1975). On the Axiomatization of fuzzy subsets theory, Eguipe de Rech. Associee Au CNRS, Doc. de Trav., Univ. of Dijon, Dijon, France.

2025 Ponsard, C. (1976). Alea flou, Note Interne, Inst. Math. Econ., Univ. of Dijon, Dijon, France.

2026 Ponsard, C. (1977). Hierarchies des places centrales et graphes phi-flous, *Environ. Plann.*, **9**, 233-252.

2027 Ponsard, C. (1977). La region en analyse spaciale, Note Interne IME No. 21, Inst. Math. Econ., Univ. of Dijon, Dijon, France, May.

2028 Ponsard, C. (1978). An application of fuzzy sets theory to the analysis of the consumer's spatial preferences, Colloq. Int. Theorie et Appl. des Sous-Ensembles Flous, Journ. de Biomath. et d'Inf. Med., Marseille, France, Sept.

2029 Ponsard, C. (1978). Jalons pour une theorie spatiale du comportement de consommateur, Doc. IME, Inst. Math. Econ., Univ. of Dijon, Dijon, France.

2030 Ponsard, C. (1978). Esquisse de simulation d'une economie regionale—L'apport de la theorie des systemes flous, Doc. de Trav. de l'IME, No. 18, Sept., Repris in *Melanges Economiques en Hommage a Pierre Moran*, Economica, Paris.

2031 Ponsard, C. (1980). Flou et science economique, *Bull. Stud. Exch. Fuzziness Appl.*, Printemps.

2032 Ponsard, C. (1980). Fuzzy economic spaces, Doc. IME No. 43, Univ. of Dijon, Dijon, France.

2033 Ponsard, C. (1980). L'equilibre spatial du consommateur dans un contexte imprecis, Doc. IME No. 41, Univ. of Dijon, Dijon, France.

2034 Ponsard, C. (1980). Producers spatial equilbrium with fuzzy constraints, 4th Eur. Congr. on Oper. Res, July.

2035 Poolock, J. L. (1975). Four kinds of conditionals, *Amer. Philos. Quart.*, **12**, 51–59.

2036 Pope, S. (1978). Multiobjective decision analysis of the value of test marketing research, ORSA/TIMS Meet., Los Angeles.

2037 Popper, K. R. (1963). *Conjectures and Refutations*, Routledge and Kegan Paul, London.

2038 Popper, K. R. (1972). *Objective Knowledge*, Clarendon Press, Oxford, UK.

2039 Popper, K. R. (1972). *The Logic of Scientific Discovery*, Hutchinson, London (1st ed., 1959).

2040 Popper, K. R. (1976). A note on verisimilitude, *Brit. J. Philos. Sci.*, **27**, 147–164.

2041 Popper, K. R. (1976). *Unended Quest*, Fontana, London.

2042 Poselov, D. A. (1976). Semoitic models—Sucesses and perspectives, *Cybern.*, **12** (6), 929–937.

2043 Pospisil, B. (1937). Remark on bicompact spaces, *Ann. Math.*, **38**, 845–846.

2044 Pospisil, B. (1939). On bicompact spaces, Publ. de la Fac. des Sci. de l'Univ. Masaryk, No. 270, Brno, Czechoslavakia, pp. 3–16.

2045 Pospisil, B. (1939). Primideale in Vollstandigen Ringen, *Fundam. Math.*, **33**, 66–74 (whole vol. published in Dec. 1945).

2046 Pospisil, B. (1940). Uber die messbaren funktionen, *Math. Ann.*, **117**, 327–355.

2047 Pospisil, B. (1941). Eine Bemerkung uber Funktionenfolgen, *Cas. Pestovani Math. Fys. (Prague)*, **70**, 119–121.

2048 Pospisil, B. (1941). Eine Bemerkung uber Stetige Vertailung, *Cas. Pestovani Math. Fys. (Prague)*, **70**, 68–72.

2049 Pospisil, B. (1941). Eine Bemerkung uber Vollstandige Raume, *Cas. Pestovani Math. Fys.* (Prague), **70**, 38–41.

2050 Pospisil, B. (1941). Von den Verteilungen auf Booleschen Ringen, *Math. Ann.*, **118**, 32–40.

2051 Post, E. L. (1921). Introduction to a general theory of elementary propositions, *Amer. J. Math.*, **43**, 163–185.

2052 Post, J. F. (1973). Shades of the liar, *Philos. Logic*, **2**, 370–386.

2053 Poston, T. (1971). Fuzzy geometry, *Manifold* (Univ. of Nottingham), **10**.
2054 Pracontal, M. (1980). Les "Ensembles flous"—Pas si flous, *Sci. Vie*, (756), Sept.
2055 Prade, H. (1980). Operations with fuzzy data, Fuzzy Sets, Theory and Appl. to Policy Anal. and Inf. Syst. (P. P. Wang and S. K. Chang, Eds.), Plenum, New York, pp. 155–169.
2056 Prade, H. (1977). Exemple d'approche heuristique, interactive, floue pour un probleme d'ordonnancement, *Congr. AFCET, Modelisation et Maitrise des Syst.*, Vol. 2, Editions Hommes et Tech., Versailles, France, pp. 347–355.
2057 Prade, H. (1977). Ordonnancement et temps reel, Doc. Thesis, ENSAE, Toulouse, France.
2058 Prade, H. (1978). Fuzzy logics—A survey, Higher Order System.
2059 Prade, H. (1978). Why fuzzy sets theory does not seem very useful for industrial robotic systems... But may be relevant for a lot of other applications. Internal Memo., Stanford Artif. Intell. Lab., Stanford Univ., Stanford, CA.
2060 Prade, H. (1979). A queuing problem with fuzzy service time, fuzzy service rule, 1st Int. Symp. on Policy Anal. and Inf. Syst., June.
2061 Prade, H. (1979). La theorie des sous-ensembles flous, resultats et perceptives, Lectures Ecole Sup. Aeronaut. et Espace, Toulouse, France, Jan.
2062 Prade, H. (1979). Nomenclature of fuzzy measures, in *Proc. of the Int. Semin. on Fuzzy Set Theory*, Johannes Kepler Univ., Linz, Austria.
2063 Prade, H. (1979). The application of fuzzy sets to height-level artificial intelligence, Survey Future Dev., Memo. LSI., Univ. de Toulouse, Toulouse, France, Jan.
2064 Prade, H. (1979). Using fuzzy set theory in a Scheduling problem—A case study, *J. Fuzzy Sets Syst.*, **2** (2), 153–165.
2065 Prade, H. (1979). Operations research with fuzzy data, 1st Symp. Int. Policy Anal. and Inf. Syst., Duke Univ., Durham, NC, June.
2066 Prade, H. (1980). An outline of fuzzy or possibilistic models for queuing systems, Fuzzy Sets Theory and Appl. to Policy Anal. and Inf. Syst.
2067 Prade, H. (1980). Compatibilite, qualification, modification, niveau de precision, *Bull. Sous Ensembles Flous Appl.*, Automne.
2068 Prade, H. (1980). Extension des modalities aristoteliciennes application a l'evaluation de la similarite entre sous-ensembles flous, Table Ronde du CNRS sur le Flou, Lyon, France, June.
2069 Prade, H. (1980). Fuzzy H.O.S. and fuzzy belief maintenance, Two examples of fuzzy data types, COMPSAC'80, Lab. Lang. and Syst. Inf., Toulouse, France.
2070 Prade, H. (1980). Fuzzy programming—Why and how?—Some hints and examples, 4th IEEE Int. Comput. Software and Appl. Conf., Chicago, Oct.
2071 Prade, H. (1980). L'ensemble de parato peut-il etre flou?, *Bull. Stud. Exch. Fuzziness Appl.*, Printemps.
2072 Prade, H. (1980). Model semantics and fuzzy set theory, *Proc. Int. Congr. on Appl. Syst. Res. and Cybern.*, Acapulco, Mexico, Dec.
2073 Prade, H. (1980). Possibilite—Necessite—Principe d'extensions, *Bull. Sous Ensembles Flous Appl.*, Automne.
2074 Prade, H. (1980). Possibilite et logique trivalente de Lukasiewicz—une remarque, *Bull. Stud. Exch. Fuzziness Appl.*, Printemps.
2075 Prade, H. (1980). Une approache des "mesures floues" bases sur les normes triangulaires, Table Ronde du CNRS sur le Flou, Lyon, France, June.

2076 Prade, H. (1980). Unions et intersections d'ensembles flous, *Bull. Stud. Exch. Fuzziness Appl.*, Ete.

2077 Prade, H., and Vaina, L. (1978). What fuzzy H.O.S. may mean, Tech. Rep., Higher Order Software Inc., Cambridge, MA.

2078 Prelecki, M. (1958). W Sprawie Terminow Nieostrych, *Stud. Logica*, **8**.

2079 Preparata, F. P., and Yeh, R. T. (1971). On a theory of continuously valued logic, in *Conf. Rec. 1971 Symp. on Theory and Appl. of Multiple-Valued Logic Des.*, pp. 124–132.

2080 Preparata, F. P., and Yeh, R. T. (1972). Continuously valued logic, *J. Comput. Syst. Sci.*, **6**, 397–418.

2081 Prevot, M. (1975). Probability calculation and fuzzy sets theory, Doc. Trav., No. 14, Inst. de Math. Econ., Univ. de Dijon, Dijon, France, Aug.

2082 Prevot, M. (1977). Sous-ensembles flous, Une approche theorique, Collect. IME, Inst. de Math. Econ., Univ. de Dijon, Dijon, France.

2083 Prevot, M. (1978). Algorithme pour la resolution des systemes flous, Colloq. Internal Theorie et Appl. de Sous-Ensembles flous, Journ. de Biomath. et d'Inf. Med., Marseille, France, Sept.

2084 Prevot, M. (1980). Axiomatization of the fuzzy subsets theory, *Bull. Stud. Exch. Fuzziness Appl.*, Automne-Hiver.

2085 Prevot, M. (1980). Fuzzy goals under several constraints, in *Proc. Int. Congr. on Appl. Syst. Res. and Cybern.*, Acapulco, Mexico.

2086 Prevot, M. (1981). Algorithm for the solution of fuzzy relations (Short Communication), *Fuzzy Sets Syst.*, **5**(3), 319–322.

2087 Prior, A. N. (1953). On propositions neither necessary nor impossible, *J. Symb. Logic*, **18**, 105–108.

2088 Prior, A. N. (1954). The interpretation of two systems of modal logic, *J. Comput. Syst.*, **4**, 201–208.

2089 Prior, A. N. (1955). Curry's paradox and 3-valued logic, *Australas. J. Philos.*, **33**, 177–182.

2090 Prior, A. N. (1955). Many-valued and modal systems—An intuitive approach, *Philos. Rev.*, **64**, 626–630.

2091 Prior, A. N. (1957). *Time and Modality*, Clarendon Press, Oxford, UK.

2092 Prior, A. N. (1962). *Formal Logic*, (2nd ed.), Clarendon Press, Oxford, UK.

2093 Prior, A. N. (1967). *Past, Present and Future*, Clarendon Press, Oxford, UK.

2094 Prior, A. N. (1971). *Objects of Thought* (Geach and Kenny, Eds.), Clarendon Press, Oxford, UK.

2095 Proctor, C. (1977). Review of *Introduction to the Theory of Fuzzy Subsets*, Vol. 1, Kaufman, *J.A.S.A.*, Mar.

2096 Procyk, T. J. (1974). The control of systems possessing delay using fuzzy set theory, Fuzzy Logic Working Group Rep., Queen Mary College, Univ. of London, London, Dec.

2097 Procyk, T. J. (1976). A fuzzy logic learning system for a single input single output plant, Fuzzy Logic Working Group Rep. 3, Queen Mary College, Univ. of London, London.

2098 Procyk, T. J. (1976). A proposal for a learning system, Internal Rep., Queen Mary College, Univ. of London, London.

2099 Procyk, T. J. (1976). Linguistic representation of fuzzy variables, Fuzzy Logic Working Group Rep. 3, Queen Mary College, Univ. of London, London.

2100 Procyk, T. J., and Mamdani, E. M. (1979). A linguistic self-organizing process controller, *Automatica*, **15**, 15–30.

2101 Prugovecki, E. (1973). A postulational framework for theories of simutaneous measurement of several observables, *Found. Phys.*, **3**, 3–18.

2102 Prugovecki, E. (1974). Fuzzy sets in the theory of measurement of incompatible observables, *Found. Phys.*, **4**, 9–18.

2103 Prugovecki, E. (1975). Measurement in quantum mechanics as a stochastic process on spaces of fuzzy events, *Found. Phys.* **5**, 557–571.

2104 Prugovecki, E. (1976). Localizability of relativistic particles in fuzzy phase space, *J. Phys. Math.*, **9** (11).

2105 Prugovecki, E. (1976). Probability measures on fuzzy events in phase space, *J. Math. Phys.*, **17**, 517–523.

2106 Prugovecki, E. (1976). Quantum two-particle scattering in fuzzy phase space, Dept. of Math., Univ. of Toronto, Toronto, Canada, Jan.

2107 Prugovecki, E. (1977). On fuzzy spin spaces, *J. Phys. A—Math.*, **10** (4).

2108 Pu, P., and Liu, Y. (1980). Fuzzy topology I—Neighborhood structure of a fuzzy point and Moore-Smith convergence, Collect. Pap. on Fuzzy Math. (Abstr.), Huazhong Inst. of Tech., Huazhong, China, Apr.

2109 Pu, P., and Liu, Y. (1980). Fuzzy topology II—Product and Quotient spaces, Collect. Pap. on Fuzzy Math. (Abstr.), Huazhong Inst. of Tech., Huazhong, China, Apr.

2110 Pu, S. (1980). Mapping properties of induced fuzzy topological spaces (in Chinese), A List of Lit. in China, The Univ. of Sci. and Tech. of China, Hofei, Anhwei, China.

2111 Pu, S. (1980). Zero-dimensional fuzzy topological spaces (in Chinese), A List of Lit. in China, The Univ. of Sci. and Tech. of China, Hofei, Anhwei, China.

2112 Pudlak, P. (1975). Polynomially complete problems in the logic of automated discovery, in *Lecture Notes in Computer Science*, Vol. 32, (J. Becvar, Ed.), Springer-Verlag, Berlin, pp. 358–361.

2113 Pudlak, P. (1975). The observational predicate calculus and complexity of computations, *Commentat. Math. Univ. Carolinae*, **16**, 395–398.

2114 Pultr, A. (1976). Closed categories of L-fuzzy sets, *Vortrage Problemsemin. Autom. Algorithmentheorie*, Weissig, Germany, Apr.

2115 Pun, L. (1975). Experience in the use of fuzzy formalism in problems with various degrees of subjectivity, Special Interest Disc. on Fuzzy Automata and Decis. Processes, 6th IFAC World Congr., Boston, MA, Aug.

2116 Pun, L. (1977). Use of fuzzy formalism in problems with various degrees of subjectivity, in *Fuzzy Automata and Decision Processes* (M. M. Gupta, G. N. Saridis, and B. R. Gaines, Eds.), North-Holland, Amsterdam, pp. 357–378.

2117 Putnam, H. (1957). Three-valued logic, *Philos. Stud.*, **8**, 73–80.

2118 Qin, G. (1980). Fuzzy space $FL(X)$ and the commentary about selection-nearness principle, *Huazhong Inst. Tech.*, **2**, A List of Lit. in China, The Univ. of Sci. and Tech. of China, Hofei, Anhwei, China.

2119 Qu, Y. (1980). An algorithm of test generation of fuzzy states of sequential circuit, Collect. Pap. on Fuzzy Math. (Abstr.), Huazhong Inst. of Tech., Huazhong, China, Apr.

2120 Qu, Y. (1980). Sequential circuit model with fuzzy border, Collect. Pap. on Fuzzy Math. (Abstr.), Huazhong Inst. of Tech., Huazhong, China, Apr.

2121 Qu, Y. (1980). The arithmetical structure of the predicate calculus and fuzzy logic, Collect. Pap. on Fuzzy Math. (Abstr.), Huazhong Inst. of Tech., Huazhong, China, Apr.

2122 Raad, M. B. (1978). Fuzzy relations in a control setting, *Kybernetes*, **7**, 185-188.
2123 Radecki, T. (1976). Application of fuzzy sets theory to the description of information retrieval process, Rep. of the Main Library and Sci. Inf. Centre, Ser. A, No. 56. Tech. Univ. of Wroclaw, Wroclaw, Poland.
2124 Radecki, T. (1976). Mathematical model of information retrieval system based on the concept of fuzzy thesaurus, *Inf. Process. Manage.*, **12**, 313-318.
2125 Radecki, T. (1977). Fuzzy set theoretical approach to document retrieval, 6th Cranfield Int. Conf. on Mech. Inf. Storage and Retr. Syst., Cranfield.
2126 Radecki, T. (1977). Level fuzzy sets, Tech. Univ. of Wroclaw, Wroclaw, Poland, *J. Cybern.*, **7**, 189-198.
2127 Radecki, T. (1977). Outline of a fuzzy logic approach to information retrieval, Comm. of the Main Library and Sci. Inf. Centre, Ser. A, No. 74, Tech. Univ. of Wroclaw, Wroclaw, Poland.
2128 Radecki, T. (1978). A model of document retrieval system based on the concept of sem. disjoint normal form, Tech. Univ. of Wroclaw, Wroclaw, Poland.
2129 Radecki, T. (1978). An evaluation of the fuzzy set theory approach to information retrieval, *EMCSR—PCSR*, **11** to be published.
2130 Radecki, T. (1981). On the inclusiveness of information retrieval systems with documents indexed by weighted descriptors, *FSS*, 159-176.
2131 Ragade, R. K. (1973). A multiattribute perception and classification of (visual) similarities, S-001-73, Syst. Res. and Plann., Bell-Northern Res., Ottawa, Canada, Nov.
2132 Ragade, R. K. (1973). On some aspects of fuzziness in communication—I Fuzzy entropies, W-002-73, Syst. Res. and Plann., Bell-Northern Res., Ottawa, Canada, Nov.
2133 Ragade, R. K. (1973). On some aspects of fuzziness in communication—II A note on fuzzy entropies associated with a fuzzy channel, W-006-73, Syst. Res. and Plann., Bell-Northern Res., Ottawa, Canada, Nov.
2134 Ragade, R. K. (1973). On some aspects of fuzziness in communication—III Fuzzy concept communication, W-005-73, Syst. Res. and Plann., Bell-Northern Res., Ottawa, Canada, Dec.
2135 Ragade, R. K. (1974). A note on fuzzy information, in *Proc. Bell-Northern Res. Center* (Canada), Dec.
2136 Ragade, R. K. (1974). Incertitude characterization of the retriever-system communication process, in *Proc. 37th Annu. Meet. Amer. Soc. Inf. Sci.*, Atlanta, GA, Oct.
2137 Ragade, R. K. (1974). Naive users and ill-formed problems in interactive systems, Tech. Rep., Bell-Northern Res. Ottawa, Canada, Dec.
2138 Ragade, R. K. (1975). Benefit cost analysis under imprecise conditions, 1-S-040675, Dept. of Syst. Des., Univ. of Waterloo, Ontario, Canada, June.
2139 Ragade, R. K. (1975). Profile transformation in groups and consensus formation by fuzzy sets, 3) Fuzzy concepts communication Univ. of Waterloo, Ontario, Canada, Jan.
2140 Ragade, R. K. (1976). A differential game formulation of fuzzy consensus, The TIMS-ORSA Conf., Miami Beach, FL.
2141 Ragade, R. K. (1976). Fuzzy games in the analysis of options, *J. Cybern.*
2142 Ragade, R. K. (1976). Fuzzy interpretive structural modelling, *J. Cybern.*
2143 Ragade, R. K. (1976). Fuzzy models in multi-objective conflict analysis, Syst. Sci. Center and the Dept. of Math., Univ. of Kentucky, Louisville, KY; also presented at the TIMS-ORSA Conf., Miami, FL, Nov.

2144 Ragade, R. K. (1976). Fuzzy set theory and the mathematical probability theory of Kolmogorov—Some observations, unpublished note.

2145 Ragade, R. K. (1976). Fuzzy sets in communication systems and consensus formation systems, TIMS-ORSA Joint Meet., Philadelphia, PA, Apr.

2146 Ragade, R. K. (1977). A mathematical model of approximate communication in information systems, *The General System Paradigm* (J. White, Ed.), pp. 334–346.

2147 Ragade, R. K. (1977). Profile transformation algebra and group consensus formation through fuzzy sets, in *Fuzzy Automata and Decision Processes* (M. M. Gupta, G. N. Saridis, and B. R. Gaines, Eds.), North-Holland, Amsterdam, pp. 331–356.

2148 Ragade, R. K. (1977). Systems analysis—Deterministic, stochastic or fuzzy?, The TIMS-ORSA Conf., Atlanta, GA.

2149 Ragade, R. K. (1978). Towards multi-attribute models by fuzzy set theory, in *Proc. 22nd Meet. of Soc. Gen. Syst. Res.*, Washington, D.C., pp. 412–417.

2150 Ragade, R. K. (1981). Statistics of fuzzy numbers, CORS-TIMS-ORSA Joint Nat. Meet., Toronto, Canada, May.

2151 Ragade, R. K., and Gupta, M. M. (1977). Fuzzy set theory—Introduction, in *Fuzzy Automata and Decision Processes* (M. M. Gupta, G. N. Saridis, and B. R. Gaines, Eds.), North-Holland, Amsterdam, pp. 105–132.

2152 Ragade, R. K., and Gupta, M. M. (1979). Fuzzy sets theory and applications—A synthesis, in *Advances in Fuzzy Sets Theory and Applications* (M. M. Gupta, R. K. Ragade, and R. R. Yager, Eds.), North Holland, Amsterdam, pp. 19–25.

2153 Ragade, R. K., and R. Hipel (1975). Non-Quantative methods in water resource management, ASCE Spec. Conf. on Water Resour. Manage., July.

2154 Ragade, R. K., Lewis, L. W., Kimswalde, Y., and Lewis, L. F. (1980). Statistical analysis of fuzzy data, in *Proc. Int. Congr. on Appl. Syst. Res. and Cybern.*, Acapulco, Mexico, Dec.

2155 Ragade, R. K., and Womak, B. F. (1977). Fuzzy graphs in societic modeling, Symp. on Fuzzy Set Theory and Appl., IEEE Conf. on Decis. and Control, New Orleans.

2156 Rajasethupathy, K. S., and Lakshmivarahan, S. (1977). Connectedness in fuzzy topology. Dept. of Math., Vivekanamdha College, Madras, India.

2157 Rajasethupathy, K. S., and Lakshmivarahan, S. (1974), Connectedness in fuzzy topology, *Kybernetika* (Prague), 13 (3).

2158 Rajeck, R. K. (1975). Benefit cost analysis under imprecise conditions, Univ. of Waterloo, Ontario, Canada, June.

2159 Ralescu, D. (1979). A survey of the representation of fuzzy concepts and its applications, in *Advances in Fuzzy Set Theory and Applications* (M. M. Gupta, R. K. Ragade, and R. R. Yager, Eds.), North-Holland, Amsterdam.

2160 Ralescu, D. (1979). Possibility theory and optimization with inexact constraints, in *Proc. 18th IEEE Conf. on Decis. and Control*, Fort Lauderdale, FL, Dec.

2161 Ralescu, D. (1980). Estimation based on subjective evaluation, in *Proc. Int. Congr. on Appl. Syst. Res. and Cybern.*, Acapulco, Mexico, Dec.

2162 Ralescu, D. A. (1974). *Fuzzy Sets and Their Applications*, Ed. Tehnica, Bucharest (in Rumanian).

2163 Ralescu, D. A. (1974). On fuzzy characters and subobjects, Semin. de Sist. Fuzzy, Dept. Econ. Cybern., Acad. of Econ. Stud., Bucharest.

2164 Ralescu, D. A. (1975). Decomposition theorems for fuzzy automata, Semin. de Sist. Fuzzy, Dept. Econ. Cybern., Acad. of Econ. Stud., Bucharest.

2165 Ralescu, D. A. (1976). *L*-fuzzy sets and *L*-flou sets, *Elektron. Inf. Kybern.*, **12**, 599–605.
2166 Ralescu, D. A. (1976). On fuzzy systems, in *Proc. 3rd Int. Congr. on Cybern.*
2167 Ralescu, D. A. (1977). Inexact solutions for large scale control problems, in *Proc. 1st Int. Congr. on Math. at the Service of Man*, Barcelona, Spain.
2168 Ralescu, D. A. (1978). Fuzzy subobjects in a category and the theory of *C*-sets, *Fuzzy Sets Syst.*, **1**, 193–202.
2169 Ralescu, D. A. (1978). Ordering, preferences and fuzzy optimization, in *Proc. 4th Int. Conf. Cybern. and Syst.*, Amsterdam, pp. 377–378.
2170 Ralescu, D. A. (1978). The interface between orderings and fuzzy optimization, The ORSA/TIMS Joint Nat. Meet., Los Angeles, Nov.
2171 Ralescu, D. A. (1980). Integration on fuzzy sets, Dept. of Math., Indiana Univ., Bloomington, *J. Math. Anal. Appl.*, 1–12.
2172 Ralescu, D., and Adams, G. (1980). The fuzzy integral, *J. Math. Anal. Appl.*
2173 Rasiowa, H. (1974). *An Algebraic Approach to Non-Classical Logics*, North-Holland, Amsterdam.
2174 Rasiowa, H., and Sikorski, R. (1970). *The Mathematics of Metamathematics*, Warsaw, Poland.
2175 Rauch, J. (1975). Ein Beitrag Zu der Guha Methode in der Dreiwertigen Logik, *Kybernetika* (Prague), **11**, 101–113.
2176 Rebrova, M. P. (1976). Fuzzy sets in classification theory, *Auto. Doc. Math. Linguist.*, **10**, 4.
2177 Reddy, D. (1972). Reference and metaphor in human language, Ph.D. Thesis, Dept. of English, Univ. of Chicago, Chicago.
2178 Reiger, B. (1974). Eine "Tolerante" Lexikonstruktur. Zur Abbildung Naturlich-Sprachlicher Bedeutung auf "Unscharfe" Mengen in Toleranzraumen, *Z. Literaturwiss. Linguist.*, **16**, 31–47.
2179 Reiger, L. (1949). A note on topological representation of distributive lattices, *Cas. Pestovani Math. Fys.* (Prague), **74**, 55–61.
2180 Reisinger, L. (1974). On fuzzy thesauri, in *Proc. Comput. Stat.*, Vienna.
2181 Rescher, N. (1963). A probabilistic approach to modal logic, *Acta Philos. Fenn.*, **16**, 215–226.
2182 Rescher, N. (1964). Quantifiers in many-valued logic, *Logique Anal.*, **7**, 181–184.
2183 Rescher, N. (1967). Semantic foundations for the logic of preference, in *The Logic of Decision and Action* (N. Rescher, Ed.), Univ. of Pittsburgh Press, Pittsburgh, PA, pp. 37–79.
2184 Rescher, N. (1968). *Topics in Philosophical Logic*, D. Reidel, Dordrecht, Holland.
2185 Rescher, N. (1969). *Many-Valued Logic*, McGraw-Hill, New York.
2186 Rescher, N. (1973). *The Coherence Theory of Truth*, Clarendon Press, Oxford, UK.
2187 Rescher, N., and Manor, R. (1970). On inference from inconsistent premises, *Theory Decis.*, **1**, 179–217.
2188 Ribeyre, S. (1978). Niveaux de flou et representations monotones d'une partie floue, Semin. "Math. Floue," Lyon, France.
2189 Richardson, G. (1975). Information, fuzzy choice, and the manipulation of social decisions, Ph.D. Thesis, Cornell Univ., in *Diss. Abstr. Int.*, **36/10-A**, 6850.
2190 Rickman, S. M., and Kandel, A. (1976). Tabular minimization of fuzzy switching functions, *IEEE Trans. Syst., Man, Cybern.*, **SMC-6**, 761–769.

2191 Rickman, S. M., and Kandel, A. (1977). Column table approach for the minimization of fuzzy functions, *Inf. Sci.*, **12** (2), 111–128.

2192 Rieger, B. (1975). On a tolerance topology model of natural language meaning, Germanic Inst., TH Aachen, Aachen, Germany.

2193 Rieger, B. (1976). Fuzzy structural semantics. On a generative model of vague natural language meaning, 3rd Eur. Meet. Cybern. Syst. Res., Vienna.

2194 Rieger, B. (1976). Theorie der unscharfen mengen und empirische textanalyze, Deutscher Germanistentag 76, Dusseldorf, Apr.

2195 Rieger, B. (1976). Zum der Reprasentation und Analyse Vager Bedeutungen, Working Pap., Inst. fur Math. Empirische Systemforschung, RWTH, Aachen, Germany.

2196 Rieger, B. (1977). Analyzing and representing vague lexical meaning on a generative model of fuzzy-structural semantic, 3rd Int. Conf. on Comput. in the Humanities, Univ. of Waterloo, Ontario, Canada, Aug. 2–5.

2197 Rieger, B. (1977). Coling 76—Concepts, frames, and scripts in aid of semantic networks, knowledge systems and fantasies, in *Sprache und Datenverarbeiten*, Max Niemeyer Verlag, Tubingen.

2198 Rieger, B. (1977). Vagheit als Problem der linguistischen semanik, in *Semantic und Pragmatik*, Akten des 11, Linguist. Kolloq. Aachen 1976, Vol. 2, Max Niemeyer Verlag, Tubingen.

2199 Rieger, B. B. (1979). Fuzzy representation systems in linguistic semantics. An empirical approach to the reconstruction of word meanings in east and west German Newspapertexts, EMCSR.

2200 Rieger, L. (1949). On the lattice theorie of Brouwerian propositional logic, *Acta Fac. Rerum Natur. Univ. Carolinae* (Prague), **189**.

2201 Rieger, L. (1967). *Algebraic Methods of Mathematical Logic*, Academia, Prague and Academic Press, New York.

2202 Riera, T. (1978). How similiarity matrices are, *Stochastics*, **II** (4), 77–80.

2203 Rine, D. (1978). Possibility theory—as a means for modeling computer security and protection, in *Inf. Sci. Program*, *Proc. 8th Int. Symp. on Multiple-Valued Logic*, 78CH1366-4C, pp. 276–286.

2204 Rine, D. (1979). Possibility theory a tool for analyzing computer security, the 9th Int. Symp. on Multi-Valued Logic, UK.

2205 Rinks, D., and Steinberg, E. (1977). Linear orderings over fuzzy preferences in a social welfare setting, in *Proc. 1977 AIDS Conf.*, Chicago.

2206 Rinks, D., and Steinberg, E. (1978). Approximate reasoning and the production scheduling problem, Joint ORSA/TIMS Meet., Los Angeles.

2207 Rinks, D. B. (1980). A heuristic approach to aggregate production planning using linguistic variables, in *Proc. Int. Congr. on Appl. Syst. Res. and Cybern.*, Acapulco, Mexico, Dec.

2208 Roberts, F. S. (1973). Tolerance geometry, *Notre Dame J. Formal Logic*, **14**, 68–76.

2209 Rocha, A. F. (1981). Neural fuzzy point processes, *FSS*, 127–140.

2210 Rocha, A. F., Francozo, E., Hadler, M. I., and Balduino, M. A. (1980). Neural languages, *Fuzzy Sets Syst.*, **3** (1), Jan.

2211 Rock, H. (1977). Interactive fuzzy decision making in a discrete mathematics program framework, Rep. Tech. Univ., Berlin.

2212 Rodabaugh, S. E. (1980). Suitability in fuzzy topological spaces, *J. Math. Anal. Appl.*

2213 Rodder, R., and Zimmermann, H. J. (1977). Analyse Beschreibung und Optimierung von Unscharf Formulierten Problemen, *Z. Oper. Res.*, **21**, 1–18.

2214 Rodder, W. (1975). On "and" and "or" connectives in fuzzy set theory, Euro 1, Lehrstuhl fur unternehmensforschig, RWTH, Aachen, Germany.

2215 Rodder, W., and Zimmermann, H. J. (1977). Duality in fuzzy programming, Int. Symp. on External Methods and Syst. Anal., Univ. of Texas, Austin, TX, Sept.

2216 Roder, W. (1975). Ein Beitrag zur Verknupfung Unscharfer Mengen, Comm. 1st Euro. Congr. on Oper. Res. (Euro 1), Brussels, Jan.

2217 Rodman, R. D. (1973). The study of fuzzy islands within the framework of transformational generative grammar, Ph.D. Thesis, Univ. of California, Los Angeles, in *Diss. Abstr. Int.*, **34/07-A**, 4234.

2218 Rongen, A., Lemke, N., and Veen, I. (1977). An autopilot for ships designed with fuzzy sets, in *Proc. 5th Conf. Int. IFAC/IFIP*, La Haye.

2219 Rosch, E. H. (1973). On the interval structure of perceptual and semantic categories, in *Cognitive Development and the Acquisition of Language* (T. E. Moore, Ed.), New York, pp. 111–144.

2220 Rosch, E. H. (1974). Universals and cultural specifics in human categorization, in *Cross-Cultural Perspectives on Learning* (R. Bristin, S. Bochnor, W. Bonnor, Eds.).

2221 Rose, A. (1950). Completeness of Lukasiewicz-Tarski Propositional Calculus, *Math. Ann.*, **122**, 296–298.

2222 Rose, A. (1951). Axiom systems for 3-valued logic, *J. London Math. Soc.*, **26**, 50–58.

2223 Rose, A. (1951). The degree of completeness of some Lukasiewicz-Tarski propositional calculi, *J. London Math. Soc.*, **26**, 47–49.

2224 Rose, A. (1952). The degree of completeness of the M-valued Lukasiewicz propositional calculus, *J. London Math. Soc.*, **27**, 92–102.

2225 Rose, A. (1953). The degree of completeness of the lamda-zero-valued Lukasiewicz propositional calculus, *J. London Math. Soc.*, **28**, 176–184.

2226 Rose, A. (1958). Many-valued logical machines, *Proc. Cambridge Philos. Soc.*, **54**, 307–321.

2227 Rose, A., and Rosser, J. B. (1958). Fragments of many-valued statement calculi, *Trans. Amer. Math. Soc.*, **87**, 1–53.

2228 Rosen, R. (1974). Planning, management policies and strategies—Four fuzzy concepts, *Int. J. Gen. Syst.*, **1**, 245–252.

2229 Rosenfeld, A. (1971). Fuzzy groups, *J. Math. Anal. Appl.*, **35**, 512–517.

2230 Rosenfeld, A. (1975). Fuzzy graphs, in *Fuzzy Sets and Their Applications to Cognitive Decision Processes* (L. A. Zadeh, K. S. Fu, K. Tanaka, and M. Shimura, Eds.), Academic Press, New York, pp. 77–95.

2231 Rosenfeld, A. (1977). Fuzzy digital topology, TR-573, Dept. of Comput. Sci. Annu. Rep., Univ. of Maryland, College Park.

2232 Rosenfeld, A., Hummel, R. A., and Zucker, S. W. (1976). Scene labeling by relaxation operations, *IEEE Trans. Syst., Man, Cybern.*, **SMC-6**, 420–433.

2233 Rosser, J. B. (1960). Axiomatization of infinite valued logics, *Logique Anal.*, **3**, 137–153.

2234 Rosser, J. B., and Turquette, A. R. (1945). Axiom schemes for M-valued propositional calculi, *J. Symb. Logic*, **10**, 61–82.

2235 Rosser, J. B., and Turquette, A. R. (1952). *Many Valued Logics*, North-Holland, Amsterdam.

2236 Rossowska, M. (1975). O Pewnim Warunku Topologisznym dla Metody, in *Metody Heuresy, Wydawnictwo Ptc, II Klajowe Symp.*, Varsovie.
2237 Roubens, M. (1976). Classification on-hierarchique (clustering) et sous-ensembles flous, Doc. 77.01, Fac, Polytech., Dept. Math. and Tech. Oper. Les, Univ. de Mons, Mons, Belgium, Nov.
2238 Roubens, M. (1977). Pattern classification problems and fuzzy sets, *Fuzzy Sets Syst.*, **1**, 239–253.
2239 Roubens, M. A. (1980). Non fuzzy clustering algorithm and its cluster validity, 4th Eur. Congr. on Oper. Res, July.
2240 Rouget, B. (1975). L'analyse spatiale en economie urbaine. Essai methodologique, 3eme partie—Morphologie imprecise de l'espace urbain et topologie floue, These Doct. Es-Sci. Econ., Univ. de Dijon, Dijon, France, pp. 221–350.
2241 Roy, B. (1975). Partial preference analysis and decision-making—The fuzzy outranking relation concept, Doc. Sema, Groupe Metra., Paris, II, ASA Workshop, Vienna, Oct.
2242 Rubin, H. (1969). A new approach to foundations of probability, in *Foundations of Mathematics, Symposium Papers Commemerating the 60th Birthday of K. Godel* (J. J. Bullof, T. C. Holyoke, and S. W. Haha, Eds.), Springer, New York.
2243 Ruitenbeek, K. (1981). Comment on "Fuzzy sets under various operations," *Bull. Sous Ensembles Flous Appl.*, **5**, *Hiver* (80–81), 8–10.
2244 Ruspini, E. (1969). A new approach to clustering, *Inf. Control*, **15**, 22–32.
2245 Ruspini, E. (1970). Numerical methods for fuzzy clustering, *Inf. Sci.*, **2**, 319–350.
2246 Ruspini, E. H. (1972). Optimization in sample descriptions—Data reduction and pattern recognition using fuzzy clustering, *IEEE Trans. Syst., Man, Cybern.*, **SMC-2**, 541.
2247 Ruspini, E. H. (1973). New experimental results in fuzzy clustering, *Inf. Sci.*, **6**, 273–284.
2248 Ruspini, E. H. (1977). A theory of fuzzy clustering, in *Proc. 1977 IEEE Conf. on Decis. and Control*, New Orleans, pp. 1378–1383.
2249 Ruspini, E. H. (1977). A theory of numerical classification, in *Proc. Symp. on Fuzzy Set Theory and Appl., IEEE Conf. on Decis. and Control*, New Orleans.
2250 Ruspini, E. H. (1980). Recent developments in fuzzy cluster analysis and its applications, in *Proc. Int. Congr. on Appl. Syst. Res. and Cybern.*, Acapulco, Mexico, Dec.
2251 Ruspini, E. H. (1980). Recent developments in fuzzy cluster analysis, Table Ronde du CNRS sur le Flou, Lyon, France, June.
2252 Russell, B. (1923). Vagueness, *Austrian J. Philos.*, **1**, 84–92.
2253 Rutherford, D. A. (1976). The implementation and evaluation of a fuzzy control algorithm for a sinter plant, in *Discrete Systems and Fuzzy Reasoning* (E. H. Mamdani and B. R. Gaines, Eds.), EES-MMS-DSFR-76, Queen Mary College, Univ. of London, London (Workshop Proc.).
2254 Rutherford, D. A., and Bloore, G. C. (1975). The implementation of fuzzy algorithms for control, Control Syst. Centre, Univ. of Manchester Inst. of Sci. and Tech., Manchester, UK.
2255 Rutherford, D. A., and Rao, G. P. (1979). Approximate reconstruction of mapping functions from linguistic descriptions in problems of fuzzy logic applied to system control, Control Syst. Centre, Univ. of Manchester Inst. of Sci. and Tech., Manchester, UK.
2256 Saaty, T. L. (1974). Measuring the fuzziness of sets, *J. Cybernet.*, **4**, 53–61.

2257 Saaty, T. L. (1977). Exploring the interface between hierarchies, multiple objectives and fuzzy sets, *Fuzzy Sets Syst.*, **1**, 57–68.

2258 Sadovskii, V. N. (1974). *Osnovanija Obscei Teorii Sistem*, Nauka, Moskow (in Russian, *On Foundations of General Systems Theories*).

2259 Safteruk, K., Staniewski, P., and Kacprzyk, J. (1980). Control of a stochastic system in a fuzzy environment in a non-Bellman-Zadeh setting, The 10th Meet. of the Eur. Working Group on Fuzzy Sets, June.

2260 Sagaama, S. (1975). A propos des sous-ensembles flous, Doc., Dept. Gestion, Ecole Nat. Sup. Mines, Saint-Etienne, France, June.

2261 Sagaama, S. (1975). Construction d'une topologie floue, Rap., Dept. Gestion, Ecole Nat. Sup. Mines, Saint-Etienne, France, Oct.

2262 Sagaama, S. (1975). Probabilites subjectives, sous-ensembles flous et aide a la decision, Rap., Dept. Gestion, Ecole Nat. Sup. Mines, Saint-Etienne, France, Oct.

2263 Sagaama, S. (1976). Subjective probabilities, fuzzy sets and decision making, 3rd Eur. Meet. Cybern. Syst. Res., Vienna.

2264 Sagaama, S. (1977). Contribution des sous-ensembles flous a l'aide a la decision et a l'analyse structurale, Thesis, Univ. Claude Bernard, Lyon, France, July.

2265 Saito, T. (1975). Chronology analysis of a social conflict, in *Summ. of Pap. on Gen. Fuzzy Problems*, The Working Group on Fuzzy Syst., Tokyo, Nov., pp. 46–48.

2266 Saitta, L. (1978). Fuzzy cluster set for medical diagnosis, in *Proc. Int. Conf. on Cybern. and Soc.*, Vol. 1, Tokyo, Nov., pp. 244–253.

2267 Saitta, L., and Torasso, P. (1978). Fuzzy characterization of coronary disease, in Colloq. Int. Theorie et Appl. des Sous-Ensembles Flous, Journ. de Biomath. et d'Inf. Med., Marseille, France, Sept.

2268 Salomaa, A. (1959). On many-valued systems of logic, *Ajatus*, **22**, 115–119.

2269 Sambuc, R. (1975). Fonction O-flous, Application a l'aide du diagnostic medical en pathologie thyroidienne, Thesis, Univ. de Marseille, Marseille, France.

2270 Sambuc, R., Sanchez, E., Gouvernet, J., and San Marco, J. L. (1980). Representation conjointe du flou et de l'indetermination dans le traitement des donnees medicales. *JIMT*.

2271 Sambuc, R., Sanchez, E., Gouvernet, J., and San Marco, J. L. (1980). Q-fuzzy functions —Application to medical diagnosis, in *Medinfo 80, Proc. 3rd World Conf. on Med. Inf.*, Tokyo, Sept., pp. 784–788.

2272 Samoylenko, S. I. (1977). Application of fuzzy heuristic techniques to computer network design, in *Proc. 5th Int. Joint Conf. Artif. Intell.*, Cambridge, MA, p. 880.

2273 San Marco, J. L., Sanchez, E., Soula, G., Sambuc, R., and Gouvernet, J. (1978). Classification de formes floues, applications au diagnostic medical, Colloq. Int. Theorie et Appl. des Sous-Ensembles Flous, Journ. de Biomath. et d'Inf. Med., Marseille, France, Sept.

2274 Sanchez, E. (1980). Resolution of composite fuzzy relations equations, *Inf. Control*.

2275 Sanchez, E. (1974). Equations de relations floues, Doct. Thesis, Biol. Humaine, Fac. de Med. de Marseille, Marseille, France, July.

2276 Sanchez, E. (1974). Fuzzy relations, Fac. de Med., Univ. de Marseille, Marseille, France.

2277 Sanchez, E. (1976). Resolution of composite fuzzy relation equations, *Inf. Control*, **30**, 38–47.

2278 Sanchez, E. (1977). Eigen fuzzy sets and fuzzy relations, ERL Memo. M77-20, Univ. of California, Berkeley.

2279 Sanchez, E. (1977). Inverses of fuzzy relations and possibility-qualification application to medical diagnosis, in *Proc. Symp. on Fuzzy Set Theory and Appl., IEEE Conf. on Decis. and Control*, New Orleans.

2280 Sanchez, E. (1977). Inverses of fuzzy relations, Application to possibility distributions and medical diagnosis, in *Proc. 1977 IEEE Conf. on Decis. and Control*, New Orleans, pp. 1384–1389.

2281 Sanchez, E. (1977). On possibilistic-qualification in natural languages, Memo. M77/28, Electron. Res. Lab., Univ. of California, Berkeley.

2282 Sanchez, E. (1977). Resolution of eigen fuzzy sets equations, *Fuzzy Sets Syst.*, **1**, 69–74.

2283 Sanchez, E. (1977). Solutions in composite fuzzy relation equations—Application to medical diagnosis in Brouwerian Logic, in *Fuzzy Automata and Decision Processes* (M. M. Gupta, G. N. Saridis, and B. R. Gaines, Eds.), North-Holland, Amsterdam, pp. 221–234.

2284 Sanchez, E. (1978). Fuzzy relations, possibility distributions, and medical diagnosis, Fac. de Med. de Marseille, Marseille, France, in *Proc. 1978 IEEE Conf. on Decis. and Control, includes the 17th Symp. on Adaptive Processes*, 78CH1392-OCS, San Diego, CA, Jan., 1979.

2285 Sanchez, E. (1978). On truth-qualification in natural languages, Lab. de Biomath., Marseille, France, in *Proc. 1978 Int. Conf. on Cybern. and Soc.*, Vol. 2, pp. 1233–1236.

2286 Sanchez, E. (1979). Compositions of fuzzy relations, in *Advances in Fuzzy Set Theory and Applications* (M. M. Gupta, R. K. Ragade, and R. R. Yager, Eds.), North-Holland, Amsterdam.

2287 Sanchez, E. (1979). Fuzzy sets with application to medical diagnosis, in *Proc. 18th IEEE Conf. on Decis. and Control*, Fort Lauderdale, FL, Dec.

2288 Sanchez, E. (1979). Medical diagnosis and composite fuzzy relations, in *Advances in Fuzzy Set Theory and Applications* (M. M. Gupta, R. K. Ragade, and R. R. Yager, Eds.), North-Holland, Amsterdam.

2289 Sanchez, E. (1980). Fuzzy logics with application to medical diagnosis, in *Proc. JACC*, San Francisco, Aug.

2290 Sanchez, E. (1980). Mesures de possibilite, qualifications de verites et classification de formes linguistics en medicine. Table Ronde du CNRS sur le Flou, Lyon, France, June.

2291 Sanchez, E. (1980). Quelques applications des ensembles flous aux sciences humaines, Table Ronde du CNRS sur le Flou, Lyon, France, June.

2292 Sanchez, E., Gouvernet, J., Bartolin, R., and Vovan, L. (1980). Linguistic approach in fuzzy logic of the WHO classification of dyslipoproteinemias, in *Proc. Int. Congr. on Appl. Syst. Res. and Cybern.*, Acapulco, Mexico, Dec.

2293 Sanchez, E., Gouvernet, J., and Joly, H. (1979). Les distributions de possibilites en classification de formes diagnostiques—Approache linguistique, *JIMT*.

2294 Sanchez, E., and Sambuc, R. (1976). Relations floues. Fonctions O-floues, Application a l'aide audiagnostic en patheologie thyroidienne, *IRIA Med. Data Process. Symp.*, Taylor and Francis, Toulouse, France.

2295 Sanchez, S. C. (1979). Contribucio a l'estudi d'entropies en la teoria des conjunts difusos, Mem. Llicenciatures en Math., Dept. Math., Univ. Politech. de Barcelona, Barcelona, Spain.

2296 Sandor, G., and Diday, E. (1974). Resultats recents concernant la methode des nuees dynamiques et applications a la recherche de profils biologiques, in *Int. Comput. Symp.* (S. Gunther, Ed.), North-Holland, Amsterdam, pp. 388–398.

Key References in Fuzzy Pattern Recognition 319

2297 Sanford, D. H. (1975). Borderline logic, *Amer. Philos. Quart.*, **12**, 29–39.
2298 Sanford, D. H. (1975). Infinity and vagueness, *Philos. Rev.*, **84**, 520–535.
2299 Sanford, D. H. (1980). Abstract of notes on logics of vagueness and some applications, Fuzzy Sets Theory and Appl. to Policy Anal. and Inf. Syst. (P. P. Wang and S. K. Chang, Eds.) Plenum, New York, pp. 123–126.
2300 Santos, E. S. (1968). Maximin automata, *Inf. Control*, **13**, 363–377.
2301 Santos, E. S. (1968). Maximin, minimax and composite sequential machines, *J. Math. Anal. Appl.*, **24**, 246–259.
2302 Santos, E. S. (1969). Maximin sequential chains, *J. Math. Anal. Appl.*, **26**, 28–38.
2303 Santos, E. S. (1969). Maximin sequential-like machines and chains, *Math. Syst. Theory*, **3**, 300–309.
2304 Santos, E. S. (1970). Fuzzy algorithms, *Inf. Control*, **17**, 326–339.
2305 Santos, E. S. (1972). Max-product machines, *J. Math. Anal. Appl.*, **37**, 677–686.
2306 Santos, E. S. (1972). On reductions of maximin machines, *J. Math. Anal. Appl.*, **40**, 60–78.
2307 Santos, E. S. (1973). Fuzzy sequential functions, *J. Cybern.*, **3**, 15–31.
2308 Santos, E. S. (1974). Context-free fuzzy languages, *Inf. Control*, **26**, 1–11.
2309 Santos, E. S. (1975). Fuzzy programs, Spec. Interest Disc. Sess. on Fuzzy Autom. and Decis. Processes, 6th IFAC World Congr., Boston, MA, Aug.
2310 Santos, E. S. (1975). Max-product grammars and languages, *Inf. Sci.*, **9**, 1–23.
2311 Santos, E. S. (1975). Realization of fuzzy languages by probabilistic max-product and maximin automata, *Inf. Sci.*, **8**, 39–53.
2312 Santos, E. S. (1976). Fuzzy automata and languages, *Inf. Sci.*, **10**, 193–197.
2313 Santos, E. S. (1977). Fuzzy and probabilistic programs, in *Fuzzy Automata and Decision Processes* (M. M. Gupta, G. N. Saridis, and B. R. Gaines, Eds.), North-Holland, Amsterdam, pp. 133–148.
2314 Santos, E. S. (1977). Regular fuzzy expressions, in *Fuzzy Automata and Decision Processes* (M. M. Gupta, G. N. Saridis, and B. R. Gaines, Eds.), North-Holland, Amsterdam, pp. 169–176.
2315 Santos, E. S., and Wee, W. G. (1968). General formulation of sequential machines, *Inf. Control*, **12**, 5–10.
2316 Saridis, G. N. (1974). Fuzzy notions in nonlinear system classification, *J. Cybern.*, **4**, 67–82.
2317 Saridis, G. N. (1975). Fuzzy decision making in prosthetic devices and other applications, Spec. Interest Disc. Sess. on Fuzzy Autom. and Decis. Processes, 6th IFAC World Congr., Boston, MA., Aug.
2318 Saridis, G. N., and Stephanou, H. E. (1977). Fuzzy decision-making in prosthetic devices, in *Fuzzy Automata and Decision Processes* (M. M. Gupta, G. N. Saridis, and B. R. Gaines, Eds.), North-Holland, Amsterdam, pp. 387–402.
2319 Saridis, G. N., and Stephanou, H. E. (1977). Fuzzy Decision-making of a prosthetic arm, *IEEE Trans. Syst., Man, Cybern.*, **SMC-7** (6), 407–420.
2320 Sasama, H. (1975). Fuzzy set model for train composition in marshalling yard, in *Summ. of Pap. on Gen. Fuzzy Problems*, The Working Group on Fuzzy Syst., Tokyo, Nov., pp. 49–54.
2321 Sasama, H. (1977). A learning model to distinguish the sex of a human name, in *Summ. of Pap. on Gen. Fuzzy Problems*, The Working Group on Fuzzy Syst., Tokyo, pp. 19–24.

2322 Sato, Y., Yamanoi, M., Miyakoshi, M., and Kawaguchi, M. (1980). An algebraic system with a fuzzy binary operation, in *Summ. of Pap. on Gen. Fuzzy Problems*, The Working Group on Fuzzy Syst., Tokyo, Dec.

2323 Savage, L. J. (1971). Elicitation of personal probabilities and expectations, *J. Amer. Stat. Assoc.*, **66**, 783–801.

2324 Scarpellini, B. (1962). Die Nicht-Axiomatisierbarkeit des Unendlichwertigen Praedikatenkalkuls von Lukasiewicz, *J. Symb. Logic*, **27**, 159–170.

2325 Schefe, P. (1979). On foundations of reasoning with uncertain facts and vague concepts, Bericht No. 56, Fachber. Inf., Univ. Hamburg, Hamburg, Germany.

2326 Schefe, P. (1980). Remarks on the concept of a possibility distribution, Table Ronde du CNRS sur le Flou, Lyon, France, June.

2327 Schek, H. J. (1977). Tolerating fuzziness in keywords by similiarity searches, *Kybernetes*, **6**, 175–184.

2328 Schock, R. (1964). On denumerably many-valued logics, *Logique Anal.*, **28**, 190–195.

2329 Schock, R. (1964). On finitely many-valued logics, *Logique Anal.*, **28**, 43–58.

2330 Schock, R. (1965). Some theorems on the relative strengths of many-valued logics, *Logique Anal.*, **30**, 101–104.

2331 Schotch, P. K. (1975). Fuzzy modal logic, in *Proc. 1975 Int. Symp. Multiple-Valued Logic*, IEEE 75CH0959-7C, May, pp. 176–182.

2332 Schuh, E. (1973). Many-valued logics and the Lewis paradoxes, *Notre Dame J. Formal Logic*, **14**, 250–252.

2333 Schutzenberger, M. P. (1962). On a theorem of R. Jungen, *Proc. Amer. Math. Soc.*, **13**, 885–890.

2334 Schwaar, M. (1977). Zweistufige Modellbilung zur Verscharfen Verhaltenbeschreibung von Linearen und Nichtlinearen Ubertragungsgliedern, in *Unscharfe Modellbildung und Steuerung*, TH Karl-Marx-Stadt, June, pp. 99–106.

2335 Schwartz, D. G. (1981). Forecasting in a fuzzy environment, in *Gen. Syst. Res. and Des. —Precursors and Futures, Proc. 25th Annu. Meet. of the Soc. for Gen. Syst. Res.*, Jan.

2336 Schwartz, D. G., Zwick, M., and Ledaris, G. G. (1978). Fuzziness and Catastrophe, *IEEE Proc. Int. Conf. on Cybern. and Soc.*, Tokyo/Kyoto, Japan, Nov.

2337 Schwartz, J. T. (1975). Optimization of very high level language, I. Value transmission and its corollaries, *Comput. Lang.*, **I**, 161–194.

2338 Schwartz, J. T. (1975). Automatic data structure choice in a language of very high level, *Comm. ACM*, **18**, 722–728.

2339 Schwartz, J. T. (1975). Optimization of very high level languages, II. Deducing relationship of inclusion and membership, *Comput. Lang.*, **I**, 197–218.

2340 Schwarz, D. (1972). Mengenlehre uber Vorgegebenen Algebraischen Systemen, *Math. Nachr.*, **53**, 365–370.

2341 Schwede, G. (1976). *N*-variable fuzzy maps with application to disjunctive decomposition of fuzzy switching functions, in *Proc. 6th Int. Symp. on Multiple-Valued Logic*, IEEE 76CH1111-4C, May, pp. 203–216.

2342 Schwede, G. W., and Kandel, A. (1977). Fuzzy maps, *IEEE Trans. Syst., Man, Cybern.*, **SMC-7**, 619–674.

2343 Schwyhla, W. (1979). Conditions for a fuzzy probability measure to be an integral, in *Proc. Int. Semin. on Fuzzy Sets Theory*, Johannes Kepler Univ., Linz, Austria, Sept.

2344 Scott, D. (1974). Completeness and axiomatizability in many-valued logic, in *Proc. Tarski Symp.* (L. Henkin, Ed.), Amer. Math. Soc., RI, pp. 412–435.

Key References in Fuzzy Pattern Recognition

2345 Scott, D. (1976). Does many-valued logic have any use?, in *Philosophy of Logic* (S. Korner, Ed.), Basil Blackwell, Oxford, UK, pp. 64–88.

2346 Scott, D., and Krauss, P. (1966). Assigning probabilities to logical formulas, in *Aspects of Inductive Logic* (J. Hintikka and P. Suppes, Eds.), pp. 219–264.

2347 Scott, L. L. (1980). Necessary and sufficient conditions for the values of a function of fuzzy variables to lie in a specified subinterval of (0, 1), Fuzzy Sets Theory and Appl. to Policy Anal. and Inf. Syst. (P. P. Wang and S. K. Chang, Eds.), Plenum, New York, pp. 35–48.

2348 Segerberg, K. (1967). Some modal logics based on a three-valued logic, *Theoria*, **33**, 53–71.

2349 Seif, A. (1978). Classification by an artificial skin using fuzzy correlation, Colloq. Int. Theorie et Appl. des Sous-Ensembles Flous, Journ. de Biomath. et d'Inf. Med., Marseille, France, Sept.

2350 Seif, A., and Aguilar-Martin, J. (1977). Classification by the artificial hand using correlated fuzzy integration, LAAS, Toulouse, France; 1st Symp. on Biophys., Le Caire, Oct.

2351 Seif, A., and Aguilar-Martin, J. (1980). Multi-group classification using fuzzy correlation, *Fuzzy Sets Syst.*, **3** (2), Mar.

2352 Seif. A. M. (1978). Classification d'objets a l'aide de la correlation floue, Application en robotique, Thesis, Univ. de Toulouse, Toulouse, France.

2353 Sekita, Y. (1975). A consideration of the fuzzy evaluation of complex social systems, *Mem. Econ. Osaka Univ.*, **25**, 312–325.

2354 Sekita, Y. (1976). A consideration of identifying fuzzy measures, *Mem. Econ. Osaka Univ.*, **25**, 133–138.

2355 Sekita, Y. (1976). A note on the identification of fuzzy measures, *Mem. Econ. Osaka Univ.*, **25** (4).

2356 Sekita, Y., and Tabaka, Y. (1979). A health status index model using a fuzzy approach, *Eur. J. Oper. Res.*, **3**, 40–49.

2357 Sembi, B. S., and Mamdani, E. H. (1979). On the nature of implication in fuzzy logic, Dept. of Elec. and Electron. Eng., Queen Mary College, Univ. of London, London; in *Proc. Int. Symp. Multi-Valued Logic*, Beth, May.

2358 Sembi, B. S., and Mamdani, E. H. (1980). Linguistic rule-based decision making using fuzzy logic, COMPSAC'80, Univ. of London, London.

2359 Serfati, M. (1974). *Algebres de Boole avec une Introtuction a la Theorie des Graphes Orientes et aux Sous-Ensembles Flous*, Edition CDU, Paris.

2360 Serika, Y., and Jabata, Y. (1979). A health status model using a fuzzy approach, *Eur. J. Oper. Res.*, **3**, 40–49.

2361 Seriwaza, M. (1973). A search technique of control rod pattern of smoothing care power distributions by fuzzy automaton, *J. Nucl. Sci. Tech.*, **10**.

2362 Shackle, G. L. S. (1949). *Expectation in Economics*, Cambridge Univ. Press, Cambridge, UK.

2363 Shackle, G. L. S. (1955). *Uncertainty in Economics and Other Reflections*, Cambridge Univ. Press, Cambridge, UK.

2364 Shackle, G. L. S. (1961). *Decision Order and Time in Human Affairs*, Cambridge Univ. Press, Cambridge, UK (2nd ed., 1969).

2365 Shafer, G. (1976). *A Mathematical Theory of Evidence*, Princeton Univ. Press, Princeton, NJ.

2366 Shaket, E. (1975). Fuzzy semantics for a natural-like language defined over a world of blocks, M.Sc. Thesis, Comput. Sci. Dept., Univ. of California, Los Angeles.

2367 Shaket, E. (1977). Fuzzy set semantics for a natural-like language, in *Proc. Symp. on Fuzzy Set Theory and Appl., IEEE Conf. on Decis. and Control*, New Orleans.

2368 Shapiro, D. I., and Torgov, J. I. (1978). Fuzzy integral games, in *Proc. 4th Int. Conf. on Cybern. and Syst.*, Amsterdam, pp. 379-380.

2369 Shaw-Kwei, M. (1954). Logical paradoxes for many-valued systems, *J. Symb. Logic*, **19**, 37-39.

2370 Shaw, M. L. G., and Gaines, B. R. (1980). Fuzzy entailment analysis, in *Proc. Int. Congr. on Appl. Syst. Res. and Cybern.*, Acapulco, Mexico, Dec.

2371 Shaw, M. L. G., and Gaines, B. R. (1980). Fuzzy semantics for personal construing, in *Proc. 24th Annu. Meet. of Soc. for Gen. Syst. Res.*, San Francisco, Jan.

2372 Shen, R. (1980). Fuzzy linear equations, Collect. Pap. on Fuzzy Math. (Abstr.), Huazhong Inst. of Tech., Huazhong, China.

2373 Shen, R. (1980). On the discrete expression of fuzzy number and its operations, News on 1980 Annu. Rep. Meet. of Beijing Working Group on Fuzzy Sets, Beijing, China.

2374 Shen, X. (1980). A philosophical approach in fuzzy mathematics, News on 1980 Annu. Rep. Meet. of Beijing Working Group on Fuzzy Sets, Beijing, China.

2375 Sheppard, D. (1954). The adequacy of everyday quantitative expressions as measurements of qualities, *Brit. J. Psychol.*, **45**, 40-50.

2376 Shi-Kuo-Chang (1972). On the execution of fuzzy algorithms using finite state machine, *IEEE Trans. Comput.*, **CF121** (3), Mar.

2377 Shibata, H. (1976). A comment on fuzzy set, in *Summ. of Pap. on Gen. Fuzzy Problems*, The Working Group on Fuzzy Syst., Tokyo, pp. 101-106.

2378 Shibata, H., and Tsutsumi, T. (1977). On logical treatment and evaluation of regulatory statements, in *Summ. of Pap. on Gen. Fuzzy Problems*, The Working Group on Fuzzy Syst., Tokyo, pp. 172-180.

2379 Shimura, M. (1972). Application of fuzzy functions to pattern classification, *Trans. IECE* (Japan), **55-D**, 218-225.

2380 Shimura, M. (1973). Fuzzy sets concept in rank ordering objects, *J. Math. Anal. Appl.*, **43**, 717-733.

2381 Shimura, M. (1975). An approach to pattern recognition and associative memories using fuzzy logic, in *Fuzzy Sets and their Applications to Cognitive and Decision Processes* (L. A. Zadeh, K. S. Fu, K. Tanaka, and M. Shimura, Eds.) Academic Press, New York, pp. 449-476.

2382 Shimura, M. (1975). Applications of fuzzy sets theory to pattern recognition, *J. Jaace*, **19**, 243-248.

2383 Shirai, T. (1937). On the pseudo-set, *Mem. College Sci. Kyoto Imp. Univ.*, **20A**, 153-156.

2384 Shortliffe, E. H. (1976). *Computer-Based Medical Consultation—Mycin*, Elsevier, New York.

2385 Shortliffe, E. H., and Buchanan, B. G. (1975). A model of inexact reasoning in medicine, *Math. Biosci.*, **23**, 351-379.

2386 Shuford, E. H., Albert, A., and Massengill, H. E. (1966). Admissible probability measurement procedures, *Psychometrika*, **31**, 125-145.

2387 Shuford, E. H., and Brown, T. A. (1975). Elicitation of personal probabilities and their assessment, *Instr. Sci.*, **4**, 137-188.

Key References in Fuzzy Pattern Recognition

2388 Si-Li, P. (1980). Mapping properties of induced fuzzy topological spaces, Rep. prepared by the Group of Topology in Sichuan Univ., Sichuan, China, (in Chinese).

2389 Si-Li, P. (1980). Zero-dimensional fuzzy topological spaces, Report prepared by the Group of Topology in Sichuan Univ., Sichuan, China, (in Chinese).

2390 Sikorski, R. (1964). *Boolean Algebras*, Springer-Verlag, Berlin.

2391 Silvert, W. (1979). Symmetric summation, A class of operations on fuzzy sets, *IEEE Trans. Syst., Man, Cybern.*, 657–659.

2392 Simon, H. A. (1967). The logic of heuristic decision making, in *The Logic of Decision and Action* (N. Rescher, Ed.), Univ. of Pittsburgh Press, Pittsburgh, PA, pp. 1–35.

2393 Simons H. W. (1976). *Persuasion*, Addison-Wesley, Reading, MA.

2394 Sinaceur, H. (1978). Logique et mathematique du flou, *Bull. Crit., Rev. Gen. Publ. Fr. Etrang.* (372), May.

2395 Singer, D. (1980). Etude des demi-groupes totalement ordonnes, en vue d'une application a la theorie des ensembles flous, Mem. de DEA, Univ. de Lyon, Lyon, France.

2396 Sinha, N. K., and Wright, J. D. (1977). Application of fuzzy control to a heat exchanger system, in *Proc. 1977 IEEE Conf. on Decis. and Control*, New Orleans.

2397 Sira-Ramierz, H. (1980). Fuzzy state estimation in linear dynamic systems, in *19th IEEE Conf. on Decis. and Control*, Albuquerque, NM, Dec, pp. 380–382.

2398 Siskos, J. (1980). A way to deal with fuzzy preferences in multicriteria decision making, 4th Eur. Congr. on Oper. Res., July.

2399 Siy, P. (1972). Fuzzy logic and handwritten character recognition, Ph.D. Diss., Dept. of Elec. Eng., Univ. of Akron, Ohio, *Diss. Abstr. Int.*, **34/04-B**, 1525.

2400 Siy, P., and Chen, C. S. (1971). Fuzzy logic approach to handwritten character recognition problem, in *Proc. IEEE Conf. Syst., Man, Cybern.*, Anaheim, CA, pp. 113–117.

2401 Siy, P., and Chen, C. S. (1971). Some properties of fuzzy logic, *Inf. Control*, **19**, 417–431.

2402 Siy, P., and Chen, C. S. (1972). Minimization of fuzzy functions, *IEEE Trans. Comput.*, **C-21**, 100–102.

2403 Siy, P., and Chen, C. S. (1974). Fuzzy logic for handwritten numerical character recognition, *IEEE Trans. Syst., Man, Cybern.*, **SMC-4**, 570–575.

2404 Skala, H. J. (1974). *On the Problem of Imprecision*, D. Reidel, Dordrecht, Holland.

2405 Skala, H. J. (1975). *Non-Archimedean Utility Theory*, D. Reidel, Dordrecht, Holland.

2406 Skala, H. J. (1976). Fuzzy concepts—Logic, motivation, application, in *Systems Theory in the Social Sciences* (H. Bossel, S. Klacko, and N. Muller, Eds.), Birkhauser Verlag, Basel, Switzerland, pp. 292–306.

2407 Skala, H. J. (1976). Not necessarily additive realizations of comparative probability relations.

2408 Skala, H. J. (1977). On many-valued logics, fuzzy sets, fuzzy logics and their applications, *Fuzzy Sets Syst.*, **1**, 129–149.

2409 Skalicka, V. (1935). *Zur Ungarischen Grammatik*, Prague.

2410 Skolem, T. (1957). Bemerkungen zum Komprehensionsaxiom, *Z. Math. Logik Grundlagen Math.*, **3**, 1–17.

2411 Skolem, T. (1960). A set theory based on a certain 3-valued logic, *Math. Scand.*, **8**, 127–136.

2412 Skolem, T. (1962). *Abstract Set Theory*, Notre Dame Press, Notre Dame, Ind.

2413 Skyrms, B. (1970). Return of the liar-three-valued logic and the concept of truth, *Amer. Philos. Quart.*, **7**, 153-161.

2414 Slack, J. M. V. (1976). A fuzzy set-theoretic approach to semantic memory—A resolution to the set-theoretic versus network model controversy, in *Discrete Systems and Fuzzy Reasoning* (E. H. Mamdani and B. R. Gaines, Eds.), EES-MMS-DSFR-76, Queen Mary College, Univ. of London, London (Workshop Proc.).

2415 Slack, J. M. V. (1976). Possible applications of the theory of fuzzy sets to the study of semantic memory, in *Discrete Systems and Fuzzy Reasoning* (E. H. Mamdani and B. R. Gaines, Eds.), EES-MMS-DSFR-76, Queen Mary College, Univ. of London, London (Workshop Proc.).

2416 Slowinski, R. (1980). Goal, parametric, interactive and fuzzy linear programming in certain resource allocation problems, 4th Eur. Congr. on Oper. Res, July.

2417 Slupecki, J. (1958). Towards a generalized mereology of Lesniewski, *Stud. Logica*, 131-154.

2418 Smets, P. (1978). Un modele mathematico-statistique simulant le processus du diagnostic medical, Mem. d'Agregation, Fac. de Med., Univ. Libre de Bruxelles, Presses Univ. de Bruxelles, Brussels.

2419 Smets, P. (1978). Medical diagnosis, fuzzy sets and degree of belief, Colloq. Int. Theorie et Appl. des Sous-Ensembles Flous, Journ. de Biomath. et d'Inf. Med., Marseille, France, Sept.

2420 Smets, P. (1978). Probability of fuzzy events—Axiomatic approach, Colloq. Int. Theorie et Appl. des Sous-Ensembles Flous, Journ. de Biomath. et Inf. Med., Marseille, France, Sept.

2421 Smets, P. (1980). Elementary operations on fuzzy sets, in *Proc. Int. Congr. on Appl. Syst. Res. and Cybern.*, Acapulco, Mexico, Dec.

2422 Smets, P. (1980). The degree of belief in a fuzzy event, Lab. Med. Stat., Brussels Free Univ., Brussels.

2423 Smets, P., Vainsel, H., Bernard, R., and Kornreich, F. (1977). Bayesian probability of fuzzy diagnosis, in *Proc. Medinfo 77* (S. Wolf, Ed.), North-Holland, Amsterdam, pp. 121-122.

2424 Smidth, F. L. (1979). Kiln control... Using fuzzy logic, *FLS-News Front*.

2425 Smith, C. A. B. (1961). Consistency in statistical inference and decision, *J. Roy. Stat. Soc.* (Ser. B), **23**, 1-37.

2426 Smith, C. A. B. (1965). Personal probability and statistical analysis, *J. Roy. Stat. Soc.* (Ser. A), **128**, 469-499.

2427 Smith, R. E. (1970). Measure theory on fuzzy sets, Ph.D. Thesis, Univ. of Saskatchewan, Saskatoon, Canada.

2428 Smithson, M. J. (1978). Measurement and probability models for fuzzy nominal data.

2429 Smullyan, R. M. (1957). Languages in which self-reference is possible, *J. Symb. Logic*, **22**, 55-67.

2430 Snyder, D. P. (1971). *Modal Logic*, Van Nostrand Reinhold, New York.

2431 Sober, E. (1975). *Simplicity*, Clarendon Press, Oxford, UK.

2432 Sobolewski, M. (1975). Zbiory rozmyte w zastosowaniu do semantyki systemov klasyfikacyjnych, in *Metody Heurezy*, Wydawnicto PTC, II Klajowe Symp., Varsovie.

2433 Sobolewski, M. (1976). Classification system semantics in terms of fuzzy sets, 3rd Eur. Meet. Cybern. Syst. Res., Vienna.

Key References in Fuzzy Pattern Recognition

2434 Sols, I. (1975). Aportaciones a la teoria de topos, al algebra universal y a las mathematics fuzzy, Ph.D. Thesis, Zaragoza, Spain.

2435 Sols, I. (1975). Fuzzy universal algebra and applications, Dept. of Geom., Fac. of Sci., Zaragoza, Spain.

2436 Sols, I. (1975). Topology in complete lattices and continuous fuzzy relations, Fac. of Sci., Zaragoza, Spain.

2437 Sols, I. (1975). Unmarco unificato para la teoria de automatas, Dept. of Geom., Fac. of Sci., Zaragoza, Spain.

2438 Sommer, G. (1976). A fuzzy programming approach to an air pollution regulation problem, 3rd Meet. Cybern. Syst. Res., Vienna.

2439 Sommer, G. (1977). An algorithm for choosing the optimal crisp solution out of optimal fuzzy sets, Lehrstuhl fur OR No. 77/06, Aachen, Germany.

2440 Sommer, G. (1978). On fuzzy information retrieval, in *Proc. 4th Int. Conf. Cybern. and Syst.*, Amsterdam, pp. 380–382.

2441 Sommer, G. (1980). Fuzzy Bayes-decision making. *EMCSR—PCSR*, **8**, to be published.

2442 Sommer, G. (1980). Fuzzy inventory scheduling, in *Proc. Int. Congr. on Appl. Syst. Res. and Cybern.*, Acapulco, Mexico.

2443 Song, D. (1980). Convex fuzzy sets and convex fuzzy mapping, Rep. Shanghai Railway Inst. (SRI), A List of Lit. in China, The Univ. of Sci. and Tech. of China, Hofei, Anhwei, China.

2444 Soula, G., Gouvernet, J., Barre, A., and San Marco, J. L. (1980). Applications of fuzzy relations to medical decision-making, *Medinfo 80, Proc. of 3rd World Conf. on Med. Inf.*, Tokyo, Sept., pp. 844–848.

2445 Soula, G., and San Marco, J. L. (1980). Application des relations floues de perception a l'aide a la decision, *JIMT*.

2446 Spillman, B., Bezdek, J., and Spillman, R. (1977). Development of an instrument for the dynamic measurement of consensus, Rep. Utah State Univ., Logan.

2447 Spillman, B., Bezdek, J., and Spillman, R. (1979). Coalition analysis with fuzzy sets, *Kybernetes*, **8** (3), 203–211.

2448 Spillman, B., Bezdek, J., and Spillman, R. (1979). A study of coalition formation in decision making groups—An application of fuzzy mathematics, *Kybernetes*.

2449 Spillman, B., Spillman, R., and Bezdek, J. (1977). New methodologies for communication research—Application of fuzzy mathematics, Western Speech Comm. Assoc. Conv., Phoenix, AZ.

2450 Spillman, B., Spillman, R., and Bezdek, J. (1979). A fuzzy analysis of consensus in small groups, 1st Int. Symp. on Policy Anal. and Inf. Syst., June.

2451 Spillman, B., Spillman, R., and Bezdek, J. (1982). Dynamic measurement of subgroup dominance and coalition formation via fuzzy preference relations, *Behav. Sci.*

2452 Srini, V. P. (1975). Realization of fuzzy forms, *IEEE Trans. Comput.*, Sept., 941–943.

2453 Stallings, W. (1977). Fuzzy set theory versus Bayesian statistics, *IEEE Trans. Syst., Man, Cybern.*, **SMC-7**, 216–219.

2454 Stalnaker, R. (1970). Probability and conditionals, *Philos. Sci.*, **37**, 64–80.

2455 Stalnaker, R. C., and Thomason, R. H. (1970). A semantic analysis of conditional logic, *Theoria*, **36**, 23–42.

2456 Staniewski, P., and Kacprzyk, J. (1980). Long-term inventory policy-making through fuzzy decision-making models, The 10th Meet. of the EWG on Fuzzy Sets, June.

2457 State, L. (1971). Quelques proprietes des algebres de Morgan, in *Logique, Automatique, Informatique* (G. C. Moisil, Ed.), Bucharest, pp. 195-207.

2458 Stefenescu, A. C. (1975). Category set $F(L)$, Semin. de Teoria Sist., Dept. Econ. Cybern., Acad. of Econ. Stud., Bucharest.

2459 Stein, E. W. (1980). Optimal stopping in a fuzzy environment, *Fuzzy Sets Syst.*, **3** (3), 253-260.

2460 Stein, W. E. (1979). Probabilistic fuzzy decision making, ORSA-TIMS, Mil., Wisconsin.

2461 Stein, W. E. (1981). Joint possibility distributions and decision making, CORS-TIMS-ORSA Joint Nat. Meet., Toronto, Canada, May.

2462 Steinacker, I. (1978). Interpretation of word association statistic in natural language text by means of fuzzy sets theory, Colloq. Int. Theorie et Appl. des Sous-Ensembles Flous, Journ. de Biomath. et d'Inf. Med., Marseille, France, Sept.

2463 Steinacker, I., and Mulhauser, G. (1970). Die Induktive Methode in der Automatischen Verarbeitung Formatierter und Naturlicher Sprache, der Aufbaueines Titel-Wort-Thessaurus fur die Nasa-Datei, Inf. Retr. Syst. and Manage. Inf. Syst., Stuttgart, Germany, Dec.

2464 Steinacker, I. (1973). Aspect of computer text processing, *Data Process.*

2465 Steinacker, I. (1973). Text processing and automatic indexing, A new approach, 1st Eur. Congr. on Doc. Syst. and Networks, Luxembourg, May.

2466 Steinberg, E. (1981). Approximate reasoning through fuzzy sets and production scheduling, CORS-TIMS-ORSA Joint Nat. Meet., Toronto, Canada, May.

2467 Steinberg, E., and Rinks, D. (1977). Linear ordering over fuzzy preferences in a social welfare setting, in *Proc. 9th Decis. Sci. Conf.*, p. 604.

2468 Steinberg, E., and Rinks, D. (1979). An application of the Blin-Whinston algorithm for resolving fuzzy group-preferences. *Manage. Sci.*, **25**.

2469 Steinlage, R. C., and Gantner, E. E. (1980). Quantification of linguistic judgements, in Proc. Int. Congr. on Appl. Syst. Res. and Cybern., Acapulco, Mexico, Dec.

2470 Steinmuller, K. (1979). Eine Fuzzy-Interpretation der Okologischen Nischentheorie, Material zur Vorlesung "Kennwertermittlung," Teil II—*Unscharfe Modellbildung und Steuerung*, TH Karl-Marx-Stadt, Oct.

2471 Stephanou, H. E., and Saridis, G. N. (1976). A hierarchically intelligent method for the control of complex systems, *J. Cybern.*, **6**, 249-261.

2472 Stephanou, H. E., and Saridis, G. N. (1976). Hierarchical control in a fuzzy environment, in *Proc. IEEE Conf. Syst., Man, Cybern.*, Washington, D.C.

2473 Stickel, M. E. (1978). Fuzzy four-valued logic for inconsistency and uncertainty, Dept. of Comput. Sci., Univ. of Arizona, Tucson, pp. 91-94.

2474 Stoica, M., and Scarlat, E. (1975). Fuzzy algorithms in economic systems, Centre of Econ. Comput. and Econ. Cybern., Bucharest, *Econ Comput. Econ. Cybern. Stud. Res.*, **3**, 239-247.

2475 Stoica, M., and Scarlat, E. (1975). Fuzzy concepts in the control of production systems, in *Proc. 3rd Int. Congr. Cybern. Syst.*, Bucharest, Aug.

2476 Stoica, M., and Scarlat, E. (1977). Some fuzzy concepts in the management of production systems, *Mod. Trends Cybern. Syst.*, **2**, 175-181.

2477 Stoica, M., Slancu-Minasian, I., and Scarlat, E. (1977). On large scale classification problems using fuzzy sets, *Econ. Comput. Cybern. Stud. Res.*, **1**, 93-100.

2478 Stone, M. H. (1938). Topological representations of distributive lattices and Brouwerian logics, *Cas. Pestovani Mat. Fys.*, **67**, 1-25.

2479 Stove, D. C. (1973). *Probability and Hume's Inductive Scepticism*, Clarendon Press, Oxford, UK.

2480 Straube, B. (1977). Ein Versuch zur Bewertung von Psychologischen Beanspruchungsztanden unter Verwendung der Theorie Unscharfer Mengen, in *Unscharfe Modellbildung und Steuerung*, Lecture Notes, TH Karl-Marx-Stadt, Germany, June, pp. 30–34.

2481 Straube, B. (1979). Unscharfe Clusterbildung und Klassifkation, Material zur Volesung "Kennwertermittlung und Modellbildung," Teil II—*Unscharfe Modellbildung und Steuerung*, Lecture Notes, TH Karl-Marx-Stadt, Oct.

2482 Straube, B. (1979). Zur Venwendung der Unscharfen Prozessrelation als Regler, Material zur Volesung "Kennwertermittlung und Modellbildung," Teil II—*Unscharfe Modellbildung und Steuerung*, Lecture Notes, TH Karl-Marx-Stadt, Germany, Oct.

2483 Su, Z. (1980). The fuzzy model in disease-diagnosis (BIAN-Zheng) of the Chinese traditional medicine, *J. Huazhong Inst. Tech.*, **2**, A List of Lit. in China, The Univ. of Sci. and Tech. of China, Hofei, Anhwei, China.

2484 Sugeno, M. (1971). On fuzzy nondeterministic problems, Annu. Conf. Rec. of SICE.

2485 Sugeno, M. (1972). Evaluation of similarity of patterns by fuzzy integrals, Annu. Conf. Rec. of SICE.

2486 Sugeno, M. (1973). Constructing fuzzy measure and grading similarity of patterns by fuzzy integrals, *Trans. SICE*, **9**, 359–367.

2487 Sugeno, M. (1974). Subjective evaluation of fuzzy objects, in *Proc. IFAC Symp. on Stochastic Control*, Budapest.

2488 Sugeno, M. (1974). Theory of fuzzy integrals and its applications, Ph.D. Thesis, Tokyo Inst. of Tech., Tokyo.

2489 Sugeno, M. (1975). Fuzzy decision-making problems, *Trans. SICE*, **11**, 709–714.

2490 Sugeno, M. (1975). Fuzzy measures and fuzzy integrals, in *Summ. of Pap. on Gen. Fuzzy Problems*, The Working Group on Fuzzy Syst., Tokyo, Nov., pp. 55–60.

2491 Sugeno, M. (1975). Inverse operation of fuzzy integrals and conditional fuzzy measures, *Trans. SICE*, **11**, 32–37.

2492 Sugeno, M. (1975). Theoretical developments of fuzzy sets, *J. JAACE*, **19**, 229–234.

2493 Sugeno, M. (1977). Fuzzy measures and fuzzy integrals—A survey, in *Fuzzy Automata and Decision Processes* (M. M. Gupta, G. N. Saridis, and B. R. Gaines, Eds.), North-Holland, Amsterdam, pp. 89–102.

2494 Sugeno, M. (1977). Fuzzy systems with underlying deterministic systems, in *Summ. of Pap. on Gen. Fuzzy Problems*, The Working Group on Fuzzy Syst., Tokyo, pp. 25–52.

2495 Sugeno, M. (1979). Application of fuzzy sets and logic to control—A survey, *J. SICE*, *18* (2).

2496 Sugeno, M. (1980). On structured set of systems, in *Summ. Pap. on Gen. Fuzzy Problems*, The Working Group on Fuzzy Syst., Tokyo, Dec.

2497 Sugeno, M., and Imaoka, H. (1978). Generalized truth value in truth qualification, Working Group on Fuzzy Syst., Rep. No. 4, Tokyo, Dec, pp. 16–25.

2498 Sugeno, M., and Takagi, T. (1980). Application of fuzzy reasoning to systems modeling, in *Proc. Int. Congr. on Appl. Syst. Res. and Cybern.*, Acapulco, Mexico, Dec.

2499 Sugeno, M., and Terano, T. (1973). An approach to the identification of human characteristics by applying fuzzy integrals, in *Proc. 3rd IFAC Symp. on Identification and Syst. Parameter Estimation*, Hague.

2500 Sugeno, M., and Terano, T. (1975). Conditional fuzzy measures and their applications,

in *Fuzzy Sets and Applications to Cognitive and Decision Processes*, Academic Press, New York, pp. 151–170.
2501 Sugeno, M., and Terano, T. (1977). A model of learning based on fuzzy information, *Kybernetes*, **6**, 157–166.
2502 Sugeno, M., and Terano, T. (1977). Analytical representation of fuzzy systems, in *Fuzzy Automata and Decision Processes* (M. M. Gupta, G. N. Saridis, and B. R. Gaines, Eds.), North-Holland, Amsterdam, pp. 177–190.
2503 Sugeno, M., Tsukamoto, Y., and Terano, T. (1974). Subjective evaluation of fuzzy objects, IFAC Symp. on Stochastic Control.
2504 Sugeno, S. (1976). Fuzzy systems and pattern recognition, Workshop on Discrete Syst. and Fuzzy Reasoning, Queen Mary College, Univ. of London, London.
2505 Sularia, M. (1977). On fuzzy programming in planning, *Kybernetes*, **6**, 230.
2506 Sundstrom, D. E. (1975). Regulatory control and modeling of responsible behavior in autonomous systems, Ph.D. Thesis, Southern Methodist Univ., Dallas, TX.
2507 Suzuki, T. (1976). Trip as triples manipulating computer language, in *Summ. of Pap. on Gen. Fuzzy Problems*, The Working Group on Fuzzy Syst., Tokyo, pp. 48–63.
2508 Swinburne, R. (1973). *An Introduction to Confirmation Theory*, Methuen, London.
2509 Tachi, S., Tanie, K, Komoriya, K., and Abe, M. (1980). Information transmission by electrocutaneous phantom sensation, in *Summ. of Pap. on Gen. Fuzzy Problems*, The Working Group on Fuzzy Syst., Tokyo, Dec.
2510 Tahani, V. (1971). Fuzzy sets in information retrieval, Ph.D. Thesis, Dept. of Elec. Eng. and Comput. Sci., Univ. of California, Berkeley.
2511 Tahani, V. (1977). A conceptual framework for fuzzy query processing—A step towards very intelligent data systems, *Inf. Process. Manage.*, **13**, 289–303.
2512 Tahani, V. A. (1976). A fuzzy model of document retrieval systems, *Inf. Process. Manage.*, **12**, 177–187.
2513 Tai, H. (1980). An approach to analyse the harm of low temperature in florescence of intercross rice, News on 1980 Annu. Rep. Meet. of Beijing Working Group on Fuzzy Sets, Beijing, China.
2514 Takahara, Y. (1977). Some topological considerations for general systems theory, in *Summ. of Pap. on Gen. Fuzzy Problems*, The Working Group on Fuzzy Syst., Tokyo, pp. 53–69.
2515 Takahashi, W. (1978). Minimax theorems for fuzzy sets, Tokyo Inst. of Tech., Tokyo.
2516 Takeda, E., and Nishida, T. (1976). An application of fuzzy graphs to the problem concerning group structure, *J. Oper. Res. Soc.* (Japan) **19** (3).
2517 Takeda, E., and Nishida, T. (1980). Multiple criteria decision problems with fuzzy domination structures, *Fuzzy Sets Syst.*, **3** (2), Mar.
2518 Takeda, E., and Nishida, T. (1980). Multiple criteria decision problems with fuzzy domination structures, *Fuzzy Sets Syst.*
2519 Takeuti, G., and Zaring, W. M. (1973). *Axiomatic Set Theory*, Springer-Verlag, Berlin.
2520 Tamura, S. (1971). Fuzzy pattern classification, in *Proc. Symp. on Fuzziness in Syst. and its Process.*, Prof. Group of Syst. Eng. of SICE.
2521 Tamura, S., Higuchi, S., and Tanaka, K. (1971). Pattern classification based on fuzzy relations, *IEEE Trans., Syst., Man, Cybern.*, **SMC-1**, 61–66.
2522 Tamura, S., and Tanaka, K. (1973). Learning of fuzzy formal language, *IEEE Trans. Syst., Man, Cybern.*, **SMC-3**, 98–102.

2523 Tanaka, H., and Asai, K. (1980). Fuzzy linear programming based on fuzzy function, *Bull. Univ. Osaka Prefect.*, Ser. A, **29** (2).

2524 Tanaka, H., and Kaneku, S. (1974). On a fuzzy decoding procedure for cyclic codes, *Trans. IECE* (Japan), **57-A**, 505–510.

2525 Tanaka, H., Okuda, T., and Asai, K. (1974). Decision-making and its goal in a fuzzy environment, US-Jap. Semin. on Fuzzy Sets and their Appl., Berkeley, CA, July.

2526 Tanaka, H., Okuda, T., and Asai, K. (1974). On fuzzy mathematical programming, *J. Cybern.*, **3**, 37–46.

2527 Tanaka, H., Okuda, T., and Asai, K. (1976). A formulation of fuzzy decision problems and its application to an investment problem. *Kybernetics*, **5**, 25–30.

2528 Tanaka, H., Okuda, T., and Asai, K. (1977). On decision-making in fuzzy environment, fuzzy information and decision making, *Int. J. Prod. Res.*, **15**, 623–635.

2529 Tanaka, H., Okuda, T., and Asai, K. (1979). Fuzzy information and decision in statistical model, in *Advances in Fuzzy Set Theory and Applications* (M. M. Gupta, R. K. Ragade, and R. R. Yager, Eds.), North-Holland, Amsterdam.

2530 Tanaka, H., and Sommer, G. (1977). On posterior probabilities concerning a fuzzy information, Lehrstuhl fur Oper. Res. No. 77/02, Aachen, Germany.

2531 Tanaka, H., Uejima, S., and Asai, K. (1980). Fuzzy linear regression model, in *Proc. Int. Congr. on Appl. Syst. Res. and Cybern.*, Acapulco, Mexico, Dec.

2532 Tanaka, K. (1972). Analogy and fuzzy logic, *Math. Sci.*

2533 Tanaka, K. (1975). Fuzzy sets theory and its application, *J. JAACE*, **19**, 227–228.

2534 Tanaka, K. (1976). Fuzzy concept and intellectual information processing, 1976 Joint Conv. of Four Inst. of Elec. Eng. of Japan.

2535 Tanaka, K. (1976). Learning in fuzzy machines, in *Computer Oriented Learning Processes* (J. C. Simon, Ed.), Noordhoff, pp. 109–148.

2536 Tanaka, K. (1980). Dealing with credibility or uncertainty in conjunction with knowledge engineering, in *Proc. Int. Congr. on Appl. Syst. Res. and Cybern.*, Acapulco, Mexico, Dec.

2537 Tanaka, K., and Asai, K. (1973). Fuzzy mathematical programming, *Trans. SICE*, **9**, 109–115.

2538 Tanaka, K., and Mizumoto, M. (1975). Fuzzy programs and their execution, in *Fuzzy Sets and their Applications to Cognitive and Decision Processes* (L. A. Zadeh, K. S. Fu, K. Tanaka, and M. Shimura, Eds.), Academic Press, New York, pp. 41–76.

2539 Tanaka, K., Okuda, T., and Asai, K. (1972). On the fuzzy mathematical programming, Ann. Conf. Rec. of SICE.

2540 Tanaka, K., Toyoda, J., Mizumoto, M., and Tsuji, H. (1970). Fuzzy automata theory and its application to automatic controls, *J. JAACE*, **14**, 541–550.

2541 Tanie, K., Tachi, S., and Abe, M. (1977). Study on the subjective magnitude in electrocutaneous stimulation, in *Summary Pap. on Gen. Fuzzy Problems*, The Working Group on Fuzzy Syst., Tokyo, pp. 79–86.

2542 Tangwen, G. S., and King-Sun-Fu (1977). An application of learning to robotic planning, TREE 77-7, School of Elec. Eng., Purdue Univ., Lafayette, IN.

2543 Tarany, C. (1976). Fuzzy aspects in cost theory, 3rd Eur. Meet. Cybern, Syst. Res., Vienna.

2544 Taranu, C. (1977). The economic efficiency—A fuzzy concept, in *Modern Trends in Cybernetics and Systems*, Vol. 2 (J. Rose and C. Bilciu, Eds.), Springer-Verlag, Berlin and New York, pp. 163–173.

2545 Tarski, A. (1956). *Logic, Semantics, Metamathematics*, Clarendon Press, Oxford, UK.

2546 Tashiro, T. (1977). Method of solution to inverse problem of fuzzy correspondence model, in *Summ. of Pap. on Gen. Fuzzy Problems*, The Working Group on Fuzzy Syst., Tokyo, pp. 70–78.

2547 Tashiro, T., Terano, T., and Tsukamoto, Y. (1978). Inverse of fuzzy coorespondence and evolutionary diagnosis, Tokyo Inst. of Tech., *Proc. 1978 Int. Conf. on Cybern. and Soc.*, Tokyo, Japan, pp. 938–941.

2548 Tasnadi, A. (1978). Fuzzy systems and learning models, in *Proc. 4th Int. Conf. Cybern. and Syst.*, Amsterdam, pp. 382–383.

2549 Tazaki, E. (1975). Heuristic synthesis in a class of systems by using fuzzy automata, in *Summ. of Pap. on Gen. Fuzzy Problems*, The Working Group on Fuzzy Syst., Tokyo, pp. 61–66.

2550 Tazaki, E., and Amagasa, M. (1979). Heuristic structure synthesis in a class of systems by using fuzzy automata, *IEEE Trans. Syst., Man, Cybern.*, Vol. 9, No. 2, pp. 73–79.

2551 Tazaki, E., and Amagasa, M. (1977). Structural modelling in a class of systems using fuzzy set theory, in *Proc. 1977 IEEE Conf. on Decis. and Control*, New Orleans.

2552 Tazaki, E., and Amagasa, M. (1978). Fuzzy algorithm and its applications, in *Proc. 22th JAACE Conf.*

2553 Tazaki, E., Amagasa, M., and Takizawa, M. (1977). Fuzzy structural modeling, *Trans. Oper. Res. Soc. Jap.*

2554 Tazaki, E., Amagasa, M., and Takizawa, M. (1977). Structural modeling in a class of systems by fuzzy sets theory, *J. Oper. Res. Soc. Jap.*, **20**, 285–310.

2555 Tchorayan, O. G. (1979). Identifying elements of the probabilistic neuronal ensemble from the standoint of fuzzy sets theory, *J. Physiol.*, **55** (in Russian).

2556 Terano, T. (1971). Fuzziness and its concept, in *Proc. Symp. on Fuzziness in Syst. and its Proc.*, Prof. Group of Syst. Eng. of SICE.

2557 Terano, T. (1972). *Fuzziness of Systems*, Nikka-Giren Eng., pp. 21–25.

2558 Terano, T. (1975). Methodology of fuzzy systems, *J. IECE*, **58**, 875–876.

2559 Terano, T. (1976). Application of fuzzy concept to controls, 1976 Joint Conv. of Four Inst. of Elec. Eng.

2560 Terano, T. (1976). Structural model for complex social problems, in *Summ. of Pap. on Gen. Fuzzy Problems*, The Working Group on Fuzzy Syst., Tokyo, pp. 70–78.

2561 Terano, T. et al. (1978). Diagnosis of engine trouble by fuzzy logic, *IFAC 7th World Congr.*, Helsinki, pp. 1621–1628.

2562 Terano, T., Kurosu, K., Murayama, Y., and Inasaka, F. (1977). Principal component analysis of engine trouble, in *Summ. of Pap. on Gen. Fuzzy Problems*, The Working Group on Fuzzy Syst., Tokyo, pp. 103–108.

2563 Terano, T., and Sugeno, M. (1975). Conditional fuzzy measures and their application, in *Fuzzy Sets and their Applications to Cognitive and Decision Processes* (L. A. Zadeh, K. S. Fu, K. Tanaka, and M. Shimura, Eds.), Academic Press, New York, pp. 151–170.

2564 Terano, T., and Sugeno, M. (1977). Macroscopic optimization using conditional fuzzy measures, in *Fuzzy Automata and Decision Processes* (M. M. Gupta, G. N. Saridis, and B. R. Gaines, Eds.), North-Holland, Amsterdam.

2565 Terano, T., and Tsukamoto, Y. (1977). Failure diagnosis by using fuzzy logic, in *Proc. on Fuzzy Set Theory and Appl., IEEE Conf. on Decis. and Control*, New Orleans.

2566 Terasaka, H. (1937). Theorie der Topologischen Verbande, *Proc. Imp. Acad.* (Tokyo), **13**.

2567 Tharp, L. (1975). Which logic is the right logic?, *Synthese*, **31**, 1–21.
2568 Thole, U., Zimmermann, H. J., and Zysno, P. (1978). The connective "and" in fuzzy decision making, An empirical study, Unpublished.
2569 Thole, U., Zimmermann, H. J., and Zysno, P. (1979). On the suitability of minimum and product operators for the intersection of fuzzy sets, *Fuzzy Sets Syst.*, **2**, No. 2, 167–180.
2570 Thole, U., Zimmermann, H. J., and Zysno, P. (1982). An empirical study on the connective "and," *Fuzzy Sets Syst.*
2571 Thole, U., Zimmermann, H. J., and Zysno, P. (1982). The "compensatory and," A new connective in fuzzy set theory, *Fuzzy Sets Syst.*
2572 Thomason, M. G. (1973). New theoretical results in fuzzy automata and their application to error control in regular languages, Ph.D. Thesis, Duke Univ., Durham, NC, in *Diss. Abstr. Int.*, **34/07-B**, 3249.
2573 Thomason, M. G. (1974). Fuzzy syntax-directed translations, *J. Cybern.*, **4**, 87–94.
2574 Thomason, M. G. (1974). The effect of logic operations on fuzzy logic distributions, *IEEE Trans. Syst., Man, Cybern.*, **SMC-4**, 309–310.
2575 Thomason, M. G. (1975). Finite fuzzy automata, Regular languages and pattern recognition, *Pattern Recogn.*, **5**, 383–390.
2576 Thomason, M. G. (1977). Convergence of powers of a fuzzy matrix, *J. Math. Anal. Appl.*, **57**, 476–480.
2577 Thomason, M. G., and Marinos, P. N. (1972). Fuzzy logic relations and their utility in role theory, in *Proc. 1972 IEEE Int. Conf. on Cybern. and Soc.*, Washington, D.C.
2578 Thomason, M. G., and Marinos, P. N. (1974). Deterministic acceptors of regular fuzzy languages, *IEEE Trans. Syst., Man, Cybern.*, **SMC-4**, 228–230.
2579 Tichy, P. (1969). Intension in terms of turing machines, *Stud. Logica*, **24**, 7–25.
2580 Tichy, P. (1974). On Popper's definition of verisimilitude, *Brit. J. Philos. Sci.*, **25**, 155–160.
2581 Tichy, P. (1976). Verisimilitude redefined, *Brit. J. Philos. Sci.*, **27**, 25–42.
2582 Tinant, B. (1976). Probleme de Taxonomie et theorie des sous-ensembles flous, Fac. Polytech. de Mons, Mem. de Fin d'Etudes, Mons, Belgium.
2583 Tong, R. M. (1976). An assessment of a fuzzy control algorithm for a nonlinear multivariable plant, in *Discrete Systems and Fuzzy Reasoning* (E. H, Mamdani and B. R. Gaines, Eds.), EES-MMS-DSFR-76, Queen Mary College, Univ. of London, London (Workshop Proc.).
2584 Tong, R. M. (1976). Analysis of fuzzy control algorithms using the relation matrix, *Int. J. Man-Mach. Stud.*, **8**.
2585 Tong, R. M. (1976). Some problems with the design and implementation of fuzzy controllers, CUED/F-CAMS/TR127(1976), Control Eng. Dept., Cambridge Univ., Cambridge, UK.
2586 Tong, R. M. (1977). A control engineering review of fuzzy systems, *Automatica*, **13**, 559–569.
2587 Tong, R. M. (1977). Analysis and control of fuzzy systems using finite discrete relations, *Int. J. Control*, **27**, 431–440.
2588 Tong, R. M. (1978). An analysis of fuzzy models and a discussion of their limitations, Control Management Syst. Div., Eng. Dept., Cambridge Univ., Cambridge, UK.
2589 Tong, R. M. (1978). Synthesis of fuzzy models for industrial processes—Some recent results, *Int. J. Gen. Syst.*, **4**, 143–162.

2590 Tong, R. M. (1979). Analysis of closed loop fuzzy systems, in *Proc. 18th IEEE Conf. Decis. and Control*, Fort Lauderdale, FL, Dec.

2591 Tong, R. M. (1979). The construction and evaluation of fuzzy models, in *Advances in Fuzzy Set Theory and Applications* (M. M. Gupta, R. K. Ragade, and R. R. Yager, Eds.), North-Holland, Amsterdam.

2592 Tong, R. M. (1980). Some properties of fuzzy feedback systems, *IEEE Trans. Syst., Man, Cybern.*, **SMC-10** (6), June.

2593 Tong, R. M. (1980). The evaluation of fuzzy models derived from experimental data, *Fuzzy Sets Syst.*, **4** (1), July, 1–12.

2594 Tong, R. M., Beck, M. B., and Latten, A. (1982). Fuzzy control of the activated sludge waste-water treatment process, Prof. Pap. Int. Inst. for Appl. Syst. Anal., Laxenburg, Austria.

2595 Tong, R. M., and Bonissone, P. P. (1979). Linguistic decision analysis using fuzzy sets, ERL Res. Rep. UCB/ERL-M79/72, Univ. of California, Berkeley, Nov.

2596 Tong, R. M., and Efstathiou, J. (1980). Rule-based decomposition of fuzzy relational models, JACE, Univ. of California, Berkeley.

2597 Tong, Z. (1980). A new Definition of Fuzzy Numbers, Rep. Sichuan Railroad Inst., A List of Lit. in China, The Univ. of Sci. and Tech. of China, Hofei, Anhwei, China.

2598 Tong, Z. (1980). A problem of Mizumoto and Tanaka, Rep. Sichuan Railroad Inst., A List of Lit. in China, The Univ. of Sci. and Tech. of China, Hofei, Anhwei, China.

2599 Tong, Z. (1980). Fuzzy number system and finite fuzzy probability model, Rep. Sichuan Railroad Inst., A List of Lit. in China, The Univ. of Sci. and Tech. of China, Hofei, Anhwei, China.

2600 Tran Qui, P. (1977). Regionalisation de l'economie et taxinomie numerique floue, Application au cas de la France, Sc. Thesis, Dept. de Econ. et de Gestion, Univ. de Dijon, Dijon, France, Dec.

2601 Tran Qui, P. (1978). La regionalisation de l'economie Francaise par une methode de taxinomie numerique floue, Colloq, Int., Theorie et Appl. des Sous-Ensembles Flous, Journ. de Biomath. et d'Inf. Med., Marseille, France, Sept.

2602 Tran Qui, P. (1978). Les regions economiques floues. Application au cas de la France. *Colloq. IME*, **16**.

2603 Tribus, M. (1979). Comments on "Fuzzy sets, fuzzy algebra and fuzzy statistics," Reply by A. Kandel, *Proc. IEEE*, **67** (8).

2604 Tribus, M. (1980). Fuzzy sets and Bayesian methods applied to the problem of literature search, *IEEE Trans. Syst., Man, Cybern.*, **SMC-10** (8), Aug.

2605 Trillas, E. (1978). Fonction caracteristique generalisee et sous-ensembles flous, Doc. Dept. Math., Escuela Tech. Sup. de Arquit., Univ. Politech., Barcelona, Spain, Aug.

2606 Trillas, E. (1979). Funciones de negation en F.S.T., Doc. Escuela Tech. Sup. de Arquit., Univ. Politech., Barcelona, Spain.

2607 Trillas, E. (1979). Sobre functiones de negacion in la teoria de conjutos difusos, *Stochastica* (Barcelona), **III** (1), 47–59.

2608 Trillas, E. (1979). Sobre una matodologica para pruebas de sigma-medibilidad?, *Prob. Est., Ens. de la Matematica y Analysis*, Univ. de Granada, Granada, Spain, pp. 67–74.

2609 Trillas, E. (1980). Do we need $\max(X, Y)$, $\min(X, Y)$ and $1 - X$ in FST?, in *Proc. Int. Congr. on Appl. Syst. Res. and Cybern.*, Acapulco, Mexico, Dec.

2610 Trillas, E., and Alsina, C. (1979). Sur les mesures de degre de flou, *Stochastica* (Barcelona), **III** (1), 81–84.

Key References in Fuzzy Pattern Recognition

2611 Trillas, E., and Batle, N. (1979). Entropy and fuzzy integral, Dept. of Math., Tech. Univ. of Barcelona, Barcelona, Spain.

2612 Trillas, E., and Riera, T. (1977). Note for the study of a concept of multi-valued entropy for a finite fuzzy set, 1st World Conf. Math. Service of Man., Barcelona, Spain, July.

2613 Trillas, E., and Riera, T. (1978). Entropies in finite fuzzy sets, *Inf. Sci.*, **15**, 159–168.

2614 Trillas, E., and Riera, T. (1978). Sobre algunas entropias los subjunctos difusos de N, Jornadas Mat. Luso-Hispanas, Aveiro, Portugal, Mar.

2615 Trillas, E., and Riera, T. (1980). On a special kind of variables in fuzzy environments, *ISMVL*, 149–152.

2616 Trillas, E., and Riera, T. (1981). Towards a representation of "synonyms" and "antonyms" by fuzzy sets, *Bull. les Sous Ensembles Flous Appl.*, **5**, Hiver (80–81).

2617 Trnkova, V. (1979). L-fuzzy functional automata, in *Lecture Notes in Computer Sciences*, Vol. 74.

2618 Tsichritzis, D. (1969). Fuzzy computability, in *Proc. Princeton Conf. Inf. Sci. and Syst.*, Princeton, NJ, pp. 157–162.

2619 Tsichritzis, D. (1969). Fuzzy properties and almost solvable problems, Tech. Rep. 70, Comput. Sci. Lab., Dept. of Elec. Eng., Princeton Univ., Princeton, NJ.

2620 Tsichritzis, D. (1969). Measures on countable sets, Tech. Rep. 8, Dept. of Comput. Sci., Univ. of Toronto, Toronto, Canada.

2621 Tsichritzis, D. (1971). Approximation and complexity of functions on the integers, *Inf. Sci.*, 70–86.

2622 Tsichritzis, D. (1971). Participation measures, *Math. Anal. Appl.*, **36**, 60–72.

2623 Tsichritzis, D. (1973). A model for iterative computation, *Inf. Sci.*, **5**, 187–197.

2624 Tsichritzis, D. (1973). Approximate logic, in *Proc. Symp. Multivalued Logic*, May.

2625 Tskumoto, Y., and Terano, T. (1977). Failure diagnosis by using fuzzy logic, in *Proc. IEEE Conf. Decis. Control*, Vol. 2, New Orleans, pp. 1390–1395.

2626 Tsuji, H., Mizumoto, M., Toyoda, J., and Tanaka, K. (1972). Interaction between random environments and fuzzy automata with variable structures, *Trans. IECE* (Japan), **55-D**, 143–144.

2627 Tsuji, H., Mizumoto, M., Toyoda, J., and Tanaka, K. (1973). Linear fuzzy automaton, *Trans. IECE* (Japan), **56-A**, 256–257.

2628 Tsukamoto, T., and Tashiro, T. (1979). Method of solution to fuzzy inverse problem, *Trans. SICE*, **15** (1).

2629 Tsukamoto, T., and Terano, T. (1977). Fault diagnosis by using fuzzy logic, in *Summ. of Pap. on Gen. Fuzzy Problems*, The Working Group on Fuzzy Syst., Tokyo, pp. 109–117.

2630 Tsukamoto, Y. (1972). Identification of preference measure by means of fuzzy integral, Preprint on JORS (in Japanese).

2631 Tsukamoto, Y. (1975). A subjective evaluation on attractivity of sightseeing zones, in *Summ. of Pap. on Gen. Fuzzy Problems*, The Working Group on Fuzzy Syst., Tokyo, Nov., pp. 73–76.

2632 Tsukamoto, Y. (1979). An approach to fuzzy reasoning method, in *Advances in Fuzzy Set Theory and Applications* (M. M. Gupta, R. K. Ragade, and R. R. Yager, Eds.), North-Holland, Amsterdam.

2633 Tsukamoto, Y. (1979). Fuzzy logic on Lukasiewicz logic and its applications to diagnosis and control, Doc. Thesis, Tokyo Inst. of Tech., Tokyo.

2634 Tsukamoto, Y., and Gupta, M. M. (1980). Characterization of some fuzzy concepts, in *Proc. Int. Congr. on Appl Syst. Res. and Cybern.*, Acapulco, Mexico, Dec.

2635 Tsukamoto, Y., and Iida, H. (1973). Evaluation models of fuzzy systems, *Annu. Conf. Rec. of SICE.*

2636 Tsukamoto, Y., Takagi, T., and Sugeno, M. (1978). Fuzzification of L aleph-1 and its application to control, in *Proc. of the 1978 Int. Conf. on Cybern. and Soc.*, Tokyo Inst. of Tech., Tokyo, Vol. 2, pp. 1217–1221.

2637 Tsukamoto, Y., and Terano, T. (1977). Failure diagnosis by using fuzzy logic, in *Proc. 1977 Conf. on Decis. and Control*, New Orleans.

2638 Tsukamoto, Y., Terano, T., Kurusu, K., and Murayama, Y. (1978). Diagnosis of engine trouble by fuzzy logic, Rep. No. 4, Working Group on Fuzzy Syst., Tokyo, Dec., pp. 16–25.

2639 Tsumura, Y., and Onisawa, T. (1980). A structure model of imprecise concepts using fuzzy theory, in *Summ. of Pap. on Gen. Fuzzy Problems, The Working Group on Fuzzy Syst., Tokyo, Dec.*

2640 Tsutsumi, T. (1976). Engineering code processing by computer, *Summ. of Pap. on Gen. Fuzzy Problems*, The Working Group on Fuzzy Syst., *Tokyo*, pp. 64–69.

2641 Turksen, I. B. (1978). Measurement of linguistis variables in medical diagnosis systems, Working Pap. 1978/02, Dept. Ind. Eng., Univ. of Toronto, Toronto, Dec.

2642 Turksen, I. B., and Martin, J. K. (1976). Decision-information systems, A conceptual framework, 76-010, Dept. of Ind. Eng., Univ. of Toronto, Toronto, Canada.

2643 Turquette, A. R. (1954). Many-valued logics and systems of strict implication, *Philos. Rev.*, **63**, 365–379.

2644 Turquette, A. R. (1963). Independent axioms for infinite-valued logic, *J. Symb. Logic*, **28**, 217–221.

2645 Twareque, A. S., and Prugovecki, E. (1977). Systems of Imprim. and representation of quantum mechanics on fuzzy phase spaces, *J. Math Phys.*, **18** (2).

2646 Ueckert, H. (1978). Nicht-Deterministisches, Nicht-Probabilistisches Denken und Erkennen—Possibilistische Kognitive Progresse, in *Tagungsband zur Arbeitstagung, Kognitive Psychologie*, Vol. 2, Hamburg, Germany.

2647 Uhr, L. (1975). Toward integrated cognitive systems which must make fuzzy decisions about fuzzy problems, in *Fuzzy Sets and their Application to Cognitive and Decision Processes* (L. A. Zadeh, K. S. Fu, K. Tanaka, and M. Shimura, Eds.), Academic Press, New York, pp. 353–393.

2648 Umano, M. (1976). Implementation of a fuzzy set-theoretic data structure system, M.S. Thesis, Fac. of Eng. Sci., Osaka Univ., Osaka, Japan (in Japanese).

2649 Umano, M. (1979). Representation and manipulation of fuzzy data, Doc. Thesis, Graduate School of Eng. Sci., Osaka Univ., Osaka, Japan, Feb.

2650 Umano, M., Mizumoto, M., and Tanaka, K. (1976). A system for fuzzy reasoning, *IJCAI.*

2651 Umano, M. Mizumoto, M., and Tanaka, K. (1976). Implementation of fuzzy sets manipulation system, Tech. Rep. on Automaton and Language of IECE, AL76-26.

2652 Umano, M., Mizumoto, M., and Tanaka, K. (1977). Application of alpha expressions to fuzzy relations, AL77-54, Tech. Rep. of IECE of Japan, Vol. 77, No. 203, on Automata and Languages, pp. 17–26 (in Japanese).

2653 Umano, M., Mizumoto, M., and Tanaka, K. (1978). FSTDS—A fuzzy set manipulation system, *Inf. Sci.*, **14**, 115–159.

2654 Umano, M., Mizumoto, M., and Tanaka, K. (1978). Implementation of approximate reasoning system using FSTDS system, Colloq. Int. Theorie et Appl. des Sous-Ensembles Flous, Journ. de Biomath. et d'Inf. Med., Marseille, France, Sept.

2655 Umano, M., Mizumoto, M., and Tanaka, K. (1978). Toward fuzzy database systems, Nat. Conv. Rec. of IECE of Japan, No. 1168 (in Japanese).

2656 Umano, M., Mizumoto, M., Toyoda, J., and Tanaka, K. (1975). On interpretation and execution of fuzzy programs, AL75-27, Tech. Rep. on Automaton and Language of IECE.

2657 Uno, K, Itakura, H., Sannomiya, N., Nishikawa, Y. (1976). Learning controls that use a fuzzy controller, *Syst. Control*, **20**, 262–268.

2658 Uragami, M., Mizumoto, M., and Tanaka, K. (1976). Fuzzy robot controls, *J. Cybern.*, **6**, 39–64.

2659 Uragami, M., Mizumoto, M., Toyoda, J., and Tanaka, K. (1975). Robot control by fuzzy programs, AL75-51, Tech. Rep. on Automaton and Language of IECE.

2660 Urquhart, A. (1973). An interpretation of many-valued logic, *Z. Math. Logik Grundlagen Math.*, **19**, 111–114.

2661 Vachet, J. (1964). On some basic principles of "Classical" Phonology, *Z. Ponetik, Sprachwiss. U. Kommunikationsforsch.* (Berlin), **17**, 409–431.

2662 Vachet, J. (1964). Prague phonological studies today, *Trav. Linguist. Prague*, **1**, 7–20.

2663 Vachet, J. (1966). On the integration of the peripheral elements into the system of language, *Trav. Linguist. Prague*, **2**, 23–37.

2664 Vachet, J. (1966). *The Linguistic School of Prague*, Indiana Univ. Press, Bloomington.

2665 Vagin, V. N., Pospelov, D. A., and Papke, W. (1977). Application of fuzzy logic in control systems, *Found. Control Eng.*, **2**, 153–160.

2666 Vaina, L. (1977). *Fuzzy Sets in the Semiotic of Text*, M.I.T., Cambridge, MA.

2667 Vainer, L. (1980). Neighborliness—A measure of similarity, in *Proc. Int. Congr. on Appl. Syst. Res. and Cybern.*, Acapulco, December, Mexico.

2668 Vallee, R. (1980). Sur la localisation de l'etat d'un systeme differential dans le cas de conditions initiales probabilistes ou floues, Table Ronde du CNRS sur le Flou, Lyon, France, June.

2669 Vallet, C., Leguyader, H., and Moulin, T. (1980). Fuzziness in arithmetical models of natural systems, in *Proc. Int. Congr. on Appl. Syst. Res. and Cybern.*, Acapulco, Mexico, Dec.

2670 Van Amerongen, J., Van Nauta Lemke, H. R., and Vander Veen, J. C. T. (1977). An autopilot for ships designed with fuzzy sets, in *Conf. 5th IFAC/IFIP Digital Comput. Appl. Process Control*, Lahaye.

2671 Van Fraassen, B. C. (1968). Presuppositions, supervaluations and self-reference, *J. Philos.*, **65**, 136–152.

2672 Van Fraassen, B. C. (1973). Lakeoff's fuzzy propositional logic, Unpublished.

2673 Van Fraassen, B. C. (1974). Hidden variables in conditional logic, *Theoria*, **40**, 176–190.

2674 Van Fraassen, B. C. (1975). Comments—Lakeoff's fuzzy propositional logic, *Contemporary Research in Philosophical Logic and Linguistic Semantics*, (D. Hockney, W. Harper, and B. Freed, Eds.), D. Reidel, Dordrecht, Holland.

2675 Van Gigch, J. P., Murray, T. J., and Pipino, L. L. (1979). Fuzzy sets and the Delphi method, ORSA-TIMS Nat. Meet., New Orleans; School of Business Adm., Univ. of Missouri, St. Louis.

2676 Van Gigch, J. P., and Pipino, L. L. (1980). From absolute to probable and fuzzy in decision making, *Kybernetes*, **9**(1).

2677 Van Heijencort, J. (ed.) (1967). *From Frege to Godel—A Source Book in Mathematical Logic 1879–1937*, Harvard Univ. Press, Cambridge, MA.

2678 Van Nauta-Lemke, H. R. (1974). Control with fuzziness, Lecture Notes, Dept. of Elec. Eng. Lab. of Autom. Control, Delft Univ. of Tech., Delft, Holland.

2679 Van Nauta-Lemke, H. R., and Kickert, W. J. M. (1976). The application of fuzzy sets theory to control a warm water, Delft Univ. of Tech., Delft, Holland, *J. A.*, **17**(1).

2680 Van Velthoven, G. D. (1974). Onderzoek naar toepasbaarheid van de theorie der vage verzamelingen op het parametrisch onderzoek inzake criminaliteit, Dec.

2681 Van Velthoven, G. D. (1975). Application of fuzzy sets theory to criminal investigation, in *Proc. 1st Eur. Congr. on Oper. Res.*, Brussels, Jan.

2682 Van Velthoven, G. D. (1975). Fuzzy models in personnel management, in *Proc. 3rd Int. Congr. of Cybern. and Syst.*, Bucharest, Aug., p. 15.

2683 Van Velthoven, G. D. (1975). Quelques applications de la taxonomie floue, Semin. sur la Contrib. des Syst. Flous a l'Autom.—Processus Humain et Ind., Centre d'Autom., Univ. des Sci. et Tech. de Lille, Lille, France, June.

2684 Vanderheydt, L., Oosterlinck, A., Van Daele, J., Van den Berghe, H. (1980). Design of a graph-representation and a fuzzy-classifier for human chromosomes. *Pattern Recogn.*, **12**(3).

2685 Vaneeden, D. (1976). Fuzzy random variables in decision problems, Rep., TH Twente.

2686 Varela, F. J. (1975). A calculus for self-reference, *Int. J. Gen. Syst.*, **2**, 5–24.

2687 Varela, F. J. (1976). The arithmetic of closure, 3rd Eur. Meet. Cybern., Syst. Res., Vienna, Apr.

2688 Varela, F. J. (1976). The extended calculus of indications interpreted as a three-valued logic, *Notre Dame J. Formal Logic*, **17**.

2689 Vayer, A. (1974). Ebauche d'une etude theorique des faits administratifs, Doc. Centres d'entrainement aux methodes d'education actives, Dec.

2690 Verma, R. R. (1970). Vagueness and the principle of the excluded middle, *Mind*, **79**, 66–77.

2691 Vickers, J. M. (1965). Some remarks on coherence and subjective probability, *Philos. Sci.*, **32**, 32–38.

2692 Villegas, C. (1964). On quantitative probability sigma-algebras, *Ann. Math. Stat.*, **35**, 1787–1796.

2693 Vincke, P. (1973). La theorie des ensembles flous, Mem., Fac. de Sci., Univ. Libre de Bruxelles, Brussels, Belgium.

2694 Vincke, P. (1973). Une application de la theorie des graphes flous, *Cah. Cent. Etud. Rech. Oper.*, **15**(3), 375–395.

2695 Von Wright, G. H. (1957). *Logical Studies*, Routledge and Kegan Paul, London.

2696 Von Wright, G. H. (1962). Remarks on the epistemology of subjective probability, in *Logic, Methodology and Philosophy of Science*, E. Nagel, P. Suppes, and A. Tarski, Eds., Stanford Univ. Press, Stanford, CA, pp. 330–339.

2697 Von Wright, G. H. (1963). *Norm and Action*, Routledge and Kegan Paul, London.

2698 Von Wright, G. H. (1963). *The Logic of Preference*, Edinburgh Univ. Press, Edinburgh, UK.

2699 Von Wright, G. H. (1972). *An Essay in Deotic Logic and the General Theory of Action*, North-Holland, Amsterdam.

Key References in Fuzzy Pattern Recognition

2700 Vopenka, P., and Hajak, P. (1972). *Theory of Semisets*, North-Holland, Amsterdam.
2701 Vossen, P. H. (1974). Fuzzy set convolution with respect to a group operation, Memo. 1974.06.20, Dept. of Psychol., Nijmegen Univ., Nijmegen, Holland.
2702 Vossen, P. H. (1974). Notes for a theory of fuzziness. The emergence of a basic concept in mathematics, science and technology, SSRG-74-01, Dept. of Psychol., Nijmegen Univ., Nijmegen, Holland.
2703 Vossen, P. H. (1975). Vertaling van Voorwoord, Voorbericht en inhoudsopgave van kaufmann 1973, Memo. 75-08, Dept. of Psychol., Nijmegen Univ., Nijmegen, Holland (in Dutch).
2704 Vossen, P. H., and Klabbers, J. H. G. (1973). In vogelvlucht over algemene systemleer en vage verzamlingenleer, SSRG-73-00, Dept. of Psychol., Nijmegen Univ., Nijmegen, Holland (in Dutch).
2705 Vossen, P. H., and Klabbers, J. H. G. (1974). A formal and experimental inquiry into the applicability of nonstandard set theory to the analysis of valuation processes in social systems, SSRG 74-11, Dept. of Psychol., Nijmegen Univ., Nijmegen, Holland.
2706 Wahlster, W. (1978). Die Simulation vager Inferenzen auf Unscharfen Wissen, eine Anwendung der Mehrwertigen Programmiersprache Fuzzy, in *Komplexe Menschliche Informations, Verarbeitung, Beitrage Zur Tagung Kognitive Psychologie*, (H. Ueckert and D. Rhenius, Eds.), Hamburg.
2707 Wahlster, W. (1979). Algorithmen zur Beactwortung von "WARUM," Fragen in Diaglog System, Ber. Nr 9, Projektgruppe Simulation von Spracheverstehen, Univ. Hamburg, Hamburg, Germany, Jan.
2708 Wahlster, W. (1980). Implementing fuzziness in dialogue systems. Ber. 14, Projecktgruppe Simulation von Sprachverstehen. Germanisches Semin., Von-Melle-Park, 6, Hamburg, Germany, to be published in *Empirical Semantics* (B. Rieger, Ed.), Brockmeyer.
2709 Wahlster, W., Jameson, A., and Hoeppner, W. (1979). Dialogue system HAM-RPM, *Amer. J. Comput. Linguist.*
2710 Wajsberg, M. (1967). Axiomatization of the three-valued propositional calculus, in *Polish Logic 1920–1939*, (S. McCall, Ed.), Clarenden Press, Oxford, UK, pp. 264–284.
2711 Wang, D., Shen, X., and Liu, Y. (1980). The distance between fuzzy numbers and its limit property, News on 1980 Annu. Rep. Meet. of Beijing Working Group on Fuzzy Sets, Beijing, China.
2712 Wang, G. (1979). Topological molecular lattices, Sanshi Sida Xueboa, A List of Lit. in China, The Univ. of Sci. and Tech. of China, Hofei, Anhwei, China (in Chinese).
2713 Wang, G. (1980). Some properties of induced fuzzy topological space, Collect. Pap. on Fuzzy Math. (Abstr.), Huazhong Inst. of Tech., Huazhong, China, Apr.
2714 Wang, H., and Che, Y. (1980). The discrete mathematical model for the pollution materials flow in a region and the regulation for it, Collect. Pap. on Fuzzy Math. (Abstr.), Huazhong Inst. of Tech., Huazhong, China, Apr.
2715 Wang, H., and Che, Y. (1980). The fuzzy classification and the identification of the environmental quality assessment of a region, Collect. Pap. on Fuzzy Math. (Abstr.), Huazhong Inst. of Tech., Huazhong, China, Apr.
2716 Wang, J., and Wang, P. (1980). On the problem of integral recognition (1) Method of cophase relations, Rep., Beijing Normal Univ., 11, A List of Lit. in China, The Univ. of Sci. and Tech. of China, Hofei, Anhwei, China.
2717 Wang, K. (1980). A more reasonable fuzzy compactness, A List of Lit. in China, The Univ. of Sci. and Tech. of China, Hofei, Anhwei, China (in Chinese).

2718 Wang, M. (1980). On the B-set theory, News on 1980 Annu. Rep. Meet. of Beijing Working Group on Fuzzy Sets, Beijing, China.
2719 Wang, M. (1980). N-fold fuzzy subsets and extension principle, Collect. Pap. on Fuzzy Math. (Abstr.), Huazhong Inst. of Tech., Huazhong, China, Apr.
2720 Wang, P. (1979). Fuzzy sets and its simple applications, Rep. Beijing Normal Univ., 12, A List of Lit. in China, The Univ. of Sci. and Tech. of China, Hofei, Anhwei, China.
2721 Wang, P. (1979). Mathematical model of synthetic decision, Rep. Beijing Normal Univ., A List of Lit. in China, The Univ. of Sci. and Tech. of China, Hofei, Anhwei, China.
2722 Wang, P. (1979). Near choice principle in the recognition of fuzzy model, Rep. Beijing Normal Univ., A List of Lit. in China, The Univ. of Sci. and Tech. of China, Hofei, Anhwei, China.
2723 Wang, P. (1980). Fuzzy contactability and fuzzy variables, in Selected Pap. on Fuzzy Subsets, Beijing Normal Univ., Beijing China, Mar.
2724 Wang. P. (1980). Fuzzy meet-grade and fuzzy distributions, *J. Beijing Normal Univ.*, **1**, A List of Lit. in China, The Univ. of Sci. and Tech. of China, Hofei, Anhwei, China.
2725 Wang, P. (1980). Fuzzy sets and the categories of fuzzy sets, Rep., Beijing Normal Univ., 11, A List of Lit. in China, The Univ. of Sci. and Tech. of China, Hofei, Anhwei, China.
2726 Wang, P. (1980). The neartude and near-choice principle, Collect. Pap. on Fuzzy Math. (Abstr.), Huazhong Inst. of Tech., Huazhong, China, Apr.
2727 Wang, P., and Bo, L. S. (1980). The reactability of fuzzy controller, Collect. Pap. on Fuzzy Math. (Abstr.), Huazhong Inst. of Tech., Huazhong, China, Apr.
2728 Wang, P., and Chuenren, G. (1980). The model of multifactorial evaluations, Collect. Pap. on Fuzzy Math. (Abstr.), Huazhong Inst. of Tech., Huazhong, China, Apr.
2729 Wang, P., and Huadong, W. (1980). Multifactorial evaluation and an application in environmental sciences, in Selected Papers on Fuzzy Subsets, Beijing Normal Univ., Beijing, China, Mar.
2730 Wang, P., and Lou, S. (1980). The reactability of fuzzy controllers, in Selected Pap. on Fuzzy Subsets, Beijing Normal Univ., Beijing, China, Mar.
2731 Wang, P., and Meng, Y. (1980). Relation equation and relation inequalities, in Selected Pap. on Fuzzy Subsets, Beijing Normal Univ., Beijing, China, Mar.
2732 Wang, P., and Meng, Y. (1980). Relation equations and relation inequalities, Collect. Pap. on Fuzzy Math. (Abstr.), Huazhong Inst. of Tech., Huazhong, China, Apr.
2733 Wang, P., and Xinggou, C. (1980). Notes on the Negoita-Ralescu flou set representation theorem, Collect. Pap. on Fuzzy Math. (Abstr.), Huazhong Inst. of Tech., Huazhong, China, Apr.
2734 Wang, P. P., and Chang, S. K. (1980). *Fuzzy Sets Theory and Applications to Policy Analysis and Information Systems*, Plenum Press, New York and London.
2735 Wang, P. P., and Togai, M. (1980). Sensitivity analysis of dynamic systems via fuzzy set theory, in *Proc. Int. Congr. on Appl. Syst. Res. and Cybern.*, Acapulco, Mexico, Dec.
2736 Wang, P. P., and Wang, C. Y. (1979). Experiment on character recognition using fuzzy filters, 1st Int. Symp. on Policy Anal. and Inf. Syst., Durham, NC, June.
2737 Wang, P. P., and Wang, C. Y. (1979). Fuzzy relations algorithms and fuzzy filter for pattern recognition, 1st Int. Symp. on Policy Anal. and Inf. Syst., Durham, NC, June.
2738 Wang, X. (1980). A problem on fuzzy classify, Rep., Beijing, Normal Univ., 11, A List of Lit. in China, The Univ. of Sci. and Tech. of China, Hofei, Anhwei, China.

2739 Wang, X. (1980). One of the problems of fuzzy cluster, News on 1980 Annu. Rep. Meet. of Beijing Working Group on Fuzzy Sets, Beijing, China.

2740 Wang, Y. (1980). Fuzzy logic and some theoretical problems, News on 1980 Annu. Rep. Meet. of Beijing Working Group on Fuzzy Sets, Beijing, China.

2741 Wang, Z. (1980). A comparison between the fuzzy topological space $(X, W(T))$ and the topological space (X, T), Collect. Pap. on Fuzzy Math. (Abstr.), Huazhong Inst. of Tech., Huazhong, China, Apr.

2742 Wang, Z. (1980). Notes on the fuzzy measures, Collect. Pap. on Fuzzy Math. (Abstr.), Huazhong Inst. of Tech., Huazhong, China, Apr.

2743 Wang, Z. (1980). Separation axioms in fuzzy topological space, Collect. Pap. on Fuzzy Math. (Abstr.), Huazhong Inst. of Tech., Huazhong, China, Apr.

2744 Warren, R. H. (1974). Closure operator and boundary operator for fuzzy topological spaces, Appl. Math. Res. Lab., Wright-Patterson, AFB, Ohio.

2745 Warren, R. H. (1974). Neighborhoods, bases and continuity in fuzzy topological spaces, Appl. Math. Res. Lab., Wright-Patterson AFB, Ohio.

2746 Warren, R. H. (1976). Optimality in fuzzy topological polysystems, *J. Math. Anal. Appl.*, **54**, 309–315.

2747 Warren, R. H. (1977). Boundary of fuzzy set, *Indiana Univ. Math. J.*, **26**(2), 191–197.

2748 Warren, R. H. (1979). Fuzzy topological properties, in *Proc. Int. Semin. on Fuzzy Sets Theory*, Johannes Kepler Univ., Linz, Austria, Sept.

2749 Warren, R. H. (1979). Fuzzy topological results, in *Proc. Int. Semin. On Fuzzy Set Theory*, Johannes Kepler Univ., Linz, Austria, Sept.

2750 Warren, R. H. (1979). Fuzzy topologies characterized by neighborhood systems, *Rocky Mountain J. Math.*, **9**(4), Fall.

2751 Warren, R. H. (1980). Neighborhoods, bases and continuity in fuzzy topological spaces, *Rocky Mountain J. Math.*

2752 Washiowski, R., and Welk, T. (1978). Application of fuzzy sets to decision making in active systems, Memo.

2753 Watada, J., Tanaka, H., and Asai, K. (1980). Alternative method for finding fuzzy transitive relations, in *Proc. Int. Congr. on Appl. Syst. Res. and Cybern.*, Acapulco, Mexico, Dec.

2754 Watanabe, S. (1961). A model of mind-body relation in terms of modular logic, *Synthesis*, 3–26.

2755 Watanabe, S. (1969). Modified concepts of logic, probability and information based on generalized continuous characteristic function, *Inf. Control*, **15**, 1–21.

2756 Watanabe, S. (1975). Creative learning and propensity automata, *IEEE Trans. Syst., Man, Cybern.*, **SMC-5**, 603–609.

2757 Watanabe, S. (1978). A generalized fuzzy-set theory, *IEEE Trans. Syst., Man, Cybern.*, **SMC-8** (10), 756–759.

2758 Watanabe, S. (1978). Fuzzification and invariance, in *Proc. 1978 Int. Conf. on Cybern. and Soc.*, Sophia Univ., Tokyo, pp. 947–951.

2759 Watson, S. R., and Weiss, J. J. (1978). Applications of fuzzy set theory to naval command and control, Rep. PR 78-13-83, Decision and Design, Inc., McLean, VA.

2760 Watson, S. R., Weiss, J. J., and Donnell, M. L. (1976). Fuzzy decision and analysis, *IEEE Trans. Syst., Man, Cybern.*, **9** (1), 1–9.

2761 Webb, D. L. (1936). The algebra of N-valued logic, C.R. *Seances Soc. Sci. Lett. Varsovie*, **29**, 153–168.

2762 Wechler, W. (1974). Analyse und Synthes Zeitvariabler *R*-Fuzzy Automaten, *Zki-Inf.* (Akad D. Wiss., Germany), **1**, 32–366.

2763 Wechler, W. (1975). Automaten uber Inputkategorien, *J. EIK*, **11**, 681–685.

2764 Wechler, W. (1975). Gesteuerte *R*-Fuzzy Automaten, *Zki-Inf.* (Akad. D. Wiss., Germany), **1**, 9–13.

2765 Wechler, W. (1975). *R*-Fuzzy automata with a time structure, in *Mathematical Foundations of Computer Science*, Vol. 28 (A. Blikle, Ed.), Springer-Verlag, Berlin, pp. 73–76.

2766 Wechler, W. (1975). The concept of fuzziness in the theory of automata, in *Proc. 3rd Int. Congr. of Cybern. and Syst.*, Bucharest, Aug.

2767 Wechler, W. (1975). Zur Verallgemeinergung des Theorems von Kleene-Schutzen-Berger auf Zeitvariable Automaten, *J. EIK*, **11**, 439–445.

2768 Wechler, W. (1976). Hierarchy of *N*-rational languages, Tech. Univ., Dresden, Germany.

2769 Wechler, W. (1976). *R*-fuzzy grammars, in "Mathematical Foundations of Computer Science, Vol. 32 (J. Bečvář, Ed.), Springer-Verlag, Berlin, pp. 450–456.

2770 Wechler, W. (1976). Zum Verhalten Gesteuerter *R*-Fuzzy Automaten, Tech. Univ., Dresden, Germany.

2771 Wechler, W. (1977). Families of *R*-fuzzy languages, *Lecture Notes Comput. Sci.*, **56**, 117–186.

2772 Wechler, W. (1977). Modelling humanistic systems by fuzzy logic, in *Proc. Symp. on Fuzzy Set Theory and Appl., IEEE Conf. on Decis. and Control*, New Orleans.

2773 Wechler, W. (1978). *The Concept of Fuzziness in Automata and Language Theory*, Akademie-Verlag, Berlin.

2774 Wechler, W. (1979). Fuzzy sets and languages, in *Advances in Fuzzy Set Theory and Applications* (M. M. Gupta, R. K. Ragade, and R. R. Yager, Eds.), North-Holland, Amsterdam.

2775 Wechler, W., and Agasandyan, G. A. (1974). Automata with a variable structure and metaregular languages, *Izv. Akad. Nauk. SSSR Tehn. Kibernet.*, **1W**, 146–148.

2776 Wechler, W., and Dimitrov, V. (1974). *R*-fuzzy automata, in *Inf. Process. 74, Proc. IFIP Congr.*, North-Holland, Amsterdam, pp. 657–660.

2777 Wechsler, H. (1975). Applications of fuzzy logic to medical diagnosis, in *Proc. 1975 Int. Symp. Multiple-Valued Logic*, IEEE 75CH0959-7C, May.

2778 Wechsler, H. (1976). A fuzzy approach to medical diagnosis, *Int. J. Bio-Med. Comput.*, **7**, 191–203.

2779 Wee, W. G. (1967). On generalizations of adaptive algorithms and applications of the fuzzy sets concept to pattern classification, Ph.D. Thesis, Purdue Univ. Lafayette, IN, *Diss. Abstr. Int.*, **28/11-B**, 4587.

2780 Wee, W. G., and Fu, K. S. (1969). A formulation of fuzzy automata and its application as a model of learning systems. *IEEE Trans. Syst., Man, Cybern.*, **SMC-5**, 215–223.

2781 Weidner, A. J. (1974). An axiomatization of fuzzy set theory, Ph.D. Thesis, Univ. of Notre Dame, *Diss. Abstr. Int.*, **35/07-B**, 3463.

2782 Weiss, J. (1980). Fuzzy methods to aid military decision making, in *Proc. Int. Congr. on Appl. Syst. Res. and Cybern.*, Acapulco, Mexico, Dec.

2783 Weiss, J. J., and Donnell, M. L. (1979). A general purpose policy capturing device using fuzzy production rules, in *Advances in Fuzzy Set Theory and Applications* (M. M. Gupta, R. K. Ragade, R. R. Yager, Eds.), North-Holland, Amsterdam.

2784 Weiss, J. J., and Donnell, M. L. (1979). Fuzzy policy capturing as a decision aid, in *Proc. 18th IEEE Conf. On Decis. and Control*, Fort Lauderdale, FL, Dec.

2785 Weiss, M. D. (1975). Fixed points separation and induced topologies for fuzzy sets, *J. Math. Anal. Appl.*, **50**, 142–150.

2786 Weiss, S. E. (1973). The Sorites antinomy—A study in the logic of vagueness and measurement, Ph.D. Thesis, Univ. of North Carolina, Chapel Hill.

2787 Weiss, S. E. (1976). The Sorites fallacy—What difference does a peanut make?, *Synthese*.

2788 Wenstop, F. (1975). Application of linguistic variables in the analysis of organizations, Ph.D. Thesis, School of Business Admin., Univ. of California, Berkeley.

2789 Wenstop, F. (1975). Evaluation of verbal organizational models, NOAK 75, Oslo, Norway.

2790 Wenstop, F. (1976). Deductive verbal models of organizations, *J. Man-Mach. Stud.*, **8**, 293–311.

2791 Wenstop, F. (1976). Fuzzy set simulation models in a systems dynamic perspective, *Kybernetes*, **6**, 290–218.

2792 Wenstop, F. (1977). Fuzzy sets and decision making, *California Eng.*, **16**, 20–24.

2793 Wenstop, F. (1978). Verbal formulation of fuzzy dynamic systems, Oslo, Inst. of Business Admin., Oslo, Norway.

2794 Wenstop, F. (1979). Exploring linguistic consequences of assertions in social sciences, in *Advances in Fuzzy Set Theory and Applications* (M. M. Gupta, R. K. Ragade, and R. R. Yager, Eds.), North-Holland, Amsterdam.

2795 Wenstop, F. (1980). Quantitative analysis with linguistic values, *Fuzzy Sets Syst.*, **4** (2), Sept., 99–116.

2796 Whalen, T. (1980). Risk minimization using *L*-fuzzy sets, Int. Conf. on Cybern. and Soc., Cambridge, MA, Oct.

2797 Whalster, W. (1977). Die Reprasentation von Vagem Wissen in Naturlichsprachen Systemen der Kunstllichen Intelligenz, Working Pap. 38, Inst. fur Inf., Univ. Hamburg, Hamburg, Germany, July.

2798 White, A. R. (1975). *Modal Thinking*, Basil Blackwell, Oxford, UK.

2799 Wiedey, G., and Zimmermann, H. J. (1982). Fuzzy sets applied to marketing, *Fuzzy Sets Syst.*

2800 Wiedey, G., and Zimmermann, H. J. (1978). Media selection and fuzzy linear programming, *J. Oper. Res. Soc.*, **29** (11), 1071–1084.

2801 Wierzshon, S. T., and Zalewski, J. (1980). Application of fuzzy integral to the control of a model of a nuclear reactor, The 10th Meet. of the Evr. Working Group on Fuzzy Sets, June.

2802 Wierzshon, S. T. (1980). Parameter identification by using fuzzy integral, *EMCSR—PCSR*, **8**.

2803 Wierzshon, S. T., and Zalewski, J. (1978). A set of subroutines for applications of fuzzy conc. to decision making problems, Int. Conf. on Syst. Sci., Wroclaw, Poland.

2804 Wierzshon, S. T., (1980). On some properties of fuzzy systems, *EMCSR—PCSR*, **8**.

2805 Wierzshon, S. T., and Zalewski, J. (1980). The application of fuzzy sets to automatic control and decision making in system-modelling-control, Zakopane, Poland.

2806 Wierzshon, S. T., Zaleswki, J., and Soltowski, T. (1979). The application of fuzzy algorithm to control the extraction process in nuclear fuel reprocessing, in *Proc. 2nd IFAC-IFIP Symp. on Software for Comput. Control, SOCOCO79*, Prague.

2807 Wierzshon, S. T., and Zalewski, J. (1978). On some equivalence between control theory and decision making theory, 4th Int. Congr. in Cybern. and Syst., Amsterdam, Aug.

2808 Wierzshon, S. T., and Zalewski, J. (1978). On some equivalence between control, Tech. Univ. of Warzawa, Warzawa, Poland, 4th Int. Congr. on Cybern. and Syst., Amsterdam.

2809 Wilkinson, J. (1973). Archetypes, language, dynamic programming and fuzzy sets, in *The Dynamic Programming of Human systems* (J. Wilkinson, R. Bellman, and R. Garaudy, Eds.) MSS Inf. Corp., New York, pp. 44–53.

2810 Wilkinson, J. (1973). Retrospective futurology, in *The Dynamic Programming of Human Systems* (J. Wilkinson, R. Bellman, and R. Garaudy, Eds.), MSS Inf. Corp., New York, pp. 19–33.

2811 Wilkinson, J., Bellman, R., and Garaudy, R. (Eds.) (1973). *The Dynamic Programming of Human Systems*, MSS Inf. Corp., New York.

2812 Wilks, Y. (1975). Preference semantics, in *Formal Semantics of Natural Language* (E. L. Keenan, Ed.), pp. 320–348.

2813 Willaeys, D. (1980). Contribution a l'etude de la theorie des sous-ensembles flous en vue de son application a l'automatique, These d'Etat, Univ. de Valenciennes, Valenciennes, France.

2814 Willaeys, D. (1980). Contribution des modeles flous a l'analyse des systemes, Table Ronde du CNRS sur le Flou, Lyon, France, June.

2815 Willaeys, D. (1981). Contribution to the study of fuzzy sets theory with a view to its application to automatic control (Abstr.), *Bull. Sous Ensembles Flous Appl.*, **5**, Hiver (80–81), 90.

2816 Willaeys, D., Hammad, P., and Malvache, H. (1975). Representation et minimisation de fontions floues, in *Proc. 3rd Congr. IFAC, IFIP*, Madrid.

2817 Willaeys, D., and Malvache, N. (1976). Utilisating referential of fuzzy sets.; Application to fuzzy algorithm, *Int. Conf. on Syst. Sci.*, Wroclaw, Poland.

2818 Willaeys, D., and Malvache, N. (1977). Utilisation of fuzzy sets for system modelling and control, Congr. Int. IEEE, New Orleans.

2819 Willaeys, D., and Malvache, N. (1978). Use of fuzzy model for process control, in *Proc. 1978 Int. Conf. On Cybern. and Soc.*, Tokyo, Japan, pp. 942–946.

2820 Willaeys, D., and Malvache, N. (1978). Utilisation de la discretisation floue pour le traitement d'informations flous par calculateur. in *Proc. Colloq. Int. sur la Theorie et les Appl. des Sous-Ensembles Flous*, Marseille, France.

2821 Willaeys, D., and Malvache, N. (1978). Use of fuzzy discrete space for fuzzy automatic control, 7th Trienn. World Congr. IFAC, Sess.—Fuzzy decision-making and appl., RT No. 15, Helsinki, June.

2822 Willaeys, D., and Malvache, N. (1979). Contribution of the fuzzy sets theory to man-machine system, in *Advances in Fuzzy Set Theory and Applications*, (M. M. Gupta, R. K. Ragade, R. R. Yager, Eds.), North-Holland, Amsterdam.

2823 Willaeys, D., and Malvache, N. (1980). Optimal control of fuzzy systems, in *Proc. Int. Congr. on Appl. Syst. Res. and Cybern.*, Acapulco, Mexico, Dec.

2824 Willaeys, D., and Malvache, N. (1981). The use of fuzzy sets for the treatment of fuzzy information by computer (short comm.), *Fuzzy Sets Syst.*, **5** (3), 323–328.

2825 Willaeys, D., Malvache, N., and Hammad, P. (1975). Minimisation et formulation des systemes a fonctions propositionnelles de variables floues, Congr. Nat. d'Inf. et d'Autom., Madrid, Oct.

2826 Willaeys, D., Mangin, P., and Manvache, N. (1977). Use of fuzzy sets for systems modelizing and control—Application to the speed regulation of a strongly perturbed motor, 5th Conf. Int. IFAC/IFIP, La Haye.

2827 Willaeys, D., Mangin, P., and Malvache, N. (1977). Contribution au concept de commande d'un processus a partir d'une identification des donnees subjectives d'un conducteur humain de processus, 1st World Conf. on Math. at the Service of Man, Barcelona, Spain, July.

2828 Willmott, R. (1979). Mean measures in fuzzy power-set theory, Rep. No. FRP-6, Dept. of Math., Univ. of Essex, Colchester, UK.

2829 Willmott, R. (1979). On the transitivity of implication and equivalence in some many-valued logics, Rep. No. FRP-5, Dept of Math., Univ. of Essex, Colchester, UK.

2830 Willmott, R. C. (1978). Two fuzzier implication operators in the theory of fuzzy power sets, Rep. No. FRP-2, Dept. of Math., Univ. of Essex, Colchester, UK.

2831 Willmott, R. C. (1979). Mean measures in fuzzy power-set theory, Rep. No. FRP-6, Dept. of Math., Univ. of Essex, Colchester, UK.

2832 Wilson, D. (1975). *Presuppositions and Non-Truth-Conditional Semantics*, Academic Press, London.

2833 Windecker, R. C. (1979). Stochastic multiple-valued combinational networks, The 9th Int. Symp. on Multi-Valued Logic, Beth, UK.

2834 Windham, M. P. (1981). Cluster validity for fuzzy clustering algorithms, *Fuzzy Sets Syst.* FSS, **5** (2), 177–186.

2835 Winkler, R. L. (1974). Probabilistic prediction—Some exact experimental results, *J. Amer. Stat. Assoc.*, **66**, 625–688.

2836 Winkler, R. L., and Murphy, A. H. (1968). Good probability assessors, *J. Appl. Meteorol.*, **7**, 751–758.

2837 Winograd, T. (1974). Lakoff on hedges, Artif. Intell. Lab., Comput. Sci. Dept., Stanford Univ., Stanford, CA. Sept.

2838 Wiredu, J. E. (1975). Truth as a logical constant, with an application to the principle of the excluded middle, *Philos. Quart.*, 305–317.

2839 Witten, I. H. (1975). Learning to control sequential and non-sequential environments, EES-MMS-CON.75, Dept. of Elec. Eng. Sci., Univ. of Essex, Colchester, UK.

2840 Wojcicki, R. (1966). Semantical criteria of empirical meaningfulness, *Stud. Logica*, **19**, 75–107.

2841 Wolf, R. G. (1975). A critical survey of many-valued logics 1966–1974, in *Proc. 6th Int. Symp. Multiple-Valued Logic*, Logan, Utah, IEEE 76CH1111-4C, pp. 468–474.

2842 Wolniewicz, B. (1970). Four notions of independence, *Theoria*, **36**, 161–164.

2843 Wong, C. K. (1973). Covering properties of fuzzy topological spaces, *J. Math. Anal. Appl.*, **43**, 697–704.

2844 Wong, C. K. (1974). Fuzzy points and logical properties of fuzzy topology, *J. Math. Anal. Appl.*, **46**, 316–328.

2845 Wong, C. K. (1974). Fuzzy topology—Product and quotient theorems, *J. Math. Anal. Appl.*, **45**, 512–521.

2846 Wong, C. K. (1975). Fuzzy topology, in *Fuzzy Sets and their Applications to Cognitive and Decision Processes*, (L. A. Zadeh, K. S. Fu, K. Tanaka, and M. Shimura, Eds.), Academic Press, New York, pp. 171–190.

2847 Wong, C. K. (1976). Categories of fuzzy sets and fuzzy topological spaces, *J. Math. Anal. Appl.*, **53**, 704–714.

2848 Wong, G. (1980). Measure on fuzzy sets, Collect Pap. on Fuzzy Math. (Abstr.), Huazhong Inst. of Tech., Huazhong, China, Apr.

2849 Wong, G. A., and Sheng, D. C. (1975). On the learning behaviour of fuzzy automata, in *Advances in Cybernetics and Systems*, Vol. 2, (J. Rose, Ed.), Gordon and Breach, London, pp. 885–896.

2850 Wong, J. (1980). Solving the fuzzy relation equations, Collect Pap. on Fuzzy Math. (Abstr.), Huazhong Inst. of Tech., Huazhong, China, Apr.

2851 Wong, J. (1980). The representation of fuzzy quantities, Collect Pap. on Fuzzy Math. (Abstr.) Huazhong Inst. of Tech., Huazhong, China, Apr.

2852 Wong, P., and Long, S. (1980). A mobile rule of fuzzy controller, News on 1980 Annu. Rep. Meet. of Beijing Working Group on Fuzzy Sets, Beijing, China.

2853 Wong, P., and Wang, J. (1980). On the isophase relation used in the problems of recognition, News on 1980 Annu. Rep. Meet. of Beijing Working Group on Fuzzy Sets, Beijing, China.

2854 Woodbury, M. A., and Clive, J. (1974). Clinical pure types as a fuzzy partition, *J. Cybern.*, **4**, (3), 111–121.

2855 Woodbury, M., Clive, J., and Garson, A. (1974). A generalized ditto algorithm for initial fuzzy clusters, *J. Cybernet.*, **4**.

2856 Woodhead, R. G. (1972). On the theory of fuzzy sets to resolve ill-structured marine decision problems, Dep. of Naval Archit. and Shipbuild., Univ. of Newcastle Upon Tyne, Newcastle, UK.

2857 Woodruff, P. W. (1974). A modal interpretation of three-valued logic, *J. Philos. Logic*, **3**, 433–439.

2858 Wright, C. (1975). On the coherence of vague predicates, *Syntheses*, **30**, 325–365.

2859 Wu, K. (1980). Pansystems analysis and mechanics models, *J. Wuhan Water Conserv. Power*, **1** (2), A List of Lit. in China, The Univ. of Sci. and Tech. of China, Hofei, Anhwei, China.

2860 Wu, K. (1980). Some problems concerning pansystems analysis and fuzziness of vision, *J. Huazhong Inst. Tech.*, **2**, A List of Lit. in China, The Univ. of Sci. and Tech. of China, Hofei, Anhwei, China.

2861 Wu, K. (1980). Transforming analysis of variation, *Appl. Math. Mech.*, **1**, A List of Lit. in China, The Univ. of Sci. and Tech. of China, Hofei, Anhwei, China.

2862 Wu, S. (1980). A method for semantics inference of natural language, News on 1980 Annu. Rep. Meet. of Beijing Working Group on Fuzzy Sets, Beijing China.

2863 Wu, S. (1980). A reduction of fuzzy transformable function by logics, News on 1980 Annu. Rep. Meet. of Beijing Working Group on Fuzzy Sets, Beijing, China.

2864 Wu, S. (1980). Fuzzy concept of orientation and the transformable technique of method, News on 1980 Annu. Rep. Meet. of Beijing, Working Group on Fuzzy Sets, Beijing, China.

2865 Wu, W. (1980). Fuzzy graph theory and its applications in the fuzzy clustering analysis, Collect. Pap. on Fuzzy Math. (Abstr.), Huazhong Inst. of Tech., Huazhong, China, Apr.

2866 Wu. W. (1980). Normal fuzzy subgroup, Collect. Pap. on Fuzzy Math. (Abstr.), Huazhong Inst. of Tech., Huazhong, China, Apr.

2867 Wu, W. (1981). Fuzzy graph, Shanghai Normal Inst., A List of Lit. in China, The Univ. of Sci. and Tech. of China, Hofei, Anhwei, China.

2868 Wu, X. (1978). A sketch on pansystems analysis, *J. Wuhan Univ.*, **3**, A List of Lit. in China, The Univ. of Sci. and Tech. of China, Hofei, Anhwei, China.

2869 Wu, X. (1979). Fuzziness, reliability and pansystems analysis, *J. Huazhong Inst. Tech.*, **1** (4), A List of Lit. in China, The Univ. of Sci. and Tech. of China, Hofei, Anhwei, China.

Key References in Fuzzy Pattern Recognition 345

2870 Wu, X. (1980). Fuzziness, reliability and pansystems analysis (III), (IV), Collect. Pap. on Fuzzy Math. (Abstr.), Huazhong, Inst. of Tech., Huazhong, China, Apr.

2871 Wu, X. (1980). Operation research in control and recognition—Fuzziness and pansystems analysis, Collect. Pap. on Fuzzy Math. (Abstr.), Huazhong Inst. of Tech., Huazhong, China, Apr.

2872 Wu, X. (1980). Reliability pansystems analysis—Fuzziness reliability and pansystems analysis (II), Collect. Pap. on Fuzzy Math. (Abstr.), Huazhong Inst. of Tech., Huazhong, China, Apr.

2873 Wu, X. (1980). A sketch on pansystems analysis—Investigations and applications of pansystems analysis (I), Collect. Pap. on Fuzzy Math. (Abstr.), Huazhong Inst. of Tech., Huazhong, China, Apr.

2874 Wu, X. (1980). Fuzziness, reliability and pansystems analysis (I), Collect. Pap. on Fuzzy Math. (Abstr.), Huazhong Inst. of Tech., Huazhong, China, Apr.

2875 Wu, X. (1980). Pansystem observocontrollability, pansystem logic and fuzziness—Investigation and applications of pansystems analysis, *J. Huazhong Inst. Tech.*, **2**, A List of Lit. in China, The Univ. of Sci. and Tech. of China, Hofei, Anhwei, China.

2876 Wu, Y. (1976). Fuzzy dynamical programming and its applications, Ph.D. Thesis, State Univ. of New York, Stony Brook, in *Diss. Abstr. Int.*, **37/07-B**, 3548.

2877 Xu, R. (1980). Analysis of fuzzy decisions process by similarity index, News on 1980 Annu. Rep. Meeting of Beijing Working Group on Fuzzy Sets, Beijing, China.

2878 Yager, R. R. (Ed.) (1982). *Recent Developments in Fuzzy Set and Possibility Theory*, Pergamon Press, New York.

2879 Yager, R. R. (1976). Comparing fuzzy constraints, in *Proc. 5th Northeast AIDS Conf.*, Philadelphia, PA.

2880 Yager, R. R. (1977). A foundation for a theory of possibility, Iona Tech. Rep. RRY 77-25, Iona College, New Rochelle, NY.

2881 Yager, R. R. (1977). Building fuzzy decision models, in *Proc. 1st Int. Conf. on Math. Modeling*, St. Louis, MO.

2882 Yager, R. R. (1977). Decisions under uncertainty using fuzzy sets, in *Proc. 6th Northeast Decis. Sci. Conf.* (T. J. Hindelang, ed.), pp. 17–20.

2883 Yager, R. R. (1977). Fuzzy decision making including unequal objectives, *Fuzzy Sets Syst.*, **1**, 87–95.

2884 Yager, R. R. (1977). Fuzzy equations, in *Proc. 7th Int. Conf. Cybern. and Soc.*, Washington D.C.

2885 Yager, R. R. (1977). Mathematical programming in a fuzzy environment, Int. Symp. on Extr. Methods and Syst. Anal., Austin, TX.

2886 Yager, R. R. (1977). Multiple objective decision-making using fuzzy sets, *Int. J. Man-Mach. Stud.*, **9**, 375–382.

2887 Yager, R. R. (1977). On validity and building fuzzy models, Rep. on the IEEE Symp. on Fuzzy Set Theory and Appl., held at the 1977 IEEE Control and Decis. Conf.

2888 Yager, R. R. (1978). A note on fuzziness in a standard uncertainty logic, Tech. Rep. No. RRY 78-19, Iona College, New Rochelle, NY.

2889 Yager, R. R. (1978). Building fuzzy systems models, in *Applied General Systems Research—Recent Developments and Trends* (G. J. Klir, Ed.), Nata Conf. Ser., Ser. II, Syst. Sci., Vol. 5, pp. 313–321.

2890 Yager, R. R. (1978). Fuzzy modeling in an urban society, in *Proc. 22nd Meet. of Soc. Gen. Syst. Res.*, Washington, D.C., pp. 451–455.

2891 Yager, R. R. (1978). Fuzzy sets over the same space, in *Proc. 1978 JACC*, Vol. 3, p. 362.
2892 Yager, R. R. (1978). Linguistic models and fuzzy truths, *Int. J. Man-Mach. Stud.*
2893 Yager, R. R. (1978). On a general class of fuzzy connectives, Tech. Rep. No. RRY 78-18, Iona College, New Rochelle, NY.
2894 Yager, R. R. (1978). On the measure of fuzziness and negation, Part II—Membership in lattices, Tech. Rep. No. RRY 78-CC, Iona College, New Rochelle, NY.
2895 Yager, R. R. (1978). On the need for membership grades, in fuzzy sets, Iona College, New Rochelle, NY.
2896 Yager, R. R. (1978). Possibilistic decision making, Tech. Rep. No. RRY 78-05B, School of Business Admin., Iona College, New Rochelle, NY.
2897 Yager, R. R. (1978). Ranking fuzzy subsets over the unit interval, *Proc. 1978 IEEE Conf. on Decis. and Control, includes the 17th Symp. on Adaptive processes*, 78CH1392-OCS, Jan. 1979, San Diego, CA, Jan.
2898 Yager, R. R. (1978). Theory of signal responses based on fuzzy sets, in *Proc. Int. Conf. Cybern. and Soc.*, Vol. II, Tokyo, Nov., pp. 921-925.
2899 Yager, R. R. (1978). Validation of fuzzy linguistic models, *J. Cybern.*, **8**, 17-30.
2900 Yager, R. R. (1978). Competitiveness and compensation in decision making—A fuzzy set based interpretation, Tech. Rep. RRY 78-14B, Iona College, New Rochelle, NY.
2901 Yager, R. R. (1979). A measurement-informational discussion of fuzzy union and intersection, *Int. J. Man-Mach. Stud.*
2902 Yager, R. R. (1979). Fuzzy sets probabilities and decision, *J. Cybern.*
2903 Yager, R. R. (1979). Fuzzy subsets of type II in decisions, *J. Cybern.*
2904 Yager, R. R. (1979). Mathematical programming with fuzzy constraints and a preference ordering on the object, *Kybernetes*.
2905 Yager, R. R. (1979). Measuring tranquility and anxiety in decision making, An application of fuzzy sets, Tech. Rep. No. RRY 79-03, Iona College, New Rochelle, NY.
2906 Yager, R. R. (1979). On choosing between fuzzy subsets, *Kybernetes*.
2907 Yager, R. R. (1979). On modeling interpersonal communications, 1st Int. Symp. on Policy Anal. and Inf. Syst., June.
2908 Yager, R. R. (1979). On solving fuzzy mathematical relationships, *Inf. Control.*
2909 Yager, R. R. (1979). On the lack of inverses in fuzzy arithmetic, *Fuzzy Sets Syst.*
2910 Yager, R. R. (1979). Properties of connectives useful in local logics, in *General Systems Research—A Science, A Methodology, A Technology* (B. R. Gaines, Ed.), SGSR.
2911 Yager, R. R. (1979). Some procedures for selecting operators for fuzzy operations, Tech. Rep. RRY-79-05, Iona College, New Rochelle, NY.
2912 Yager, R. R. (1979). Level sets for membership evaluation of fuzzy subsets, Tech. Rep. RRY 79-14, Iona College, New Rochelle, NY.
2913 Yager, R. R. (1979). On the measure of fuzziness and negation. Part 1—Membership in the unit interval. *Int. J. Gen. Syst.*, **5**, Aussi Tech. Rep. RRY 79-01B. Iona College, New Rochelle, NY.
2914 Yager, R. R. (1979). Prototypical values for fuzzy subsets, Tech. Rep. RRY 79-13, Iona College, New Rochelle, NY.
2915 Yager, R. R. (1979). Satisfaction and fuzzy decision functions, in *Fuzzy Sets, Theory and Applications to Policy Analysis and Information Systems* 171-194 (P. P. Wang and S. K. Cheng, Eds.), Plenum, New York.

Key References in Fuzzy Pattern Recognition   347

2916 Yager, R. R. (1979). Using level sets to extract fuzzy membership grades, *EMCSR—PCSR*, **8**, to be published.

2917 Yager, R. R. (1980). A logical on-line bibliographic searcher an application of fuzzy sets, *IEEE Trans. Syst., Man. Cybern.*, **10** (1), Jan, 51–53.

2918 Yager, R. R. (1980). An approach to multiobjective decisions using fuzzy sets, Res. Memo., Iona College, New Rochelle, NY.

2919 Yager, R. R. (1980). A foundation for a theory of possibility, *J. Cybern.*, **10**.

2920 Yager, R. R. (1980). A linguistic variable for importance of fuzzy sets, *J. Cybern.*, **10**.

2921 Yager, R. R. (1980). Aspects of possibilistic uncertainty, *Int. J. Man-Mach. Stud.*, **12**.

2922 Yager, R. R. (1980). Finite linearly ordered fuzzy sets with applications to decisions, *Int. J. Man-Mach. Stud.*, **12**.

2923 Yager, R. R. (1980). Fuzzy sets in decisions in complex systems, in *Proc. of JACC*, San Francisco, Aug.

2924 Yager, R. R. (1980). Generalized "and-or" operators for multivalued and fuzzy logic, Proc. Int. Symp. Multivalued Logic.

2925 Yager, R. R. (1980). On the implication operator in fuzzy logic. Tech. Rep. RRY 80-03, Iona College, New Rochelle, NY.

2926 Yager, R. R. (1980). Some observations of linguistic approximation, in *Proc. Int. Congr. on Appl. Syst. Res. and Cybern.*, Acapulco, Mexico, Dec.

2927 Yager, R. R., and Basson, D. (1975). Decision making with fuzzy sets, *Decis. Sci.*, **6**, 590–600.

2928 Yamashita, T., and Hara, F. (1977). Classification of facial expression, in *Summ. of Pap. on Gen. Fuzzy Problems*, The working group on fuzzy syst., Toyko, pp. 167–171.

2929 Yamashita, T., and Hara, F. (1980). The use of face graphs in the malfunction diagnosis of a rotary machine group, in *Summ. of Pap. on Gen. Fuzzy Problems*, The Working Group on Fuzzy Syst., Tokyo, Dec.

2930 Yang, K. (1980). Fuzzy classical probability scheme, Collect. Pap. on Fuzzy Math. (Abstr.), Huazhong Inst. of Tech., Huazhong, China, Apr.

2931 Yang, K. (1980). On the entropy of fuzzy relation, News on 1980 Annu. Rep. Meet. of Beijing Working Group on Fuzzy Sets, Beijing, China.

2932 Yang, T., Jia, S., and Zhang, B. (1980). Applications of fuzzy sets to the comprehensive evaluation of human tolerance to hypoxia, News on 1980 Annu. Rep. Meet. of Beijing Working Group on Fuzzy Sets, Beijing, China.

2933 Yeh, R. T. (1974). Toward an algebraic theory of fuzzy relational systems, in *Proc. Int. Congr. Cybern.*, Namur, Belgium.

2934 Yeh, R. T., and Bang, S. Y. (1975). Fuzzy relations, fuzzy graphs and their applications to clustering analysis, in *Fuzzy Sets and their Applications to Cognitive and Decision Processes* (L. A. Zadeh, K. S. Fu, K. Tanaka, and M. Shimura, Eds.), Academic Press, New York, pp. 125–149.

2935 Yeong, L. C. and Terano, T. (1978). Story formulation by computers, Rep. No. 4, Working Group on Fuzzy Syst., Dec., pp. 44–47.

2936 Yezhjova, I. V., and Pospelov, D. A. (1977). Prinyatie resheniy pri nechyotkikh osnovaniyakh (Decision making in fuzzy bases), *Univers. Nayashkola, Tekh. Kaya Kibern.*, **6**, 3–11.

2937 Yoeli, M. (1961). A note on a generalization of Boolean matrix theory, *Amer. Monthly Math.*, **68**, 552–557.

2938 Yuan, M. (1980). Countable fuzzy functions, Collect. Pap. on Fuzzy Math. (Abstr.), Huazhong, Inst. of Tech., Huazhong, China, Apr.
2939 Yuan, M. (1980). Evaluable fuzzy functions, Rep. A.E.P.S., A List of Lit. in China, The Univ. of Sci. and Tech. of China, Hofei, Anhwei, China.
2940 Yuan, M. (1980). Fuzzy analysis, Collect. Pap. on Fuzzy Math. (Abstr.), Huazhong Inst. of Tech., Huazhong, China, Apr.
2941 Yuan, M. (1890). Fuzzy arithmetic, Collect. Pap. on Fuzzy Math. (Abstr.), Huazhong, Inst. of Tech., Huazhong, China, Apr.
2942 Yuan, M. (1980). On the family of fuzzy systems, News on 1980 Annu. Rep. Meet. of Beijing Working Group on Fuzzy Sets, Beijing, China.
2943 Yun, T. (1980). A note on the solution of statical problem of elastic bodies with fuzzy boundary conditions, *J. Huazhong Inst. Tech.*, **2**, A List of Lit. in China, The Univ. of Sci. and Tech. of China, Hofei, Anhwei, China.
2944 Zadeh, L. A. (1965). Fuzzy sets and systems, in *System Theory*. (J. Fox, Ed.), Microwave Res. Inst. Symp. Ser. XV, Polytechnic Press, Brooklyn, NY., pp. 29–37.
2945 Zadeh, L. A. (1965). Fuzzy sets, *Inf. Control*, **8**, 338–353.
2946 Zadeh, L. A. (1966). Shadows of fuzzy sets, Probl. *Transm. Inf.*, **2**, 37–44 (in Russian).
2947 Zadeh, L. A. (1968). Fuzzy algorithms, *Inf. Control*, **12**, 94–102.
2948 Zadeh, L. A. (1968). Probability measures of fuzzy events, *J. Math. Anal. Appl.*, **23**, 421–427.
2949 Zadeh, L. A. (1969). Biological applications of the theory of fuzzy sets and systems, in *Biocybernetics of the Central Nervous System* (L. D. Proctor, Ed.), Little, Brown and Co., Boston, pp. 199–212.
2950 Zadeh, L. A. (1969). The concepts of system, aggregate and state in system theory, in *System Theory* (L. A. Zadeh and E. Polak, Eds.), McGraw-Hill, New York, 3–42.
2951 Zadeh, L. A. (1971). Fuzzy languages and their relation to human and machine intelligence, Memo. ERL-M 302, Electron. Res. Lab., Univ. of California, Berkeley.
2952 Zadeh, L. A. (1971). Human intelligence vs. machine intelligence, in *Proc. Int. Conf. on Sci. and Soc.*, Belgrade, Yugoslavia, pp. 127–133.
2953 Zadeh, L. A. (1971). Quantitative fuzzy semantics, *Inf. Sci.*, **3**, 159–716.
2954 Zadeh, L. A. (1971). Similarity relations and fuzzy orderings, *Inf. Sci.*, **3**, 177–200.
2955 Zadeh, L. A. (1971). Towards a theory of fuzzy systems, in *Aspects of Networks and Systems Theory* (R. E. Kalman and R. N. Declairis, Eds.), Holt, Rinehart and Winston, New York.
2956 Zadeh, L. A. (1971). Towards fuzziness in computer systems—Fuzzy algorithms and languages, in *Architecture and Design of Digital Computers* (G. Boulaye, Ed.), Dunod, Paris, pp. 9–18.
2957 Zadeh, L. A. (1972). A fuzzy-set-theoretic interpretation of linguistic hedges, *J. Cybern.*, **2**, 4–34.
2958 Zadeh, L. A. (1972). A rationale for fuzzy control, *J. Dyn. Syst., Meas. Control*, **G94**, 3–4.
2959 Zadeh, L. A. (1972). Fuzzy languages and their relation to human intelligence, in *Proc. Int. Conf. on Man and Comput.*, S. Karger, Basel, Switzerland, pp. 130–165.
2960 Zadeh, L. A. (1972). On fuzzy algorithms, Memo. ERL-M325, Univ. of California, Berkeley.
2961 Zadeh, L. A. (1973). A system-theoretic view of behaviour modification, in *Beyond the Punitive Society* (H. Wheeler, Ed.), W. H. Freeman, San Francisco, pp. 160–169.

Key References in Fuzzy Pattern Recognition

2962 Zadeh, L. A. (1973). On the analysis of very large systems, Memo. ERL-M418, Electron. Res. Lab., Univ. of California, Berkeley, Jan.

2963 Zadeh, L. A. (1973). Outline of a new approach to the analysis of complex systems and decision processes, *IEEE Trans. Syst., Man, Cybern.*, **2**, 28–44.

2964 Zadeh, L. A. (1974). A new approach to system analysis, in *Man and Computer* (M. Marois, Ed.), North-Holland, Amsterdam, pp. 55–94.

2965 Zadeh, L. A. (1974). Fuzzy logic and its application to approximate reasoning, in *Inf. Process. 74, Proc. IFIP Congr. 74*, Vol. 3, North-Holland, Amsterdam, pp. 591–594.

2966 Zadeh, L. A. (1974). Numerical versus linguistic variables, *Newspaper Circuits Syst. Soc.*, **7**, Feb., 3–4.

2967 Zadeh, L. A. (1974). The concept of a linguistic variable and its application to approximate reasoning, in *Learning Systems and Intelligent Robots*, (K. S. Fu and J. T. Tou, Eds.), Plenum Press, New York, pp. 1–10.

2968 Zadeh, L. A. (1975). A bibliography on fuzzy sets and their application to decision processes in Zadeh, Fu, Etc. (Eds.) *Fuzzy Sets and their Applications to Cognitive and Decision Processes* (L. A. Zadeh, K. S. Fu, K. Tanaka, and M. Shimura, Eds.), Academic Press, New York.

2969 Zadeh, L. A. (1975). A relational model for approximate reasoning, IEEE Int. Conf. on Cybern. and Soc., San Francisco, Sept.

2970 Zadeh, L. A. (1975). Calculus of fuzzy restrictions, in *Fuzzy Sets and Their Applications to Cognitive and Decision Processes* (L. A. Zadeh, K. S. Fu, K. Tanaka, and M. Shimura, Eds.), Academic Press, New York, pp. 1–39.

2971 Zadeh, L. A. (1975). Fuzzy logic and approximate reasoning, *Synthese*, **30**, 407–428.

2972 Zadeh, L. A. (1975). Linguistic cybernetics, in *Advances in Cybernetics and Systems*, Vol. 3, (J. Rose, Ed.), Gordon and Breach, London, pp. 1607–1615.

2973 Zadeh, L. A. (1976). A fuzzy approach to decision analysis, ORSA/TIMS Conf., Miami, FL, Nov.

2974 Zadeh, L. A. (1976). A fuzzy-algorithmic approach to the definition of complex or imprecise concepts, *Int. J. Man-Mach. Stud.*, **8**, 249–291.

2975 Zadeh, L. A. (1976). Semantic inference from fuzzy data by mathematical programming, *IEEE Man, Syst. Cybern. Conf.*

2976 Zadeh, L. A. (1976). Semantic inference from fuzzy premises, in *Proc. 6th Int. Symp. Multiple-Valued Logic*, IEEE 76CH1111-4C, May, pp. 217–218.

2977 Zadeh, L. A. (1976). The linguistic approach and its application to decision analysis, in *Directions in Large-Scale Systems* (Y. C. Ho, and S. K. Mitter, Eds.), Plenum Press, New York, 339–370.

2978 Zadeh, L. A. (1977). A theory of approximate reasoning (AR), Memo, UCB/ERL M77/58, Electron. Res. Lab., College of Eng., Univ. of California, Berkeley, pp. 1–71.

2979 Zadeh, L. A. (1977). Fuzzy logic as a model for human reasoning, Int. Conf. on Appl. Gen. Syst. Res., Binghamton, NY.

2980 Zadeh, L. A. (1977). Fuzzy set theory—A perspective, in *Fuzzy Automata and Decision Processes*, (M. M. Gupta, G. N. Saridis, and B. R. Gaines, Eds.), North-Holland, Amsterdam, pp. 3–4.

2981 Zadeh, L. A. (1977). Fuzzy sets and their application to classification and clustering, in *Classification and Clustering* (J. Van Ryzin, Ed.), Academic Press, New York, pp. 251–299.

2982 Zadeh, L. A. (1977). Information analysis applications in a possibilistic framework and its meaning in natural languages, Colloq. CNRS, Les Dev. Recents de la Theorie de l'Inf. et leurs Appl., Cachan, France.

2983   Zadeh, L. A. (1977). Linguistic characterization of perference relations as a basis for choices in social systems, Memo. UCB/ERL M77/24, Electron. Res. Lab., College of Eng., Univ. of California, Berkeley. Also: Erkenntnis, 11, 383–410.

2984   Zadeh, L. A. (1977). Possibility theory versus probability theory in decision analysis, in *Proc. Symp. on Fuzzy Set Theory and Appl., 1977 IEEE Conf. on Decis. and Control*, New Orleans.

2985   Zadeh, L. A. (1977). PRUF and its application to inference from fuzzy propositions, in *Proc. 1977 IEEE Conf. on Decis. and Control*, New Orleans, pp. 1359–1360.

2986   Zadeh, L. A. (1977). PRUF—A meaning representation language for natural languages, Memo. ERL-M77/61, Electron. Res. Lab., College of Eng., Univ. of California, Berkeley, Oct. Also: *Int. J. Man-Machine Studies*, 10 (1978), 395–460.

2987   Zadeh, L. A. (1977). Theory of fuzzy reasoning and probability theory vs. possibility theory in decision making, in *Proc. Symp. on Fuzzy Set Theory and Appl., IEEE Conf. on Decis. and Control*, New Orleans.

2988   Zadeh, L. A. (1977). Theory of fuzzy sets, Memo. UCB/ERL M77/1, Electron. Res. Lab., College of Eng., Univ. of California, Berkeley.

2989   Zadeh, L. A. (1978). Application of fuzzy sets to knowledge representation and approximate reasoning, Colloq. Int Theorie et Appl. des Sous-Ensembles Flous, Journ. de Biomath. et d'Inf. Med., Marseille, France, Sept.

2990   Zadeh, L. A. (1978). Fuzzy logic and its applications to decision and control analysis, in *Proc. 1978 IEEE Conf. on Decis. and Control, includes the 17th Symp. on Adaptive Processes*, 78CH1392-OCS, San Diego, CA, Jan. 1979.

2991   Zadeh, L. A. (1982). Possibility theory as a basis for meaning representation, *Proc. Sixth Wittgenstein Symp.*, Kirchberg.

2992   Zadeh, L. A. (1978). Fuzzy system theory—A framework for the analysis of humanistic systems, *IEEE Conf. on Circuit and Syst.*

2993   Zadeh, L. A. (1978). On translation of imperatives, Seminary Dept. Elec. Eng. and Comput. Sci., Univ. of California, Berkeley, June.

2994   Zadeh, L. A. (1978). Fuzzy sets as a basis for a theory of possibility, *Int. J. Fuzzy Sets Syst.* **1** (1), 3–28.

2995   Zadeh, L. A. (1978). Possibility theory and its applications to information analysis, in *Proc. Int. Colloq. on Inf. Theory*, CNRS, Paris.

2996   Zadeh, L. A. (1979). Approximate reasoning based on fuzzy logic, Memo. UCB-ERL M 79 32, Electron. Res. Lab., College of Eng., Univ. of California, Berkeley.

2997   Zadeh, L. A. (1979). Fuzzy sets and information granularity, in *Advances in Fuzzy Set Theory and Applications* (M. M. Gupta, R. K. Ragade, and R. R. Yager, Eds.), North-Holland, Amsterdam.

2998   Zadeh, L. A. (1979). Liar's paradox and truth qualification principle, UCB-ERL Memo. M 79-34, Electron. Res. Lab., College of Eng., Univ. of California, Berkeley.

2999   Zadeh, L. A. (1979). On the validity of Dempster's rule of combination of evidences, Memo. UCB-ERL M79-24, Electron Res. Lab., College of Eng., Univ. of California, Berkeley.

3000   Zadeh, L. A. (1979). Possibility theory with applications to decision analysis, in *Proc. 18th IEEE Conf. on Decis. and Control*, Dec. Ft. Lauderdale, Florida.

3001   Zadeh, L. A. (1979). Possibility theory and its application to the representation and manipulation of uncertain data, in *Proc. Workshop on Image Understanding*, NESC, Washington, D.C.

3002 Zadeh, L. A. (1979). Possibility theory and soft data analysis, ERL Memo. M79/66, Proc. of the AAAS Symp. on Soft Data Analysis.

3003 Zadeh, L. A. (1980). Fuzzy sets versus probability, *Proc. IEEE*, **68** (3), 421.

3004 Zadeh, L. A. (1980). Inference from fuzzy knowledge via fuzzy and ultra fuzzy logics, in *Proc. Int. Congr. on Appl. Syst. Res. and Cybern.*, Acapulco, Mexico, Dec.

3005 Zadeh, L. A. (1980). Inference in fuzzy logic, in *Proc. 10th Int. Symp. on Multiple-Valued Logic*, Northwestern Univ., Evanston, IL.

3006 Zadeh, L. A. (1980). Possibility theory as a basis for a mathematical model of human communication, UNESCO, Vienna.

3007 Zadeh, L. A. (1980). Possibility theory as a basis for information processing and knowledge representation, in *Proc. 4th IEEE Comput. Software and Appl. Conf.*, Chicago, Oct., pp. 842–849.

3008 Zadeh, L. A. (1980). Possibility theory as a basis for man-machine communication, COMPSAC'80, Univ. of California, Berkeley.

3009 Zadeh, L. A. (1980). Soft data analysis in decision making, in *Proc. of JACC*, San Francisco, Aug.

3010 Zadeh, L. A. (1981). Fuzzy probabilities in decision analysis, CORS-TIMS-ORSA Joint Nat. Meet., Toronto, Canada, May.

3011 Zadeh, L. A., Fu, K. S., Tanaka, K., and Shimura, M. (Eds.) (1975). *Fuzzy Sets and their Applications to Cognitive and Decision Processes*, Academic Press, New York.

3012 Zeleny, M. (1974). A concept of compromise solutions and the method of displaced ideal, computers and O.R., **1** (4), 479–496.

3013 Zeleny, M. (1976). Fuzzy assessment of intensities of preferences ORSA/TIMS Conf., Miami, FL, Nov.

3014 Zeleny, M. (1976). The theory of displaced ideal, in *Multiple Criteria Decision Making, Kyoto 1975*, Lecture Notes in Econ. and Math. Syst., Vol. 123 (M. Zeleny, Ed.), Springer-Verlag, Berlin, pp. 153–206.

3015 Zeleny, M. (1978). Membership functions and their assessment, Columbia Univ., New York, 4th Int. Congr. on cybern. and Syst., Amsterdam.

3016 Zeleny, M. (1980). Fuzzy sets and precision—Relationship theoretical, in *Proc. Int. Congr. on Appl. Syst. Res. and Cybern.*, Acapulco, Mexico, Dec.

3017 Zhang, C. (1980). An approach to describe the fuzziness, News on 1980 Annu. Rep. Meet. of Beijing Working Group on Fuzzy Sets, Beijing, China.

3018 Zhang, C. (1980). An approach to fuzzy math, Rep. Beijing Normal Univ. 11, A List of Lit. in China, The Univ. of Sci. and Tech. of China, Hofei, Anhwei, China.

3019 Zhang, J. (1980). A family of construction on the lattice value fuzzy sets, News on 1980 Annu. Rep. Meet. of Beijing Working Group on Fuzzy Sets, Beijing, China.

3020 Zhang, J. (1980). A family of models of lattice-valued sets, Working Pap., Inst. of Comput. Tech., Acad. Sinica Beijing, Beijing, China.

3021 Zhang, J. (1980). Normal fuzzy set structures, *Rep. on the Int. Congr. on Appl. Syst. Res. and Cybern.*, 12, A List of Lit. in China; The Univ. of Sci. and Tech. of China, Hofei, Anhwei, China.

3022 Zhang, J. (1980). Some basic properties of the normal fuzzy set structures, *J. Huazhong Inst. Tech.*, **2** (1), (English Ed.).

3023 Zhang, J. (1980). Survey of fuzzy sets in the Peoples Republic of China, in *Proc. Int. Congr. on Appl. Syst. Res. and Cybern.*, Acapulco, Mexico, Dec.

3024 Zhang, J. W. (1979). A unified treatment of fuzzy sets theory and Boolean valued set theory, Fuzzy sets and structures, Doc. Inst. of Comput. Tech., Acad. Sinica, Peking, China, Aug.

3025 Zhang, J. W. (1979). The normal fuzzy set structure and the Boolean-valued models, 6, *J. Huazhong Inst. Tech.*, (Wukan, Peking, China), **2**, 1–7.

3026 Zhang, J. W. (1980). Fuzzy set structures and normal fuzzy set structure, in *Proc. Int. Congr. on Appl. Syst. Res. and Cybern.*, Acapulco, Mexico, Dec.

3027 Zhang, W. (1980). A method based on fuzzy clustering for clinical data reduction, Collect. Pap. on Fuzzy Math. (Abstr.), Huazhong Inst. of Tech., Huazhong, China, Apr.

3028 Zhang, W. (1980). The generalizations of fuzzy measure and fuzzy integration, Collect. Pap. on Fuzzy Math. (Abstr.), Huazhong Inst. of Tech., Huazhong, China, Apr.

3029 Zhang, W. (1981). Possibility degree spaces, Dept. of Math. Xi-An Jiaotong Univ., A List of Lit. in China, The Univ. of Sci. and Tech. of China, Hofei, Anhwei, China.

3030 Zhao, C. (1980). Applications of fuzzy sets to the arrangement of railway transport, News on 1980 Annu. Rep. Meet. of Beijing Working Group on Fuzzy Sets, Beijing, China.

3031 Zhao, H., and Wong, P. (1980). Fuzzy mathematics and the science of science, News on 1980 Annu. Rep. Meet. of Beijing Working Group on Fuzzy Sets, Beijing, China.

3032 Zhao, R. (1980). The construction of transitive closure and the net-making method for fuzzy clustering analysis, Collect. Pap. on Fuzzy Math. (Abstr.), Huazhong Inst. of Tech., Huazhong, China, Apr.

3033 Zheng, W., Ren, S., Wu, C., and Tsuei, T. (1980). The intersection of fuzzy subsets and the robustness of fuzzy control, News on 1980 Annu. Rep. Meet. of Beijing Working Group on Fuzzy Sets, Beijing, China.

3034 Zhong, C. (1980). A new definition of the fuzzy sets, Collect. Pap. on Fuzzy Math. (Abstr.), Huazhong Inst. of Tech., Huazhong, China, Apr.

3035 Zhou, K. (1980). Relations between topological spaces and fuzzy topological spaces, a List of Lit. in China, The Univ. of Sci. and Tech. of China, Hofei, Anhwei, China.

3036 Zhou, K. (1980). Fuzzy complete subgroups, Rep. Beijing Normal Univ., 11, A List of Lit. in China, The Univ. of Sci. and Tech. of China, Hofei, Anhwei, China.

3037 Zhou, K. (1980). An example of the relation between fuzziness and unfuzziness, Rep., Beijing Normal Univ., 11, A List of Lit. in China, The Univ. of Sci. and Tech. of China, Hofei, Anhwei, China.

3038 Zhou, K. (1980). Fuzzy ring, Rep. Beijing Normal Univ., 11, A List of Lit. in China, The Univ. of Sci. and Tech. of China, Hofei, Anhwei, China.

3039 Zhou, K. (1980). Fuzzy topological algebra, Rep. Beijing Normal Univ., 11, A List of Lit. in China, The Univ. of Sci. and Tech. of China, Hofei, Anhwei, China.

3040 Zhu, Y., and Wu, X. (1980). Pansystem analysis of dynamic yin-yang logic and fuzziness, *J. Huazhong Inst. Tech.*, **2**, A List of Lit. in China, The Univ. of Sci. and Tech. of China, Hofei, Anhwei, China.

3041 Zied, A. M. (1980). Fuzzy clustering in a partitioned Karhunen-Loeve transform domain-application to characterization of multiple-diagnosis VCG's, Ph.D. Thesis, The Ohio State Univ., in *Diss. Abstr. Int.*, **41/04-B**, 1457.

3042 Zimmermann, H. J. (1974). Optimization in fuzzy environments, Tech. Rep., Inst. for Oper. Res., TH, Aachen, Germany.

Key References in Fuzzy Pattern Recognition

3043 Zimmermann, H. J. (1975). Bibliography—Theory and applications of fuzzy sets, Lehrstuhl fur Unternehmensforschung, RWTH, Aachen, Germany, Oct.

3044 Zimmermann, H. J. (1975). Description and ecoptimization of fuzzy systems, *Int. J. Gen. Syst.*, **2**, 209–215.

3045 Zimmermann, H. J. (1975). Description and optimization of fuzzy systems, *Int. J. Gen. Syst.*, **2** (4).

3046 Zimmermann, H. J. (1975). Fuzzy decisions, fuzzy algorithms—A promising approach to problem solving, NOAK75, Oslo, Norway, Oct.

3047 Zimmermann, H. J. (1975). Optimale Entscheidungen Bei Unscharfen Problembeschreibungen, Lehrstuhl fur Unternehmensforsch., RWTH, Aachen, Germany.

3048 Zimmermann, H. J. (1975). The potential of fuzzy decision making in the private and public sector, SOAK-75, Lidingo, Sweden.

3049 Zimmermann, H. J. (1976). Fuzzy programming vs. multiple projective functions, ORSA/TIMS Conf., Miami, FL.

3050 Zimmermann, H. J. (1976). Un Scharfe Entscheiden und Multi-Criteria-Analyse, in *Proc. Oper. Res.*, (Wurzburg), **6**, 99–109.

3051 Zimmermann, H. J. (1977). Duality in fuzzy programming, Int. Symp. on External Methods and Syst. Anal., Austin, TX.

3052 Zimmermann, H. J. (1977). Fuzzy programming and linear programming with several objective functions, *Fuzzy Sets Syst.*, **1**, 45–55.

3053 Zimmermann, H. J. (1977). Results of empirical studies in fuzzy set theory, in *Applied General Systems Research, Recent Development* (G. Klir, Ed.), Plenum Press, New York,

3054 Zimmermann, H. J. and Gehring, H. (1975). Fuzzy information profiles for information selection, *Congr. Book*, Vol. II, 4th Int. Congr., AFCET, Paris.

3055 Zimmermann, H. J., and Pollarscheck, M. A. (1976). APL routines for fuzzy 0-1 linear programs and their performance, Aachen Working Pap. No. 76/10, Aachen Univ., Aachen, Germany.

3056 Zimmermann, H. J., and Rodder, W. (1975). Analyse, Beschreibung und optimierung von unscharf Formulierten Problemen, Lehrstuhul fur Unternehmensforsch., RWTH, Aachen, Germany.

3057 Zimmermann, H. J., and Zysno, P. (1982). Human decision making in different connectives, *Fuzzy Sets Syst.*

3058 Zimmermann, H. J., and Zysno, P. (1980). Latent connectives in human decision making, *Fuzzy Sets Syst.*, **4** (1), July, 37–52.

3059 Zinovev, A. A. (1963). Philosophical problems of many-valued logic, D. Reidel, Dordrecht, Holland.

3060 Zu-Wei, L. (1981). Mutually inverse properties of fuzzy matrix and its inverse, *Bull. Sous Ensembles Flous Appl.*, **5**, Hiver (80–81), 13–19.

3061 Zu-Wei, L. (1981). Necessary and sufficient conditions for the existence of solutions of fuzzy equations, *Bull. Sous Ensembles Flous Appl.*, **5**, Hiver (80–81), 20–29.

3062 Zwick, M., Schwartz, D. G., and Lendaris, G. G. (1978). Fuzziness and catastrophe, in *Proc. 1978 Int. Conf. on Cybern. and Soc.*, Tokyo, Vol. 2, pp. 1237–1241.

3063 Zysno, P. (1979). One class of operators for the aggregation of fuzzy sets, Eur. III Congr., Amsterdam.

3064 Zysno, P. (1980). The aggregation of subjective categories within judgmental and evaluative processes, The 9th Meet. of the Eur. Working Group on Fuzzy Sets, Vienna, Apr.

Index

α-composite fuzzy relation, 137
Adjoint of matrix, 143
Agglutination, 117
Atom, 48

Botryology, 2
Bounded-difference, 35
Bounded-sum, 35
Branch-bound-back track (BBB) algorithm, 112
Brouwrian lattice, 137

Cartesian product, 35
Clause, 46
Cluster, 5
Clustering performance measure, 206
Complement, 33
Compositional rule of inference, 58
Composition of fuzzy relations, 60, 61, 135
Concentration, 34
Conjunctive rule, 55
Constant fuzzy matrix, 144
Convex combination, 35
Convex decomposition, 168
Crossover point, 26, 28
Cut-point languages, 97

Decision path, 111
Degree of belief, 65
Dendrogram, 149
Dilation, 35
Disjunctive rule, 55
Dissimilarity relation, 63
Distance:
  absolute, 12
  Euclidean, 12
  functional, 13
  fuzzy, 42
  Mahalanobis, 12
  Minkowski, 12
  probabilistic, 12
  weighted, 12
Distance function, 152
Dynamic fuzzy relation (DFR), 165

Entropy, 49
Extension principle, 39

Factorization, 14
Feature extraction, 6, 94
Fractionally fuzzy grammar (FFG), 98, 104
Fuzziness, 23
Fuzzy α-cuts, 31, 41
Fuzzy covariance matrix, 193, 199
Fuzzy coverage, 115
Fuzzy decision tree, 110
Fuzzy expected value (FEV), 67
Fuzzy filter, 182
Fuzzy formal language, 95
Fuzzy grammar, 96
Fuzzy graph, 131, 158, 182
Fuzzy ISODATA, 125
Fuzzy linguistic variable, 174
Fuzzy logic, 43
Fuzzy mean, 197
Fuzzy measure, 66
Fuzzy partition, 113, 155, 187
Fuzzy r-cluster, 114
Fuzzy relation, 59, 131, 134
Fuzzy restriction, 63
Fuzzy sets, 23
Fuzzy syntax-directed translation scheme (FSDTS), 97

Height, 26, 62
Hierarchical clustering, 149

Implication rule, 55

Inconsistent formula, 46
Inexact matrix, 142
Inexact transmission matrix, 146
Inf-Max composition, 141
Intersection, 32
ISODATA, 119

λ-compatible, 165
λ-complementation, 34
λ-degree connected graph, 132
Left square, 35
Level sets, 31
L-fuzzy sets, 183
Literal, 46
Local maximum, 190

Masking, 183
Max-Product composition, 161
Max-Δ transitivity, 168
Measure of cluster validity, 200
Membership function, 24
Min-Max composition, 159
Min-Max operators, 38
Modifier rule, 54
M-partition, 203

n-ary fuzzy relation, 133
n-step fuzzy relation, 139

Opaque, 92

Parallel partitioning, 17
Parallel threshold partitioning, 16
Phrase, 46
Possibility assignment equation, 52
Possibility distribution, 52
 function, 52
Possibility postulate, 52
Possibility/probability consistency
 principle, 59
Precategorical classification feature (PCF),
 173

Primitive connection matrix, 146
Probability space, 86
Product, 34
Projection, 56, 62
Propinquity, 140
Proximity relation, 151
Pseudocomplemented distributive lattice, 45
Pseudometric, 42
Pseudosimilarity relation, 169

Quantification rule, 55

Reflexive, 62, 139

Sequential threshold partitioning, 15
Similarity:
 graph, 150
 measure, 11, 13
 region, 62
 relation, 139
 vector, 179
Single-linkage algorithm, 150
Squared error cluster programs, 203
Subtypical, 71
Supertypical, 71
Support, 26, 62
Symmetric, 63, 139
Symmetrical difference, 38

Taxonomy, 2
Transitive, 63, 139
Transparent, 92
Transpose of matrix, 142
Triangle product, 136
Truth quantification rule, 56
Typical, 71, 92

Unimodal fuzzy sets, 186
Union, 32

Valid formula, 46
Validity, 200